Deep Learning for Advanced X-ray Detection and Imaging Applications

Krzysztof (Kris) Iniewski • Liang (Kevin) Cai
Editors

Deep Learning for Advanced X-ray Detection and Imaging Applications

Springer

Editors
Krzysztof (Kris) Iniewski
Redlen Technologies
Port Moody, BC, Canada

Liang (Kevin) Cai
Canon Medical Research USA
Vernon Hills, IL, USA

ISBN 978-3-031-75652-8 ISBN 978-3-031-75653-5 (eBook)
https://doi.org/10.1007/978-3-031-75653-5

This Springer imprint is published by the registered company Springer Nature Switzerland AG
The registered company address is: Gewerbestrasse 11, 6330 Cham, Switzerland

Contents

About the Editors

Krzysztof (Kris) Iniewski is Director of detector architecture and applications at Redlen Technologies Inc. in British Columbia, Canada. During his 18 years at Redlen, he managed the development of highly integrated CdZnTe detector products in medical imaging and security applications. Prior to Redlen, Kris held various management and academic positions at PMC-Sierra, University of Alberta, Simon Fraser University (SFU), University of British Columbia (UBC), and University of Toronto. Dr. Iniewski has published over 150+ research papers in international journals and conferences. He holds 25+ international patents granted in USA, Canada, France, Germany, and Japan. He wrote and edited 75+ books for Wiley, Cambridge University Press, Mc-Graw Hill, CRC Press, and Springer. He is a frequent invited speaker and has consulted for multiple organizations internationally.

Liang (Kevin) Cai is Manager of CT Reconstruction at Canon Medical Research, USA. During his time at Canon, Dr. Cai has overseen the development of multiple industry-leading deep learning CT reconstruction algorithms, including advanced intelligent clear-IQ engine (AiCE), precise IQ engine (PIQE), and deep learning spectral CT. Dr. Cai has extensive experience in research and development for X-ray and Gamma-ray imaging systems, which can be attributed to his comprehensive expertise in both radiation detection physics and image reconstruction algorithms. He has contributed to over 50 research papers published in international journals and conferences and has been named as an inventor on more than 20 granted US patents.

Deep Learning Techniques for CT Image Denoising and Resolution Enhancement

Jian Zhou, Ting Xia, Efren Lee, and Kevin Cai

1 Introduction

Computed tomography (CT) has revolutionized clinical diagnosis for decades. Modern CT scanners can produce images with superior spatial resolution and greatly reduced noise compared to earlier generations. However, the quality of the raw data acquired by the scanner is not the sole determinant of the final image quality. Image reconstruction techniques play a critical role in transforming this raw data into clinically useful images. The choice of reconstruction algorithm substantially impacts the final image characteristics, including noise levels, spatial resolution, and the presence of artifacts.

Conventional CT reconstruction methods primarily fall into two categories: analytical methods and model-based iterative reconstruction (MBIR). Analytical methods, such as filtered backprojection (FBP), remain prevalent due to their computational efficiency and are the default option in most commercial scanners. However, FBP offers limited control over noise, making image quality highly dependent on scan parameters such as dose levels and scan duration. MBIR improves reconstruction by incorporating an accurate system model, enabling the reconstruction of high-quality images with controllable noise properties through regularization techniques. While MBIR offers several advantages over FBP, its computational complexity presents a significant hurdle to widespread adoption in clinical settings. Additionally, the computational demands of MBIR can far exceed the capabilities of older CT scanners, potentially limiting its applicability in resource-constrained environments.

J. Zhou · T. Xia · E. Lee · K. Cai (✉)
CMRU, Vernon Hills, IL, USA
e-mail: lcai@mru.medical.canon

K. (K.) Iniewski, L. (K.) Cai (eds.), *Deep Learning for Advanced X-ray Detection and Imaging Applications*, https://doi.org/10.1007/978-3-031-75653-5_1

The advent of deep learning (DL) has fundamentally transformed various fields, including medical imaging. Unlike traditional machine learning, it eliminates the need for manual feature extraction and uses a large and deep neural network (DNN) that directly learns from the training data to make predictions. In CT imaging, DL-based reconstruction methods offer a compelling alternative to traditional techniques, enabling the generation of high-fidelity CT images without substantial computational costs. This chapter focuses on two key areas where deep learning has significantly improved CT image quality: denoising and spatial resolution enhancement. In denoising, DNNs can be trained to remove noise artifacts while preserving anatomical details in reconstructed images. The denoising DNNs often act as a post-processing step, which can be seamlessly integrated into existing CT reconstruction pipelines. Some approaches also implement DNNs into the MBIR algorithm with a differentiable forward model, enabling noise suppression with reduced computational cost and enforced data consistency conditions. Furthermore, deep learning techniques can be tailored to address specific challenges in CT image reconstruction, such as image resolution enhancement for improved diagnostic accuracy. Therefore, this chapter also explores various super resolution (SR) techniques that leverage deep learning to enhance CT image spatial resolution, improving image detail and thus allowing for more precise diagnoses.

It is worth noting that numerous studies have been published on DL-based CT image reconstruction. While this review may not cover all of them, we hope that it will, nevertheless, provide readers with a comprehensive examination that highlights the significant advancements of DL-based applications in CT image reconstruction.

2 CT Image Denoising

Quantum noise, also referred to as X-ray photon statistical noise, is a significant source of noise in the CT imaging process. This type of noise is largely influenced by the radiation dose administered to a patient during data acquisition. Higher radiation doses increase photon statistics, thereby reducing noise levels and improving image quality. However, as the dose increases, the risk of radiation-induced damage to the patient also rises. To balance safety and effectiveness, the principle of ALARA (As Low As Reasonably Achievable) dictates that low-dose CT (LDCT) scans are preferred. Nonetheless, reducing the dose while maintaining suitable image quality under varying scan conditions remains a great challenge for CT reconstruction.

Denoising techniques have long served as valuable alternatives to MBIR in CT reconstruction, especially in LDCT applications, where they are essential for achieving high diagnostic image quality. These techniques are typically divided into two main categories: data-domain denoising and image-domain denoising. Data-domain denoising focuses on removing noise directly from the raw data before image reconstruction, whereas image-domain denoising is applied as a post-processing step on reconstructed images.

In data-domain denoising, filtering is a commonly used approach. For example, structural adaptive filters [1] and bilateral filters [2] have been employed to suppress noise in the raw data. Iterative denoising methods, such as those based on penalized weighted least-squares algorithms [3], have also been developed. These approaches leverage second-order data statistics in combination with data-domain regularization techniques to estimate clean noiseless data. More recently, DL-based data-domain denoising approaches have been investigated in [4–7] where convolutional neural networks (CNNs) and transformer networks have been trained to predict high-dose data directly from low-dose data. These networks can effectively remove noise across various scan dose levels. Notably, certain networks [5] and [6] can achieve this without requiring high-dose data as a reference in training.

While techniques in both domains can sometimes be applied interchangeably, data-domain denoising often results in inferior image reconstruction compared to image-domain denoising [8]. This is because optimal noise removal at the data level does not always translate to optimal image quality. Additionally, research on data-domain denoising is relatively limited compared to image-domain denoising, largely due to the difficulty in accessing clinical raw data. Over the past few decades, LDCT image-domain denoising has been one of the most active research areas in medical imaging. Several representative methods have emerged, including nonlocal mean filtering [9] and dictionary learning-based approaches [10]. These techniques are largely grounded in the theories of sparse representation and compressed sensing where noise and signal can be effectively separated through sparse feature transformation. Recent advancements in deep learning technologies have driven significant innovation in this field, leading to numerous learning-based denoising methods that outperform traditional approaches relying on handcrafted features.

Most DL-based denoising approaches developed to date can be classified as regression models, where neural networks are typically trained to take LDCT images as input and estimate the corresponding high-dose or normal-dose images. The training process generally involves using a regression loss function, such as mean squared error (MSE) or mean absolute error (MAE), to quantify how closely the network output matches the target image. However, many studies have observed that networks trained with conventional MSE or MAE often produce overly smooth CT images that lack natural noise texture. This is because the regression loss drives the network to predict an averaged image, which may not necessarily be sharp. As a result, some denoising networks also incorporate additional training loss functions, such as perceptual loss and generative adversarial network (GAN) loss, to better capture specific image characteristics. Some representative works are listed in Table 1.

2.1 Denoising as Conventional Regression Model

CNNs are particularly popular for building denoising neural networks due to their efficiency in processing 2D and 3D image data, closely mimicking the human visual

Table 1 A list of DL-based LDCT image denoising approaches

	Loss function	Network structure
Chen et al. [11]	MSE	Sequential CNN
Kang et al. [15, 16]	MSE	U-Net, wavelet residual network
Chen et al. [8]	MSE	RED-CNN, U-Net
Wu et al. [22]	MSE	A cascade of 3 sequential CNNs
Yang et al. [14]	MSE	Sequential CNN
Liang et al. [23]	MSE + MSP	Densely connected CNN
Yuan et al. [5]	MAE	U-Net
Gholizadeh-Ansari et al. [24]	MSE + perceptual loss	Sequential CNN
Zhang et al. [20]	MSE	Sequential CNN
Matsuura et al. [21]	MAE	U-Net
Li et al. [25]	MS-SSIM-L1+edge loss	MSFLNet
Zhang et al. [26]	MSP	MFAF-net
Luthra et al. [27]	MSE	Transformer
Wang et al. [28]	MSE	Transformer
Li et al. [29]	MSE+ MSP + Sobel-L1	Transformer
Wolterink et al. [30]	GAN + MSE	Generator: sequential 3D CNN
Choi et al. [31]	GAN + MSE + MAE	Generator: sequential CNN
Yang et al. [32]	WGAN + perceptual loss	Generator: sequential CNN
Hu et al. [33]	WGAN + MSE	Generator: CNN with cascaded residual block
Yi et al. [34]	Least-square GAN + MAE + sharpness loss	SAGAN, generator: residual U-Net
Shan et al. [35]	WGAN + perceptual loss	Generator: CPCE
You et al. [36]	WGAN + structure preservation + MAE	SMGAN 2D and 3D
Li et al. [37]	WGAN + auto-encoder perceptual loss	3D SACNN (self-attention CNN)
Bera et al. [38]	GAN + noise aware MSE loss	Generator: modified RED-CNN, discriminator: SA-SNMP
Huang et al. [39]	Image-domain GAN + gradient-domain GAN + MSE + MAE (gradient)	DU-GAN, generator: REDCNN, discriminator: 2 U-Nets
Geng et al. [40]	GAN + MAE	Generator: sequential CNNs, discriminator: patch-based
Kang et al. [41]	GAN + cyclic consistency loss + additional identity loss	CycleGAN with two sequential CNNs as generator
Gu et al. [42]	GAN + cyclic consistency loss + additional identity loss	CycleGAN, generator with AdaIN

cortex. Various denoising CNNs have been proposed, each with unique network architectures and objective functions. Chen et al. [11] introduced the first DL-based LDCT denoising method, using a lightweight CNN with just three convolutional layers. This simple, sequential CNN was trained using low-dose FBP images as input and normal-dose FBP images as labels, with the conventional MSE as the training loss function. Despite its simplicity, this three-layer CNN demonstrated effective noise removal from LDCT images.

While deeper CNNs often exhibit superior generalization capabilities and enhanced denoising performance, their training can be more challenging due to the potential vanishing gradient problem during error back-propagation. To address this issue, residual learning [12, 13] was commonly introduced, incorporating bypass connections to learn the difference between the input and target images. Yang et al. [14] proposed the use of 2D and 3D residual convolutional networks for low-dose CT imaging, noting that a 3D model can achieve better denoising performance than a 2D model, particularly in terms of edge preservation and feature continuity. Kang et al. [15, 16] suggested using a U-Net-based CNN [17] for LDCT denoising. Instead of operating directly in the spatial domain, they applied denoising in the wavelet domain. They found that utilizing a directional wavelet transform enabled the network to better identify and suppress directional components of noise streak artifacts, leading to improved denoising performance.

Another noteworthy approach is the residual encoder-decoder CNN (RED-CNN) proposed by Chen et al. [8], which features an encoder-decoder structure specifically designed for CT image denoising. In this model, the pooling and unpooling layers of the original U-Net were replaced with convolutional and deconvolutional layers, a modification reported to enhance noise suppression while preserving image details and improving lesion detectability. Inspired by the noise2noise training [18], Yuan et al. [5] proposed an LDCT denoising U-Net trained without using high-dose reference images. They developed a novel method that is able to split a high-dose scan into two independent lower-dose scans. The training then uses one of the lower-dose reconstructions as input and the other as target. The trained network is theoretically guaranteed to be identical to the conventional low-dose to high-dose denoising network. Recently, the same algorithm has been extended to photon-counting CT (PCCT) image denoising [19].

Zhang et al. [20] proposed a multifeature extraction approach based on a residual module learning framework. By combining different sizes of convolution filters with different learning capabilities, their network could more comprehensively capture image feature information during denoising. Matsuura et al. [21] argued that the choice of learning target can also be critical to denoising performance. They proposed to use CT images reconstructed from multiple reconstruction kernels as the training target. They also suggested using MBIR to obtain these reconstructions in order to preserve the best image quality for different anatomies. The resulting neural network, namely the feature-aware CNN, outperforms both FBP and standard MBIR by producing improved CT image quality potentially suitable for a broad range of clinical diagnoses. Instead of employing a more complex structure with additional layers, a cascaded model can also be considered. For example, Wu et al. [22] proposed applying a trained CNN on the training dataset to initiate new training and remove artifacts induced by the initial denoising process. This cascaded denoising approach can achieve better performance with a simple sequential CNN structure.

To improve sharpness in denoised images, Liang et al. [23] suggested using an edge enhancement–based densely connected convolutional neural network (EDCNN) for CT image denoising. This network incorporates an edge enhancement module with a trainable Sobel convolution, which adaptively extracts edge features during the optimization process. The training loss function combines MSE with a multi-scale perceptual (MSP) loss. Gholizadeh-Ansari et al. [24] introduced a residual linear sequence convolutional learning network for low-dose CT denoising. Their network includes dilated convolution and a non-trainable edge detection convolutional layer utilizing four Sobel kernels for edge extraction at the initial stage. They employed a loss function based on MSE and perceptual loss, computed using the pretrained VGG-16 model [43]. Li et al. [25] proposed a multi-scale feature fusion network (MSFLNet) that extracts multi-scale information and multichannel features from LDCT images to enhance feature fusion across different dimensions. Their approach used a composite loss function consisting of MAE loss, multi-scale structural similarity index measure (SSIM) loss, and edge loss. Recently, Zhang et al. [26] introduced a novel multi-scale feature aggregation and fusion network (MFAF-Net) for image denoising. Their method employs a multi-scale residual feature aggregation module to capture structural information at different scales in CT images, along with a cross-level feature fusion module to integrate features across various levels. They also proposed a self-supervised multilevel perceptual loss, replacing the traditional single-level perceptual loss to more effectively recover fine textures and structures in tissues and lesions within CT images.

Conventional CNN-based models often struggle to capture global contextual information due to their limited receptive fields, making them less effective at modeling structural similarities across an entire image. Recently, the transformer model [44] has demonstrated exceptional performance in computer vision tasks. Unlike CNNs, transformers excel at capturing global information and long-range feature interactions, enabling the utilization of richer contextual data. Wang et al. [28] introduced a novel transformer-based model, CTformer, specifically designed for CT image denoising. This work is among the first to apply visual transformers to the LDCT denoising task, incorporating techniques like expansion and cyclic displacement to capture longer-range interactions. Several other studies have also explored the application of transformer models in LDCT denoising. For instance, Zhang et al. [45] were pioneers in applying transformer modules to the outputs of Gaussian filters for enhanced denoising performance. Luthra et al. [27] employed Sobel-Feldman operators and the locally enhanced window (LeWin) transformer block to optimize noise residual information in LDCT images. Lastly, Li et al. [29] developed a dual-branch model featuring an overlapping-free,

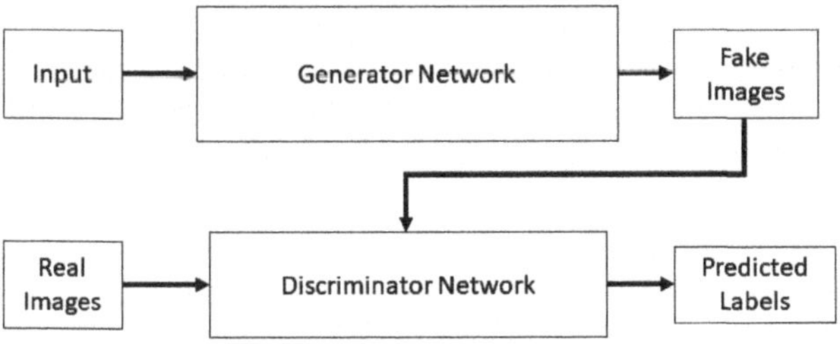

Fig. 1 A typical architecture of GAN which consists of a generator and a discriminator

window-based self-attention transformer block on the main branch, along with an enhancement module to improve edge, texture, and contextual information in LDCT images.

2.2 *Denoising with GAN*

Generative adversarial networks (GANs) [46] have achieved great success in many deep learning applications, including LDCT image denoising (see Table 1 for several examples). A typical GAN consists of two competing networks: the generative network (generator) and the discriminative network (discriminator). The generator is trained to produce samples that mimic a target distribution, while the discriminator attempts to distinguish between real and generated samples using an adversarial loss. The two networks engage in a min-max optimization, competing against each other during training until they reach an equilibrium. Figure 1 illustrates a typical GAN architecture.

Wolterink et al. [30] were among the first to introduce GANs to address the noise texture issue in low-dose cardiac CT image denoising. Their proposed GAN includes a discriminator CNN that aims to differentiate between real routine-dose CT images and denoised low-dose CT images. It is important to note that the discriminator is only used during training and not during testing. Their results showed that GAN-based training enabled the generator to produce images that were visually closer to routine-dose CT images. A similar GAN-based denoising network was presented in [31] for real-time recovery of high-quality images from noisy LDCT images. Yang et al. [32] also applied GANs to LDCT denoising except using the Wasserstein distance with an additional gradient penalty term, namely WGAN, to mitigate the mode collapse issue often associated with GAN training. Additionally, they proposed using the perceptual loss instead of the conventional MSE loss to enhance feature preservation. Their results indicated that the additional perceptual loss led to more visually appealing CT images compared to those produced using MSE or MAE loss. The same WGAN loss has also been employed in [33] for CT dental image denoising.

Yi et al. [34] further advanced GAN-based LDCT denoising by developing a residual U-Net-like generator, distinct from those used by Wolterink [30] and Yang, et al. [32]. To enhance the spatial resolution in low-contrast regions of CT images, they introduced a sharpness loss function in place of the traditional MSE loss. Their resulting network, termed SA-GAN, was evaluated on a diverse range of CT images from head to pelvis, demonstrating promising results across various dose levels. Shan et al. [35] used WGAN to train their conveying path-based convolutional encoder-decoder (CPCE) network. Unlike U-Net, their network does not have pooling or down-sampling layers, which can cause information loss. Instead, it uses conveying paths to feed feature maps from earlier layers directly into later layers, preserving high-resolution features. This idea also led to the development of their new network, referred to as the modularized adaptive processing neural network (MAP-NN) for LDCT imaging via progressive denoising [47]. This framework can be extended to any existing denoising network. To further capture subtle structural features while maintaining high visual sensitivity for LDCT, You et al. [36] proposed a 3D clinical structurally sensitive multi-scale generative adversarial network (SMGAN) under the WGAN framework. They used a 3D convolutional network as the generator and introduced a structurally sensitive loss that combines adversarial loss, perceptual structural loss, and pixel-wise MAE loss. Li et al. [37] introduced WGAN with a 3D self-attention model, combining plane attention and depth attention to capture global relationships in the spatial domain. Their proposed 3D self-attention neural network enables convolution layers to capture long-range dependencies in 3D CT images, further improving denoising performance. Instead of using VGG feature extractors, originally trained for classifying natural images, they trained a self-supervised auto-encoder neural network for perceptual loss.

Bera et al. [38] suggested using a novel nonlocal module that exploits neighborhood similarity in CT images for enhanced denoising. Within the GAN framework, they employed a modified version of the RED-CNN architecture for the generator, incorporating a nonlocal module in the middle layer of the network, a self-attentive spectral normalized Markovian patch discriminator (SA-SNMP) for the discriminator, and noise-aware MSE loss. Huang et al. [39] proposed DU-GAN, which leverages U-Net-based discriminators in the GAN framework to learn both global and local differences between denoised and normal-dose images in both the image and gradient domains. Geng et al. [40] proposed a content-noise complementary learning (CNCL) strategy for medical image denoising within the GAN framework. Their denoising strategy simultaneously learns the content and noise using two predictors in a complementary manner and then reconstructs the final content from the extracted features of both the content predictor and noise predictor.

Another advantage of using the GAN framework in deep learning is that it enables unsupervised learning, which is particularly valuable when large amounts of spatially well-aligned input and reference image pairs are not available in clinical settings. Kang et al. [41] proposed using an unsupervised CycleGAN [48] for multiphase coronary CT angiography denoising. Specifically, they trained two networks between two different domains (low-dose and routine-dose), with the goal that the networks should be inverses of each other. Although only one of the generator networks is used during the inference phase, the secondary auxiliary generator is necessary to enforce cycle consistency. However, the additional memory requirements and increased number of learnable parameters present challenges for CycleGAN training. To address the issue of network weight reduction, the same group later improved their CycleGAN network structure by introducing an adaptive instance normalization (AdaIN) layer into the generator [42]. This modification allows the network to be trained in a manner similar to CycleGAN but without the need for an additional generator, leading to greater memory efficiency during network training.

2.3 *Denoising with Diffusion Model*

Diffusion models are another type of generative model that have gained significant attention in recent years. Unlike GANs, diffusion models typically rely on an iterative procedure to gradually generate samples that follow the target data distribution. These models involve two fundamental processes: the forward diffusion process and the reverse sampling process. The forward process mimics a diffusion or Markov process by adding stochastic noise, often Gaussian noise, to an image until it becomes a noise image with known statistics. The reverse sampling process, on the other hand, attempts to invert the forward procedure, starting from random noise and progressively removing it until a clear image emerges. This noise removal is performed using a neural network, which constitutes the core of the diffusion model. A graphic illustration of diffusion model is provided in Fig. 2. Compared to GAN training, training diffusion models is relatively straightforward, as it is similar to training a denoising neural network approximating the score function of the data distribution. For more detailed information on the classic denoising diffusion probabilistic model (DDPM) and related models like score-based generative models, interested readers are referred to [49–52].

In the field of CT imaging, diffusion models have already been applied to the LDCT denoising problem. For instance, Gao et al. [53] proposed a contextual

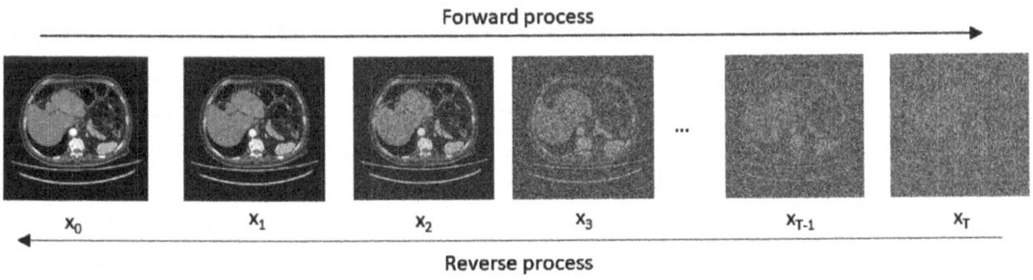

Fig. 2 A graphical illustration of the denoising diffusion probabilistic model. The forward process starts with a clean image, progressively adding noise, while the reverse process begins with a noise image, gradually removing the noise to recover the clean image

conditional diffusion model for LDCT image denoising. In their model, the low-dose CT image is used as a condition to guide the diffusion process. Additionally, they leveraged the contextual information from adjacent slices to better restore structural details that are challenging to distinguish from noise in a single 2D image. Their findings indicated that the proposed method demonstrated superior generalization performance across different dose images compared to conventional CNN- and GAN-based models. Later, in [54], the same authors introduced an improved version based on the cold diffusion method [55] which begins with the LDCT image and performs iterative denoising. This approach significantly reduces the number of sampling steps while maintaining good denoising performance. Xia et al. [56] also studied a fast ordinary differential equation solver to accelerate the reverse sampling process for LDCT image denoising, achieving a remarkable 20x speed improvement.

2.4 Commercial Products of Clinical CT Image Denoising

Thanks to their superior image quality, increased diagnostic accuracy, and faster clinical workflows, DL-based denoising approaches have been integrated into commercial image reconstruction products by several CT vendors. Currently, three major FDA-approved DL-based products are available on the market: TrueFidelity by GE Healthcare [57], Advanced Intelligent Clear-IQ Engine (AiCE) by Canon Medical Systems [58], and Precise Image by Philips Healthcare [59].

TrueFidelity (GE Healthcare) was trained on high-quality FBP datasets, allowing it to effectively distinguish between noise and signal, thus reducing noise without compromising anatomical and pathological details. By leveraging high-dose FBP as the ground truth, TrueFidelity can accurately represent the scanned object, resulting in CT images with a more desirable noise texture. Precise Image (Philips Healthcare) was trained using FBP images reconstructed from lower-dose-simulated sinogram as input, matched with routine-dose FBP images as ground truth. Precise Image achieves both dose reduction and improved low-contrast imaging while maintaining a traditional noise texture [59]. AiCE (Canon Medical Systems) utilized full-dose clinical MBIR images as ground truth during training. The use of MBIR allows for not only noise reduction but also artifact reduction due to its comprehensive modeling of system optics, physics, scanner statistical properties, and human anatomy. AiCE reconstruction enables effective dose reduction, preserves edges, maintains image details, and delivers high contrast resolution with low noise and much faster reconstruction speed than MBIR. Figure 3 shows a comparison of clinical images reconstructed using the adaptive iterative dose reduction (AIDR 3D, Canon Medical Systems) algorithm and AiCE.

In addition to the offerings from CT vendors, two independent software companies have developed vendor-agnostic DL-based denoising algorithms: ClariCT.AI [60] and PixelShine [61]. Both received FDA clearance in 2019. These DL-based commercial denoising products also tend to deliver superior image quality

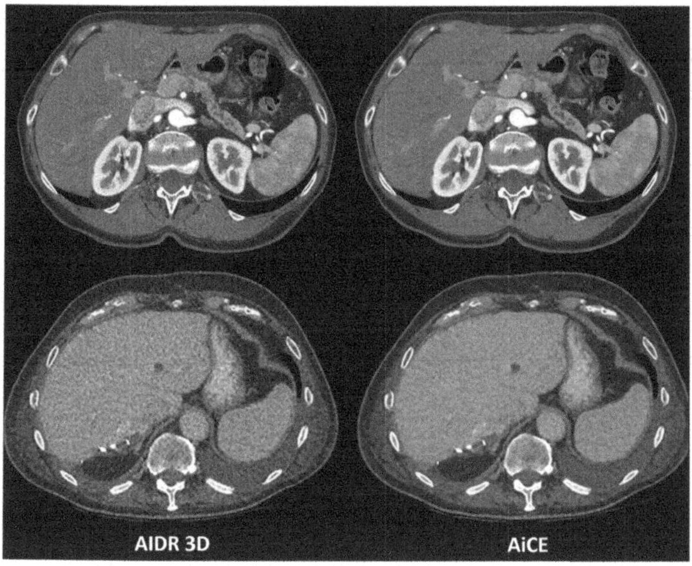

Fig. 3 A comparison of AIDR 3D and AiCE images. Top row: 120 kV, CTDI: 9.6 mGy; Bottom row: 100 kV, CTDI: 2.0 mGy. Images courtesy of Prof. Prokop, Radboudumc, Nijmegen, the Netherlands

compared to conventional FBP or even iterative reconstruction techniques while offering faster reconstruction speeds.

3 CT Image Resolution Enhancement

The resolution of clinical CT has significantly improved over the past few decades, primarily due to advancements in hardware. For instance, modern high-resolution CT can achieve a spatial resolution of approximately 125 μm [62]. However, these high-resolution CT systems are expensive and not widely accessible in routine clinical practice. Several methods have been developed to enhance CT resolution through specific algorithms. For example, in FBP, different reconstruction kernels can be selected to achieve higher spatial resolution, but this typically results in increased noise in the reconstructed images. Another approach is MBIR, which improves resolution by incorporating accurate system optics models. However, the resulting MBIR often suffers from slow convergence, leading to longer reconstruction times. The emergence of deep learning technologies has enabled a novel class of resolution enhancement methods, known as DL-based super resolution techniques. Similar to DL-based denoising techniques, most DL-based SR methods for CT work as post-processing steps that can be easily applied to rapidly produce resolution-enhanced

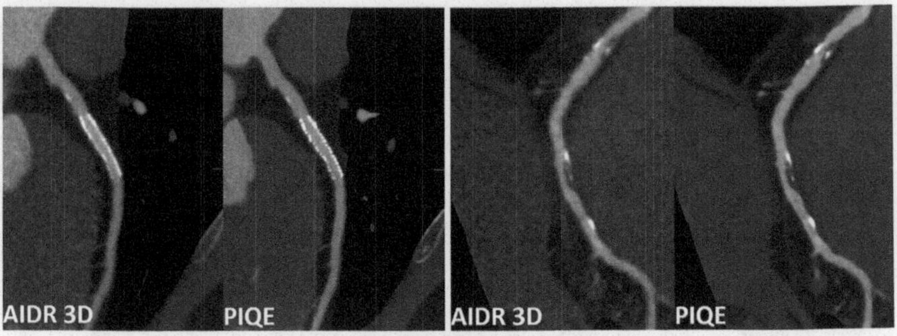

Fig. 4 A comparison of AIDR 3D and PIQE reconstructions of LAD stent (**a**) and RCA plaques. Images courtesy of Dr. Marcus Chen, MD, Director of Cardiothoracic Imaging at the NIH, USA

images. These approaches can also be viewed as regression models, where the network takes a low-resolution image and predicts a high-resolution version.

DL-based SR approaches have demonstrated significant success in enhancing image resolution. They show promise in extracting intricate patterns and features that traditional methods may overlook. Recently, one DL-enabled super resolution technique has been commercialized as the Precision IQ Engine (PIQE) in the Canon Aquilion ONE/PRISM Edition CT scanner. Figure 4 shows some PIQE images as compared to normal-resolution (NR) AIDR 3D images. It can be seen that PIQE images exhibited better noise texture, enhanced spatial resolution, and improved visualization of stents, calcification, and soft tissue features compared to other reconstruction methods.

Similar to image denoising, there exists a significant amount of research effort on image resolution enhancement, spanning from traditional regression-based DL approaches to today's popular diffusion models. In the following context, we will briefly review a few representative attempts to provide the readers with a general research landscape in this area.

Mansoor et al. [63] proposed utilizing SRGAN [64] to generate diagnostic-quality volumetric CT images from lower-dose acquisitions. Similar to the approach in [32], they employed perceptual loss to assess the difference between the network output and the target, while adversarial loss was used to distinguish between the estimated and diagnostic-quality images. The proposed framework demonstrated excellent performance, generating super resolution images with high diagnostic quality. Other similar studies can be found in the references provided [65, 66].

Tang et al. [67] developed a data-domain DL-enabled SR approach. They applied the SR network to low-resolution binned sinogram data. Additionally, their investigation extended the method by integrating this network with an iterative reconstruction method, enhancing the reconstruction of high-resolution images. In a recent study, Koons et al. [68] proposed training a super resolution CNN using high-resolution PCCT images as the target and smoothed lower-resolution images as the input. Their approach focused on cardiac imaging, and the results indicated that the network produced images with reduced noise and improved structural similarity, closely matching the high-resolution reference images.

Supervised DL-based SR methods typically require paired training data, including high-resolution target images, which are not always available in many clinical applications. To mitigate this challenge, researchers have been exploring the use of GANs with unsupervised or semi-supervised training approaches. For instance, You et al. [69] introduced GAN-CIRCLE, a semi-supervised deep learning method that employs two GAN networks linked together in a circular configuration. This approach aims to accurately recover high-resolution CT images from unmatched low-resolution inputs. In their study, the authors used cycle consistency loss based on the Wasserstein distance to create an end-to-end mapping from noisy low-resolution images to high-resolution outputs. In addition to this, they incorporated joint constraints into the loss function to ensure structural preservation across the two GAN networks. This mechanism enhances the robustness of the GAN network, potentially improving its generalizability for CT imaging applications.

Li et al. [70] introduced a novel approach that leverages micro-CT images to enhance the training of SR networks. Micro-CT imaging, which relies on specialized hardware and system geometry, offers significantly higher spatial resolution compared to conventional CT, capturing intricate details that closely resemble human anatomical structures. However, the major challenge with this approach lies in the difficulty of obtaining paired datasets, which are essential for conventional supervised training methods. The authors proposed an unsupervised training method based on CycleGAN, which effectively handles the unpaired data problem by learning the mapping between the micro-CT images and the target clinical CT images in a bidirectional manner. Furthermore, they incorporated Bayesian inference into their model, allowing them to quantify uncertainty and improve the robustness of their predictions. The combination of CycleGAN and Bayesian inference enabled the model to achieve superior performance, particularly in the imaging of inner ear bone structures, when compared to conventional SR approaches.

Like its applications for image denoising, the diffusion model holds significant potential in SR applications as recently revealed in several natural image generation studies [71, 72]. Chen et al. [73] developed a conditional DDPM for CT image resolution improvement. The low-resolution image was utilized as a condition to perform guided reverse sampling for the high-resolution image. The study showed that the proposed method had better resolution enhancement and less over-smoothing artifacts comparing to conventional supervised learning, yielding better low-contrast details and better preservation of noise texture (see Fig. 5). They also proposed a step-sharing mechanism to largely accelerate the reverse sampling process to mitigate the well-known speed bottleneck in diffusion models.

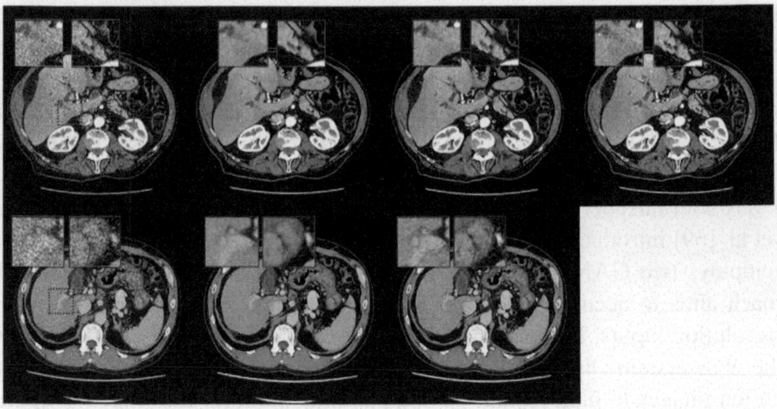

Fig. 5 Clinical CT image resolution enhancement using a conditional DDPM (images courtesy of [73]). Top row, test-on-training results (left to right): high-resolution target image, low-resolution high-noise conditional image, U-Net-based supervised learning result, and the conditional DDPM-generated result. Bottom row, real testing results (left to right): low-resolution clinical image, U-Net-based supervised learning result, and the conditional DDPM-generated result. Zoomed-in views of two selected regions of interest (ROIs) are shown from the areas marked by dotted boxes. Window level/window width: 40/300 HU

4 Discussion and Future Work

We have reviewed deep learning techniques in applications of CT image denoising and super resolution. As we have seen, CNNs have been widely used in these two applications achieving great success, but they also come with inherent limitations. One major drawback is their limited receptive field size, which restricts the network to capture long-range dependencies and complex structures in images. Additionally, the loss functions commonly used in regression-based CNNs can sometimes lead to suboptimal performance. While transformers have emerged as a powerful alternative due to their ability to model global dependencies and improve context understanding, they may require substantial computational resources, making them less practical for many large-scale medical imaging applications.

Diffusion-based generative models have gained a lot of attention recently. Unlike GANs, which have been explored for medical image generation but often face challenges such as difficult training and model collapse, diffusion models have emerged as state-of-the-art techniques. However, diffusion models also have their own challenges. For example, they are typically very slow, as the sampling process requires many iterative steps to produce a high-quality output. Additionally, like other generative models, diffusion models are prone to generating hallucinations and false positives, which can be especially problematic in the medical field, where accuracy is crucial. Several interesting ideas have been proposed to tackle these challenges, including generating uncertainty maps [74] using Bayesian approximations of the Gaussian process [75], applying conformal risk control [76] with diffusion models, and employing the concept of hallucination decomposition [77].

Another emerging category of methods includes deep unrolled models and image diffusion priors. The diffusion prior approach integrates the diffusion model with the data likelihood model to perform posterior sampling, ensuring data consistency and effectively reducing hallucination artifacts. A few pioneering studies, such as [78–81], are highly recommended. However, since these methods are focused on solving inverse problems, which are fundamentally different from traditional denoising and SR tasks, we have not reviewed them in detail here. We plan to explore these techniques in future work, as they represent a promising direction for improving both the accuracy and efficiency of medical image reconstruction.

Ongoing research and development in CT image reconstruction point to a promising future where CT imaging becomes not only safer but also more precise, driven by the continuous advancement of machine learning-based techniques. As these technologies evolve, they have the potential to revolutionize medical diagnostics, providing clinicians with tools that offer clearer insights into patient health, ultimately leading to better treatment outcomes. The synergy between deep learning and medical imaging exemplifies the remarkable progress in healthcare technology, ushering in a new era of diagnostic accuracy and patient safety.

References

1. M. Balda, J. Hornegger, B. Heismann, Ray contribution masks for structure adaptive sinogram filtering. IEEE Trans. Med. Imaging **31**(6), 1228–1239 (2012)
2. A. Manduca, et al., Projection space denoising with bilateral filtering and CT noise modeling for dose reduction in CT. Med. Phys. **36**(11), 4911–4919 (2009)
3. J. Wang et al., Penalized weighted least-squares approach to sinogram noise reduction and image reconstruction for low-dose X-ray computed tomography. IEEE Trans. Med. Imaging **25**(10), 1272–1283 (2006)
4. T.-C. Lee, J. Zhou, Z. Yu, Deep learning based adaptive filtering for projection data noise reduction in x-ray computed tomography, in *15th International Meeting on Fully Three-Dimensional Image Reconstruction in Radiology and Nuclear Medicine*. ed. by S. Matej, S.D. Metzler, vol. 11072 (International Society for Optics and Photonics. SPIE, New York, 2019), p. 110721D
5. N. Yuan, J. Zhou, J. Qi, Half2Half: deep neural network based CT image denoising without independent reference data. Phys. Med. Biol. **65**(21), 215020 (2020)
6. E. Zainulina, A. Chernyavskiy, D.V. Dylov, No-reference denoising of low-dose CT projections, in *2021 IEEE 18th International Symposium on Biomedical Imaging (ISBI)* (2021), pp. 77–81
7. L. Yang, et al., Low-dose CT denoising via sinogram inner-structure transformer. IEEE Trans. Med. Imaging **42**, 910–921 (2023)
8. H. Chen, et al., Low-Dose CT with a residual encoder-decoder convolutional neural network. IEEE Trans. Med. Imaging **36**(12), 2524–2535 (2017)
9. J. Ma, et al., Low-dose computed tomography image restoration using previous normal-dose scan. Med. Phys. **38**(10), 5713–5731 (2011)
10. Q. Xu, et al., Low-dose X-ray CT reconstruction via dictionary learning. IEEE Trans. Med. Imaging **31**(9), 1682–1697 (2012)

11. H. Chen, et al., A low-dose CT via convolutional neural network. Biomed. Opt. Express **8**(2), 679–694 (2017)
12. K. He, et al., Deep residual learning for image recognition, in *Proceedings of the IEEE Computer Society Conference on Computer Vision and Pattern Recognition*, vol. 2016-December (2016)
13. K. Zhang et al., Beyond a Gaussian denoiser: Residual learning of deep CNN for image denoising. IEEE Trans. Image Process. **26**, 3142–3155 (2017)
14. W. Yang, et al., Improving low-dose CT image using residual convolutional network. IEEE Access **5**, 24698–24705 (2017)
15. E. Kang, J. Min, J.C. Ye, A deep convolutional neural network using directional wavelets for low-dose X-ray CT reconstruction. Med. Phys. **44**(10), e360–e375 (2017)
16. E. Kang, et al., Deep convolutional framelet denosing for low-dose CT via wavelet residual network. IEEE Trans. Med. Imaging **37**(6), 1358–1369 (2018)
17. O. Ronneberger, P. Fischer, T. Brox, U-net: Convolutional networks for biomedical image segmentation. in *Lecture Notes in Computer Science (including subseries Lecture Notes in Artificial Intelligence and Lecture Notes in Bioinformatics)*, vol. 9351 (2015)
18. J. Lehtinen, et al., Noise2Noise: learning image restoration without clean data, in *Proceedings of the 35th International Conference on Machine Learning*, ed. by J. Dy, A. Krause, vol. 80. Proceedings of Machine Learning Research (2018), pp. 2965–2974
19. S. Wang, et al., Noise2Noise for denoising photon counting CT images: generating training data from existing scans, in *Medical Imaging 2023: Physics of Medical Imaging*, vol. 12463 (SPIE, New York, 2023), pp. 15–19
20. J. Zhang, et al., CNN and multi-feature extraction based denoising of CT images. Biomed. Signal Process. Control **67**, 102545 (2021)
21. M. Matsuura, et al., Feature-aware deep-learning reconstruction for context-sensitive X-ray computed tomography. IEEE Trans. Radiat. Plasma Med. Sci. **5**(1), 99–107 (2021)
22. D. Wu, et al., A cascaded convolutional neural network for X-ray low-dose CT Image Denoising. *CoRR* abs/1705.04267 (2017)
23. T. Liang, et al., EDCNN: edge enhancement-based densely connected network with compound loss for low-dose CT denoising, in *International Conference on Signal Processing Proceedings, ICSP*, vol. 2020-December (2020)
24. M. Gholizadeh-Ansari, J. Alirezaie, P. Babyn, Deep learning for low-dose CT denoising using perceptual loss and edge detection layer. J. Digit. Imaging **33**(2), 504–515 (2020)
25. Z. Li, et al., Multi-scale feature fusion network for low-dose CT denoising. J. Digit. Imaging **36**(4), 1808–1825 (2023)
26. Y. Zhang, et al., Multi-scale feature aggregation and fusion network with self-supervised multi-level perceptual loss for textures preserving low-dose CT denoising. Phys. Med. Biol. **69**(10), 105003 (2024)
27. A. Luthra, et al., Eformer: edge enhancement based transformer for medical image denoising. arXiv preprint arXiv:2109.08044 (2021)
28. D. Wang, et al., CTformer: convolution-free Token2Token dilated vision transformer for low-dose CT denoising. Phys. Med. Biol. **68**(6), 065012 (2023)
29. H. Li, et al., Transformer with double enhancement for low-dose CT denoising. IEEE J. Biomed. Health Inform. **27**(10), 4660–4671 (2023)
30. J.M. Wolterink, et al., Generative adversarial networks for noise reduction in low-dose CT. IEEE Trans. Med. Imaging **36**(12), 2536–2545 (2017)
31. K. Choi, S. Won Kim, J.S. Lim, Real-time image reconstruction for low-dose CT using deep convolutional generative adversarial networks (GANs), in *Medical Imaging 2018: Physics of Medical Imaging*, vol. 10573 (SPIE, New York, 2018), pp. 790–795
32. Q. Yang, et al., Low-dose CT image denoising using a generative adversarial network with Wasserstein distance and perceptual loss. IEEE Trans. Med. Imaging **37**(6), 1348–1357 (2018)
33. Z. Hu, et al., Artifact correction in low-dose dental CT imaging using Wasserstein generative adversarial networks. Med. Phys. **46**(4), 1686–1696 (2019)

34. X. Yi, P. Babyn, Sharpness-aware low-dose CT denoising using conditional generative adversarial network. J. Digit. Imaging **31**, 655–669 (2018)
35. H. Shan, et al., 3-D convolutional encoder-decoder network for low-dose CT via transfer learning from a 2-D trained network. IEEE Trans. Med. Imaging **37**(6), 1522–1534 (2018)
36. C. You, et al., Structurally-sensitive multi-scale deep neural network for low-dose CT denoising. IEEE Access **6**, 41839–41855 (2018)
37. M. Li, et al., SACNN: Self-attention convolutional neural network for low-dose CT denoising with self-supervised perceptual loss network. IEEE Trans. Med. Imaging **39**(7), 2289–2301 (2020)
38. S. Bera, P.K. Biswas, Noise conscious training of non-local neural network powered by self attentive spectral normalized Markovian patch GAN for low dose CT denoising. IEEE Trans. Med. Imaging **40**(12), 3663–3673 (2021)
39. Z. Huang, et al., DU-GAN: generative adversarial networks with dual-domain U-net-based discriminators for low-dose CT denoising. IEEE Trans. Instrum. Meas. **71**, 1–12 (2022)
40. M. Geng et al., Content-noise complementary learning for medical image denoising. IEEE Trans. Med. Imaging **41**(2), 407–419 (2022)
41. E. Kang, et al., Cycle-consistent adversarial denoising network for multiphase coronary CT angiography. Med. Phys. **46**(2), 550–562 (2019)
42. J. Gu, J.C. Ye, AdaIN-based tunable CycleGAN for efficient unsupervised low-dose CT denoising. IEEE Trans. Comput. Imaging **7**, 73–85 (2021)
43. K. Simonyan, A. Zisserman, Very deep convolutional networks for large-scale image recognition, in *3rd International Conference on Learning Representations, ICLR 2015—Conference Track Proceedings* (2015)
44. A. Vaswani, et al., Attention is all you need, in *Advances in Neural Information Processing Systems*, vol. 2017-December (2017)
45. Z. Zhang, et al., TransCT: Dual-path transformer for low dose computed tomography, in *Medical Image Computing and Computer Assisted Intervention—MICCAI 2021*, ed. by M. de Bruijne, et al. (Springer International Publishing, Berlin, 2021), pp. 55–64
46. I. Goodfellow, et al., Generative adversarial nets, in *Advances in Neural Information Processing Systems*, ed. by Z. Ghahramani, et al., vol. 27 (2014)
47. H. Shan, et al., Competitive performance of a modularized deep neural network compared to commercial algorithms for low-dose CT image reconstruction. Nat. Mach. Intell. **1**(6), 269–276 (2019)
48. J.Y. Zhu, et al., Unpaired image-to-image translation using cycle-consistent adversarial networks, in *Proceedings of the IEEE International Conference on Computer Vision*, vol. 2017-October (2017)
49. J. Ho, A. Jain, P. Abbeel, Denoising diffusion probabilistic models, in *Proceedings of the 34th International Conference on Neural Information Processing Systems* (2020)
50. A. Nichol, P. Dhariwal, Improved denoising diffusion probabilistic models, in *Proceedings of Machine Learning Research*, vol. 139 (2021)
51. Y. Song, et al., Score-based generative modeling through stochastic differential equations, in *International Conference on Learning Representations* (2021)
52. D.P. Kingma, et al., Variational diffusion models, in *Advances in Neural Information Processing Systems*, vol. 26 (2021)
53. Q. Gao, H. Shan, CoCoDiff: a contextual conditional diffusion model for low-dose CT image denoising, in *Developments in X-Ray Tomography XIV*, ed. by B. Müller, G. Wang, vol. 12242 (SPIE, New York, 2022), p. 122420I
54. Q. Gao, et al., CoreDiff: contextual error-modulated generalized diffusion model for low-dose CT denoising and generalization. IEEE Trans. Med. Imaging **43**(2), 745–759 (2024)
55. A. Bansal, et al., Cold diffusion: inverting arbitrary image transforms without noise. Adv. Neural Inf. Proces. Syst. **36** (2022). arXiv:2208.09392
56. W. Xia, Q. Lyu, G. Wang, Low-dose CT using denoising diffusion probabilistic model for 20 times speedup. arXiv preprint arXiv:2209.15136 (2022)

57. J. Hsieh, et al. A new era of image reconstruction: TrueFidelity™. White Paper (JB68676XX) (2019). https://www.gehealthcare.com/jssmedia/040dd213fa89463287155151fdb01922.pdf
58. K. Boedeker. AiCE deep learning reconstruction: Bringing the power of ultra-high resolution CT to routine imaging. White Paper (2019). https://global.medical.canon/publication/ct/2019WP_AiCE_Deep_Learning
59. AI for significantly lower dose and improved image quality. White Paper (2021). https://www.documents.philips.com/assets/20240223/4f2cbfa4d0df428f94bfb11f01751a87.pdf
60. J.H. Kim, Apparatus and method for denoising CT images. US Patent 9,852,527B2 (2015)
61. What's That Noise? White Paper. https://www.algomedica.com/medical-imaging-resources#white-papers
62. C.H. McCollough, P.S. Rajiah, Milestones in CT: Past, present, and future. Radiology **309**(1), e230803 (2023)
63. A. Mansoor, T. Vongkovit, M.G. Linguraru, Adversarial approach to diagnostic quality volumetric image enhancement, in *2018 IEEE 15th International Symposium on Biomedical Imaging (ISBI 2018)* (2018), pp. 353–356
64. C. Ledig, et al., Photo-realistic single image super-resolution Using a generative adversarial network, in *2017 IEEE Conference on Computer Vision and Pattern Recognition (CVPR)* (2017), pp. 105–114
65. H. Yu, et al., Computed tomography super-resolution using convolutional neural networks, in *2017 IEEE International Conference on Image Processing (ICIP)* (2017), pp. 3944–3948
66. K. Umehara, J. Ota, T. Ishida, Application of super resolution convolutional neural network for enhancing image resolution in chest CT. J. Digit. Imaging **31**(4), 441–450 (2018)
67. C. Tang, et al., Generative adversarial network-based sinogram super-resolution for computed tomography imaging. Phys. Med. Biol. **65**(23), 235006 (2020)
68. E.K. Koons, et al., Learned high-resolution cardiac CT imaging from ultra-high-resolution PCD-CT, in *Proceedings of SPIE–the International Society for Optical Engineering*, vol. 12925 (2024), p. 129252Q
69. C. You, et al., CT super-resolution GAN constrained by the identical, residual, and cycle learning ensemble (GAN-CIRCLE). IEEE Trans. Med. Imaging **39**(1), 188–203 (2020)
70. H. Li, et al., Micro-CT synthesis and inner ear super resolution via generative adversarial networks and Bayesian inference, in *2021 IEEE 18th International Symposium on Biomedical Imaging (ISBI)* (2021), pp. 1500–1504
71. C. Saharia, et al., Image super-resolution via iterative refinement. IEEE Trans. Pattern Anal. Mach. Intell. **45**(4), 4713–4726 (2023)
72. R. Rombach, et al., High-resolution image synthesis with latent diffusion models, in *Proceedings of the IEEE/CVF Conference on Computer Vision and Pattern Recognition* (2022), pp. 10684–10695
73. X. Chen, T.-C. Lee, L. Cai, Accelerating diffusion-based CT image restoration through sampling step sharing, in *8th International Conference on Image Formation in X-Ray Computed Tomography* (SPIE, New York, 2024)
74. X. Zhang, et al., Combining physics-based models with deep learning image synthesis and uncertainty in intraoperative cone-beam CT of the brain. Med. Phys. **50**(5), 2607–2624 (2023)
75. Y. Gal, Z. Ghahramani, Dropout as a Bayesian approximation: representing model uncertainty in deep learning, in *Proceedings of the 33rd International Conference on International Conference on Machine Learning—Volume 48*, ICML'16 (2016), pp. 1050–1059
76. J. Teneggi, et al., How to trust your diffusion model: a convex optimization approach to conformal risk control, in *Proceedings of the 40th International Conference on Machine Learning*, ICML'23. JMLR.org (2023)
77. S. Bhadra, et al., On hallucinations in tomographic image reconstruction. IEEE Trans. Med. Imaging **40**(11), 3249–3260 (2021)
78. H.K. Aggarwal, M.P. Mani, M. Jacob, MoDL: Model-based deep learning architecture for inverse problems. IEEE Trans. Med. Imaging **38**(2), 394–405 (2019)
79. V. Monga, Y. Li, Y.C. Eldar, Algorithm unrolling: interpretable, efficient deep learning for signal and image processing. IEEE Signal Process. Mag. **38**(2), 18–44 (2021)

80. Y. Song, et al., Solving inverse problems in medical imaging with score-based generative models, in *International Conference on Learning Representations* (2021)
81. H. Chung, et al., Diffusion posterior sampling for general noisy inverse problems, in *The Eleventh International Conference on Learning Representations* (2023)

Physically Interpretable Deep Learning Reconstruction for Photon Counting Spectral CT

Yuxiang Xing, Hewei Gao, and Le Shen

In recent years, photon-counting detector-based spectral computed tomography (PCD-SCT) has been significantly advanced, with great potentials in medical and industrial applications and many advantages over energy-integrating detectors, such as no electronic noise, higher spatial resolution, better material discrimination, K-edge imaging capability, etc. However, its sophisticated physics and special signal acquisition mechanism make it extremely important to model the PCT-SCT with a careful consideration of the spectral response of the detector.

1 Physics Model for PCD-SCT

Considering cross talk which contains charge sharing and K-escape, and variations in energy thresholds in photon-counting detector, one incident photon will be recorded as more than one signal with lower energies. This effect can be modeled approximately by using detector response function $D(E^{'}; E)$, i.e., the overall probability of a photon with energy E reached on a detector is recorded as a photon of energy $E^{'}$. Because the response function $D(E^{'}; E)$ is generally shift-variant, we would like to denote it as a function of E and $E^{'}$, instead of a function of $E - E^{'}$. Another thing we would like to point out is that this $D(E^{'}; E)$ does not model the cross-pixel charge sharing effect. Thus, the signal collected in an energy bin of a PCD-SCT system can be formulated as:

Y. Xing (✉) · H. Gao · L. Shen
Tsinghua University, Beijing, China
e-mail: xingyx@mail.tsinghua.edu.cn; gaohewei@tsinghua.org.cn; shenle.2007@tsinghua.org.cn

K. (K.) Iniewski, L. (K.) Cai (eds.), *Deep Learning for Advanced X-ray Detection and Imaging Applications*, https://doi.org/10.1007/978-3-031-75653-5_2

$$\begin{aligned}\lambda_i &= \int_{E_L(i)}^{E_H(i)} \mathrm{d}E' \int_0^\infty S_0(E) D\left(E'; E\right) \exp\left(-\int \mu\left(E, \vec{r}\right) \mathrm{d}\vec{r}\right) \mathrm{d}E \\ &= \int_0^\infty \mathrm{d}E\, S_0(E) \exp\left(-\int \mu\left(E, \vec{r}\right) \mathrm{d}\vec{r}\right) \int_{E_L(i)}^{E_H(i)} D\left(E'; E\right) \mathrm{d}E'\end{aligned} \tag{1}$$

where $[E_L(i), E_H(i)]$ defines the ith energy bin, λ_i represents the photon counts in the bin indexed by i, $S_0(E)$ denotes the X-ray source spectrum, and $\mu\left(E, \vec{r}\right)$ is the linear attenuation coefficient with $\vec{r}$ being a vector denoting spatial location. Let $D_i(E) = \int_{E_L(i)}^{E_H(i)} D\left(E'; E\right) \mathrm{d}E'$; we have:

$$\lambda_i = \int_0^\infty \mathrm{d}E\, S_0(E) \exp\left(-\int \mu\left(E, \vec{r}\right) \mathrm{d}\vec{r}\right) D_i(E) \tag{2}$$

In a discretized form, we use $\boldsymbol{\mu}(E) \in \mathbb{R}^N$ to denote an N-dimension attenuation map at a certain energy E, $\mathbf{H} \in \mathbb{R}^{M \times N}$ a discretized projection matrix, $\mathbf{p}(E) = \mathbf{H}\boldsymbol{\mu}(E)$ an M-dimension projection at the energy E, and $\mathbf{n} \in \mathbb{R}^M$ a zero-mean noise vector. Therefore, $\boldsymbol{\lambda}_i \in \mathbb{R}^M$ is a vector of collected counts for an energy bin indexed by i. Accordingly, Eq. (2) becomes

$$\boldsymbol{\lambda}_i = \int_0^\infty \mathrm{d}E\, S_0(E) \exp\left(-\mathbf{p}(E)\right) D_i(E) + \mathbf{n}_i \tag{3}$$

with additive noise. Obviously, $\boldsymbol{\lambda}_i$ is nonlinearly related to $\mathbf{p}(E)$. Generally, we need to pre-calibrate $S_0(E)D_i(E)$ to get a discrete effective detector response matrix $\mathbf{D}^{\text{eff}}$: $D_{i,\tau}^{\text{eff}} = S_0\left(E_\tau\right) D_i\left(E_\tau\right)$, so that

$$\boldsymbol{\lambda}_i = \sum_\tau D_{i,\tau}^{\text{eff}} \exp\left(-\mathbf{p}\left(E_\tau\right)\right) + \mathbf{n}_i \tag{4}$$

with sufficient resolution on E_τ.

2 Framework for Model-Based PCD Spectral CT Reconstruction

Neural networks have demonstrated great power in various applications in all image- and language-related fields. Universal approximation theorem [1, 2] has revealed the magnificent potential of neural networks. Therefore, it is possible to construct a deep learning method with input being acquired PCD-SCT data and output being whatever we need, i.e., we simply rely on a huge-dimensionally parameterized black box of a neural network to solve our PCD-SCT reconstruction problems. However, it often leads to the difficulty to understand the prediction results hence could raise issues about the trustworthy especially in high-risk medical diagnosis applications where one failure might cause irreparable damages. Many efforts have been devoted to the interpretation of neural networks and explaining their behavior in deep learning fields. Incorporating physical model or knowledge in traditional methods is an easy-to-implement solution to improve the interpretability and predictability for deep learning-based PCD-SCT reconstruction.

We start with some notation definitions. Traditional PCD-SCT reconstruction methods are briefly overviewed before elaborating deep learning-based methods.

2.1 Notation Definitions of Quantities in PCD-SCT

The targeted quantities in PCD spectral CT reconstruction could be virtual monochromatic images (VMI), images of effective atomic number distribution, images of electron density distribution, and material component distribution such as contrast image. We denote them by vectors $\boldsymbol{\mu}(E)$, $\mathbf{z}^{\text{eff}}, \boldsymbol{\rho}_{\text{e}}, \mathbf{c}^{\text{Cntr}} \in \mathbb{R}^{N\text{x}1}$, respectively, with N being the number of image pixels. Some intermediate quantities often computed are decomposition coefficients in dual-effect/basis material decomposition denoted by $\mathbf{c} \in \mathbb{R}^N$ and their line integrals denoted by $\mathbf{p}^{\text{c}} \in \mathbb{R}^M$ with M being the dimension of projections, virtual monochromatic sinogram (VMS) denoted by $\mathbf{p}(E) \in \mathbb{R}^M$, and reconstructions from each energy bins denoted by $\boldsymbol{\mu}_i \in \mathbb{R}^N$. One may also compute attenuation ratio between two energies ($\mathbf{r} \in \mathbb{R}^N$) that is a quantity independent to density. Here, we give a summary of the relationship between these quantities, which is very useful to the design of deep learning methods for PCD spectral CT reconstruction.

After spatial discretization, the projection in CT imaging is a linear operation expressed in a matrix-vector format, and one can solve the inverse problem by various methods:

$$\mathbf{p}(E) = \mathbf{H}\boldsymbol{\mu}(E) \Longleftrightarrow \boldsymbol{\mu}(E) = \Phi^{\text{SpaRec}}\left(\mathbf{p}(E)\right) \tag{5}$$

$$\mathbf{p}_i = \mathbf{H}\boldsymbol{\mu}_i \Longleftrightarrow \boldsymbol{\mu}_i = \Phi^{\text{SpaRec}}\left(\mathbf{p}_i\right) \tag{6}$$

$$\mathbf{p}^{\text{c}} = \mathbf{H}\mathbf{c} \Longleftrightarrow \mathbf{c} = \Phi^{\text{SpaRec}}\left(\mathbf{p}^{\text{c}}\right) \tag{7}$$

Here, we use $\Phi^{\text{SpaRec}}()$ to denote a general operator of spatial reconstruction[1] for the inverse problem without limiting it to any specific reconstruction algorithm. For multi-energy or spectral CT, one basic step is to separate energy-dependent and

[1] We use the term "spatial reconstruction" to distinguish it from spectral reconstruction. It is a processor to obtain image of spatial distribution from a sinogram in projection domain. In other words, spatial reconstruction here is same as reconstruction for traditional single-energy CT.

energy-independent factor in linear attenuation coefficient curve of materials. One type of it is referred as basis material decomposition:

$$\boldsymbol{\mu}(E) = \sum_b \mu_b(E)\mathbf{c}_b \tag{8}$$

with $\mu_b(E)$ denoting the linear attenuation coefficients of some chosen materials and $\mathbf{c}_b$ the decomposition coefficient vector for b^{th} basis materials. The density and effective atomic number can be approximately computed by [3]:

$$\boldsymbol{\rho} = \sum_b \rho_b \mathbf{c}_b \tag{9}$$

$$\mathbf{z}^{\text{eff}} = \sqrt[v]{\sum_b \left(z_b^{\text{eff}}\right)^v \text{diag}^{-1}(\boldsymbol{\rho})\, \rho_b \mathbf{c}_b} \tag{10}$$

The other type is referred as basis effect decomposition:

$$\boldsymbol{\mu}(E) = f^{\text{PE}}(E)\mathbf{c}^{\text{PE}} + f^{\text{Compt}}(E)\mathbf{c}^{\text{Compt}} + f^{\text{PP}}(E)\mathbf{c}^{\text{PP}} \tag{11}$$

with $f^{\text{PE}}(E)$, $f^{\text{Compt}}(E)$, and $f^{\text{PP}}(E)$ corresponding to the energy-related photoelectric, Compton, and pair-production cross-section factors, and $\mathbf{c}^{\text{PE}}$, $\mathbf{c}^{\text{Compt}}$, $\mathbf{c}^{\text{PP}}$ the decomposition coefficient vectors, respectively. In the energy level of medical imaging, there are no pair-production effect, and Eq. (11) reduces to two terms. Since $\mathbf{c}^{\text{PE}} \sim \mathbf{Z}^v \rho_e$, $v \approx 3$ and $\mathbf{c}^{\text{Compt}} \sim \rho_e$, we can compute

$$\mathbf{z}^{\text{eff}} = \left(\kappa \text{ diag}^{-1}\left(\mathbf{c}^{\text{Comp}}\right)\mathbf{c}^{\text{PE}}\right)^{\frac{1}{v}} \tag{12}$$

with κ being some constants.

Besides the general basis material terms and basis effect terms, one may also use $\mu^{\text{Cntr}}(E)\mathbf{c}^{\text{Cntr}}$ additionally in Eqs. (8) and (11) for specific contrast imaging, because contrasts have special shapes in their attenuation curve. Thus, we can have:

$$\boldsymbol{\mu}(E) = \sum_b \mu_b(E)\mathbf{c}_b + \mu^{\text{Cntr}}(E)\mathbf{c}^{\text{Cntr}} \tag{13}$$

or

$$\boldsymbol{\mu}(E) = f^{\text{PE}}(E)\mathbf{c}^{\text{PE}} + f^{\text{Compt}}(E)\mathbf{c}^{\text{Compt}} + \mu^{\text{Cntr}}(E)\mathbf{c}^{\text{Cntr}} \tag{14}$$

In material discrimination, the ratio of linear attenuation coefficients can also be useful to remove the effect from density:

$$\mathbf{r}(E_1, E_2) = \text{diag}^{-1}(\boldsymbol{\mu}(E_2))\,\boldsymbol{\mu}(E_1) \tag{15}$$

with E_1 and E_2 being two different energies.

For convenience, we refer the process of separating energy-dependent factors and retrieving material-dependent coefficients as spectral decomposition.

2.2 Roadmaps of Two-Step Reconstruction Methods for PCD Spectral CT

In the literature, we often categorize spectral CT reconstruction according to one-step and two-step. Here, we discuss two-step methods firstly. Two-step methods are categorized by the order of spectral decomposition and spatial reconstruction. Briefly, the category of preprocessing method refers to the methods doing spatial reconstruction after spectral decomposition, and the category of postprocessing methods refers to the methods doing spatial reconstruction first and spectral decomposition afterward. In this section, we give a map of various physical quantities involved in PCD-SCT imaging and reconstruction with their relationship illustrated. Roadmaps of PCD-SCT reconstruction can be obtained by drawing some routes on this map. Different reconstruction methods for PCD-SCT are simply a cascade of the inversion operations in the Eqs. (5, 6, 7, 8, 9, 10, 11, 12, 13, and 14). For example, the orange route in Fig. 1 is a cascade of $\mathbf{p}_i = \Phi^{\mathrm{Cnts} \rightarrow \mathrm{Sino}}(\boldsymbol{\lambda}_i)$, $\boldsymbol{\mu}_i = \Phi^{\mathrm{SpaRec}}(\mathbf{p}_i)$, $\{\mathbf{c}_b, \forall b\} = \Phi^{\mathrm{dcmp} - \mu}(\{\boldsymbol{\mu}_i, \forall i\})$:

$$\psi_1^{\mathrm{SCT-rec}}\left(\{\boldsymbol{\lambda}_i, \forall i\}\right) = \Phi^{\mathrm{dcmp-\mu}} \circ \Phi^{\mathrm{SpaRec}} \circ \Phi^{\mathrm{Cnts \rightarrow Sino}}\left(\{\boldsymbol{\lambda}_i, \forall i\}\right) \tag{16}$$

Here, we use Φ as dummy function for an inversion operation with the superscription "Cnts $\rightarrow$ Sino" being the short for "counts to sinogram" estimation, "SpaRec" for "spatial reconstruction," and "dcmp-μ" for "decomposition in μ domain." The other superscriptions in the context follow similarly.

The red route in Fig. 1 is a cascade of $\mathbf{p}_i = \Phi^{\mathrm{Cnts} \rightarrow \mathrm{Sino}}(\boldsymbol{\lambda}_i)$, $\left\{\mathbf{p}_b^{\mathrm{c}}\right\} = \Phi^{\mathrm{dcmp-p}}\left(\left\{\mathbf{p}_i, \forall i\right\}\right)$,$\mathbf{c}_b = \Phi^{\mathrm{SpaRec}}\left(\mathbf{p}_b^{\mathrm{c}}\right)$:

$$\psi_2^{\mathrm{SCT-rec}}\left(\{\boldsymbol{\lambda}_i, \forall i\}\right) = \Phi^{\mathrm{SpaRec}} \circ \Phi^{\mathrm{dcmp-p}} \circ \Phi^{\mathrm{Cnts \rightarrow Sino}}\left(\{\boldsymbol{\lambda}_i, \forall i\}\right) \tag{17}$$

We also indicate a purple route in Fig. 1 that is a cascade of $\{\mathbf{p}(E)\} = \Phi^{\mathrm{Cnts} \rightarrow \mathrm{MonoSino}}(\{\boldsymbol{\lambda}_i, \forall i\})$, $\boldsymbol{\mu}(E) = \Phi^{\mathrm{SpaRec}}(\mathbf{p}(E))$, $\{\mathbf{c}_b\} = \Phi^{\mu \rightarrow \mathrm{cmp}}(\{\boldsymbol{\mu}(E)\})$.

$$\psi_3^{\mathrm{SCT-rec}}\left(\{\boldsymbol{\lambda}_i, \forall i\}\right) = \Phi^{\mu \rightarrow \mathrm{cmp}} \circ \Phi^{\mathrm{SpaRec}} \circ \Phi^{\mathrm{Cnts \rightarrow MonoSino}}\left(\{\boldsymbol{\lambda}_i, \forall i\}\right) \tag{18}$$

with the superscription "$\mu \rightarrow$ cmp" denoting the decomposition from linear attenuation coefficients to decomposition coefficients of components. Here, $\boldsymbol{\mu}(E)$ at multiple E is needed.

All these operators can be defined analytically or iteratively in the sense of optimization problems. We refer them as traditional operators that are not our focus

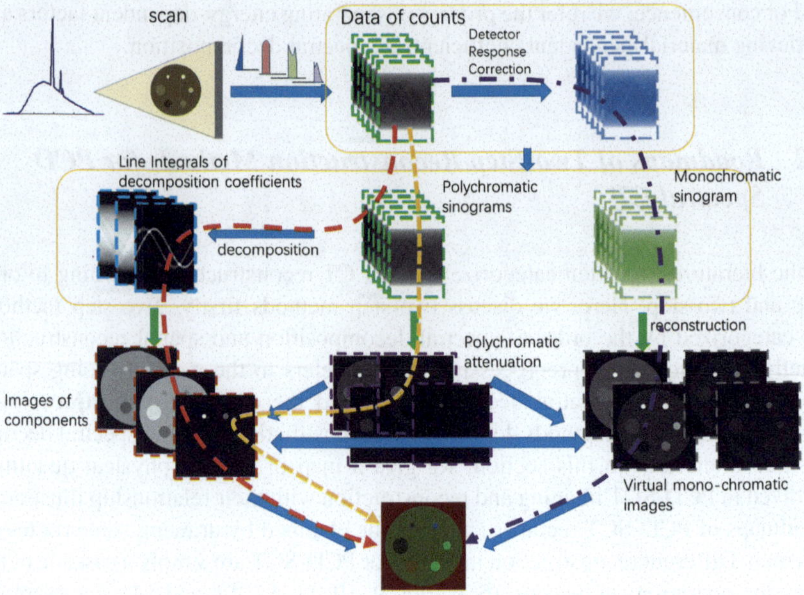

Fig. 1 Roadmap of PCD-SCT reconstruction

here. In the following sections, we focus on realizing these operators using neural networks.

3 Incorporating Deep Learning into Reconstruction Methods for PCD-SCT

One may directly adopt a black box neural network to solve the problem of PCD-SCT. However, knowledge of imaging physics is precious and could help lower error bound of solutions [4]. Here, we try to give some explanation on how to make deep learning methods for PCD-SCT not fully black.

For convenience, we denote a neural network version of these operators with an additional superscript "NN", e.g., $\Phi_{\boldsymbol{\theta}}^{\mathrm{NNdcmp-p}}$ parameterized by $\boldsymbol{\theta}$ to denote the spectral decomposition in projection domain by a neural network. In this way, we can easily construct various reconstruction methods by combining traditional and neural network operators as we will. Neural network operators are extensively studied in this field for all $\Phi_{\boldsymbol{\theta}}^{\mathrm{NNSpaRec}}$, $\Phi_{\boldsymbol{\theta}}^{\mathrm{NNdcmp-p}}$, and $\Phi_{\boldsymbol{\theta}}^{\mathrm{NNdcmp-\mu}}$. Please notice that we haven't specified any neural network architecture or training scheme for these operators.

3.1 Transform from Conventional PCD-SCT Reconstruction to Deep Learning Methods

In this section, we are to demonstrate some of these methods incorporating deep learning neural networks in more detail based on the above skeleton roadmap for PCD-SCT data processing in Sect. 2.

(a) Decompose spectral information in the image domain.

Replacing $\Phi^{\mathrm{dcmp}-\mu}$ in $\psi_1^{\mathrm{SCT-rec}}(\{\boldsymbol{\lambda}_i, \forall i\})$ with a $\Phi_{\boldsymbol{\theta}}^{\mathrm{NNdcmp}-\mu}$ is a straightforward spectral CT reconstruction method:

$$\psi_{1'}^{\mathrm{SCT-rec}}(\{\boldsymbol{\lambda}_i, \forall i\}; \boldsymbol{\theta}) = \Phi_{\boldsymbol{\theta}}^{\mathrm{NNdcmp}-\mu} \circ \Phi^{\mathrm{SpaRec}} \circ \Phi^{\mathrm{Cnts}\rightarrow\mathrm{Sino}}(\{\boldsymbol{\lambda}_i, \forall i\}) \tag{19}$$

Though data from a PCD are from a narrower energy window, it could still be unsatisfying to solve linear equations for spectral decomposition, because reconstructions from a polychromatic sinogram could be affected by beam hardening, charge sharing, etc. A neural network operator being able to catch the nonlinear property is expected to do a good job in removing these nonideal factors. According to our analysis, VMI $\boldsymbol{\mu}(E)$ is nonlinearly related to $\boldsymbol{\mu}_i$ and is a linear function of component coefficient images $\mathbf{c}_b$. Therefore, we can specifically solve the nonlinear relation between $\boldsymbol{\mu}(E)$ and $\boldsymbol{\mu}_i$ by training a neural network:

$$\hat{\boldsymbol{\mu}}(E) = \Phi_{\boldsymbol{\theta}}^{\mu}\left(\boldsymbol{\mu}_1, \cdots, \boldsymbol{\mu}_i, \cdots\right) \tag{20}$$

Feng et al. [5] set up a simple patch-based fully connected (FC) neural network (as shown in Fig. 2) to estimate VMI from polychromatic reconstructions from multiple energy bins. Besides dealing the nonlinearity, the output of the neural network incorporated the local information within its neighbors by using a patch-based neural network, which can help reduce noise and improve robustness. The pixel-wise estimation is:

$$\hat{\mu}_n(E) = \Phi_{\boldsymbol{\theta}}^{\mu}\left(\mathcal{P}_n(\boldsymbol{\mu}_1), \cdots, \mathcal{P}_n(\boldsymbol{\mu}_i), \cdots\right) \tag{21}$$

where $\mathcal{P}_n()$ denotes patch extraction around pixel n. The neural network is trained by a simple loss function of mean square error (MSE) for a train dataset:

$$\mathrm{MSE} = \frac{1}{\#\text{of samples}} \sum_{n \in \mathrm{ROIs}} \left(\hat{\mu}(n) - \mu^*(n)\right)^2 \tag{22}$$

with $\mu^*(n)$ being the label in training dataset.

With nonideal factors removed in VMI, the linear model in Eqs. (13) and (14) can be directly applied to get component coefficient images.

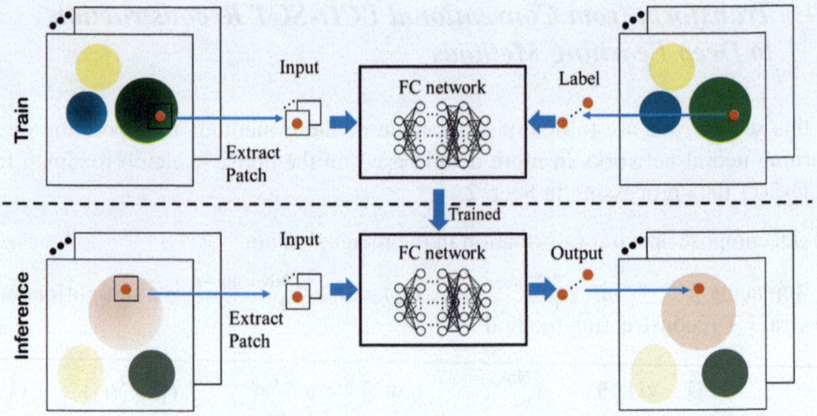

Fig. 2 Illustration of using FC neural network to estimate VMI from multiple non-monochromatic reconstruction images. Different colors are used to indicate different materials. The non-monochromatic images could have beam hardening effect and are noisy. The resulted monochromatic images are relieved from beam hardening effect and of less noise

Image domain decomposition methods for PCD-SCT have been widely explored. Some works directly output component coefficient images by a neural network without explicitly computing VMI, i.e., nonlinear and linear factors are processed as a whole. Since image-domain neural networks for spectral information decomposition are after the spatial reconstruction step, it is independent of scanning geometry, so that it has the convenience in data acquisition, implementation, and deploy.

(b) Decompose spectral information in the sinogram domain.

There are also two ways to decompose spectral information in the sinogram domain. As show in Fig. 1, one way is to compute monochromatic sinogram $\mathbf{p}(E)$ from polychromatic data as in $\psi_3^{\text{SCT-rec}}(\{\boldsymbol{\lambda}_i, \forall i\})$. Replacing $\Phi^{\text{Cnts}\rightarrow\text{MonoSino}}$ in $\psi_3^{\text{SCT-rec}}(\{\boldsymbol{\lambda}_i, \forall i\})$ with a $\Phi_{\boldsymbol{\theta}}^{\text{NNCnts}\rightarrow\text{MonoSino}}$ gives another deep learning-based spectral CT reconstruction method:

$$\psi_{3'}^{\text{SCT-rec}}(\{\boldsymbol{\lambda}_i, \forall i\}) = \Phi^{\mu\rightarrow\text{cmp}} \circ \Phi^{\text{SpaRec}} \circ \Phi_{\boldsymbol{\theta}}^{\text{NNCnts}\rightarrow\text{MonoSino}}(\{\boldsymbol{\lambda}_i, \forall i\}) \tag{23}$$

We can also simply choose to use a neural network denoted by $\Phi_{\boldsymbol{\theta}}^{\text{NNCnts}\rightarrow\text{cmpPrj}}(\{\boldsymbol{\lambda}_i, \forall i\})$ (e.g., PMS-GAN in [6]) to obtain the projection of decomposition coefficients, i.e., replacing $\Phi^{\text{dcmp-p}} \circ \Phi^{\text{Cnts}\rightarrow\text{Sino}}(\{\boldsymbol{\lambda}_i, \forall i\})$ with $\Phi_{\boldsymbol{\theta}}^{\text{NNCnts}\rightarrow\text{cmpPrj}}(\{\boldsymbol{\lambda}_i, \forall i\})$ in Eq. (17):

$$\psi_{2'}^{\text{SCT-rec}}(\{\boldsymbol{\lambda}_i, \forall i\}) = \Phi^{\text{SpaRec}} \circ \Phi_{\boldsymbol{\theta}}^{\text{NNCnts}\rightarrow\text{cmpPrj}}(\{\boldsymbol{\lambda}_i, \forall i\}) \tag{24}$$

Since monochromatic projection data or projection of component coefficients are not directly available, a direct way to get corresponding training labels is by the following four steps: (1) scan some phantoms of known materials using a spectral CT; (2) reconstruct these phantoms; (3) assign the material information to each pixel in the reconstruction; and (4) compute the line integrals of linear attenuation coefficients (or component coefficients) at a specified energy along some ray paths match to detected signals. We can see that this is difficult and costly. Firstly, making physical phantoms with known materials and sizes of great variety is expensive and not easy. Secondly, there will be error in line integral computation when pairing with scanning ray paths of collected signals because of geometrical uncertainty and partial volume effect in reconstructions. Feng et al. proposed a loss defined in image domain though neural network output in sinogram domain [7]:

$$\text{Loss} = \sum_{k} \frac{1}{N_k \mu_k^*(E)} \sum_{n(k)} \left(\hat{\mu}_{n(k)}(E) - \mu_k^*(E)\right)^2, \quad n(k) \in k^{\text{th}} \text{ region of known material} \tag{25}$$

where N_k is the number of pixels which belongs to the kth region of known material with its ground truth attenuation coefficient $\mu_k^*(E)$ in the phantoms that are scanned for training data. An analytical Φ^{SpaRec} is used in Eq. (25) as $\hat{\boldsymbol{\mu}}(E) = \Phi^{\text{SpaRec}}\left(\hat{\mathbf{p}}(E)\right) = \mathbf{H}_{\text{W}}^{\text{T}}\mathbf{F}\mathbf{W}\hat{\mathbf{p}}(E)$ where, if for a fan beam scan, $\mathbf{W}$ weights the projection data, $\mathbf{F}$ is a ramp filter, and $\mathbf{H}_{\text{W}}^{\text{T}}$ is the weighted back projection matrix so that $\mathbf{H}_{\text{W}}^{\text{T}}\mathbf{F}\mathbf{W}$ forms an FBP operator. Hence, the parameters in the sinogram-domain neural network are optimized by:

$$\begin{aligned} \boldsymbol{\theta}(E) = \underset{\boldsymbol{\theta}(E)}{\operatorname{argmin}} \sum_{k} \frac{1}{N_k \mu_k^*(E)} \sum\nolimits_{n(k)} \\ \times \left(\left[\mathbf{H}_{\text{W}}^{\text{T}}\mathbf{F}\mathbf{W}\Phi_{\boldsymbol{\theta}(E)}^{\text{NNCnts}\to\text{MonoSino}}\left(\{\boldsymbol{\lambda}_i, \forall i\}\right) \right]_{n(k)} - \mu_k^*(E) \right)^2 \end{aligned} \tag{26}$$

$$\text{Pixel } n(k) \in k^{\text{th}} \text{ region of known material.}$$

For this loss, it's easy to leverage the pixels completely within the regions of known materials and minimize the effect from partial volume or discretization error. This method of loss definition significantly eases the data preparation from real experiments. For example, it is easy to get different compounds with liquids. Liquids need to be in some a container. If labels of projections are needed, we need to know the exact material composition of containers. Now we can ignore those pixels that belong to containers in Eq. (3). For another example, we may increase data variation by using some known materials surrounded by different unknown materials, which is not usable to projection domain loss but is valuable to image domain loss.

Please note that $\Phi_{\boldsymbol{\theta}}^{\text{NNCnts}\to\text{MonoSino}}$ will be energy dependent; i.e., we will have $\Phi_{\boldsymbol{\theta}(E)}^{\text{NNCnts}\to\text{MonoSino}}$ trained at each specified monochromatic energy E. Of course, partial neural network layers for different E may be shared. Multiple monoenergetic images can be synthesized as needed. The values of energy shall be specified according to clinical tasks. At least two monoenergetic images at appropriately chosen energies are necessary for material distribution estimation afterward. Three monoenergetic images will be needed if with a K-edge contrast.

Another way to decompose spectral information in the sinogram domain can be realized by replacing $\Phi^{\mathrm{dcmp-p}}$ with $\Phi_{\boldsymbol{\theta}}^{\mathrm{NNdcmp-p}}$ in $\psi_2^{\mathrm{SCT-rec}}(\{\boldsymbol{\lambda}_i, \forall i\})$, so that we have

$$\psi_{2'}^{\mathrm{SCT-rec}}(\{\boldsymbol{\lambda}_i, \forall i\}) = \Phi^{\mathrm{SpaRec}} \circ \Phi_{\boldsymbol{\theta}}^{\mathrm{NNdcmp-p}} \circ \Phi^{\mathrm{Cnts}\to\mathrm{Sino}}(\{\boldsymbol{\lambda}_i, \forall i\}) \tag{27}$$

To supervise the training of $\Phi_{\boldsymbol{\theta}}^{\mathrm{NNdcmp-p}}$, we will need the ground truth of $\mathbf{p}^{\mathrm{c}}$ that could be available but is not easy to get in reality. Similar to Eqs. (25) and (26), we can use a loss defined in the domain of decomposition coefficient images to optimize:

$$\boldsymbol{\theta} = \underset{\boldsymbol{\theta}}{\operatorname{argmin}} \sum_b \beta_b \sum_k \frac{1}{N_k c_{k,b}^*} \sum\nolimits_{n(k)} \left(\left[\mathbf{H}_{\mathrm{W}}^{\mathrm{T}} \mathbf{F} \mathbf{W} \Phi_{\boldsymbol{\theta}}^{\mathrm{NNdcmp-p}}(\{\boldsymbol{\lambda}_i, \forall i\}) \right]_{n(k),b} - c_{k,b}^* \right)^2 \tag{28}$$

Pixel $n(k) \in k^{\mathrm{th}}$ region of known material.

with β_b being a hyperparameter adjust balancing the error in image of decomposition coefficients corresponding to b^{th} basis-component, and $c_{k,b}^*$ the ground truth.

(c) Decouple spectral information in the domain of counts or pre-log domain.

An ideal detector for spectral CT is one that can record each photon with its exact energy. Decoupling spectral information in the domain of counts is in the sense of recovering ideal response from practical detector response. Mathematically, the difference between dealing counts and projections is only the logarithm. Therefore, the operator can be formulated as:

$$\mathbf{p}(E) = \Phi^{\mathrm{Cnts}\to\mathrm{MonoSino}}(\{\boldsymbol{\lambda}_i, \forall i\}) = \log\left(\Phi^{\mathrm{Cnts}\to\mathrm{MonoCnts}}(\{\boldsymbol{\lambda}_i, \forall i\})\right) + C \tag{29}$$

with log() being an elementwise logarithm and C a constant related to blank scan. Everything else follows very straightforwardly. If we adopt a neural network version of $\Phi_{\boldsymbol{\theta}}^{\mathrm{NNCnts}\to\mathrm{MonoCnts}}(\{\boldsymbol{\lambda}_i\})$, the difficult is in how to train the neural network $\Phi_{\boldsymbol{\theta}}^{\mathrm{NNCnts}\to\mathrm{MonoCnts}}$, since it is impossible to obtain the ground truth of counts in a specific mono-energy in reality, thus not possible to do a supervised training. Hence, we need to use unsupervised learning or supervise it in another domain. Some tricks similar to Sect. II.B.2 can be used. We will discuss these in the next subsection about the loss functions for spectral CT deep reconstruction in more detail.

To summarize this section II.B, the basic idea in the above straightforward roadmap to incorporate neural networks into PCD-SCT reconstruction is using neural networks to solve the nonlinear subproblems in the data processing of PCD-SCT that could be difficult by traditional methods, so that superior performance can be expected. The rest linear operations can be easily done with traditional methods. According to this principle, we are able to construct all kinds of interpretable deep learning methods, train end-to-end or independently for each sub-network, as well as execute multiple-domain learning as [8] did.

(d) Specific deep learning tasks in PCD-CT to improve reconstruction.

In last subsection, we gave a rather general roadmap for PCD-CT reconstruction. In fact, we can obtain reasonable images using non-deep learning methods if data acquisition condition is rather ideal, i.e., if we have high counts data with low noise, sufficient views covering full object with an ideal high energy-resolution PCD. Nowadays, deep learning methods are mostly leveraged to tackle problems that are hard for traditional methods or to further improve results obtained in traditional methods. These tasks of hot topics including denoising for ultralow dose scans, sparse view reconstruction, artifact reduction, etc. These challenging tasks are not necessarily PCD-SCT but are more complicated when involving PCD and multiple energy channels. We give two exemplary tasks here.

Firstly, denoising is one typical task extensively studied in the field. Generally, deep learning neural networks favor smooth results. According to our discussion in the last subsection, no matter we decouple spectral information with deep learning methods in image domain, sinogram domain, or pre-log count domain, we will have results denoised to some extent, though further improvement might be needed. Based on the roadmap in Sect. 2.2, we know we can have an additional denoising neural network $\Phi_{\boldsymbol{\theta}}^{\text{NNdenoiseCnts}}$ for detected count data cascaded before synthesizing monochromatic projections,

$$\Phi'^{\text{NNCnts}\to\text{MonoCnts}}_{\boldsymbol{\theta}}(\{\boldsymbol{\lambda}_i, \forall i\}) = \Phi_{\boldsymbol{\theta}}^{\text{NNCnts}\to\text{MonoCnts}} \circ \Phi_{\boldsymbol{\theta}}^{\text{NNdenoiseCnts}}(\{\boldsymbol{\lambda}_i, \forall i\}) \tag{30}$$

or have a denoising neural network to post-process material coefficient images:

$$\Phi_{\boldsymbol{\theta}}^{\text{NNdenoiseCmp}} \circ \Phi^{\mu\to\text{cmp}}(\{\boldsymbol{\mu}(E)\}) \tag{31}$$

or have them both so that form a rather complicate overall algorithm:

$$\begin{aligned}\{\mathbf{c}_b, \forall b\} =& \Phi_{\boldsymbol{\theta}}^{\text{NNdenoiseCmp}} \circ \Phi^{\mu\to\text{cmp}} \circ \Phi^{\text{SpaRec}} \circ \Phi_{\boldsymbol{\theta}}^{\text{NNCnts}\to\text{MonoCnts}} \\ & \circ \Phi_{\boldsymbol{\theta}}^{\text{NNdenoiseCnts}}(\{\boldsymbol{\lambda}_i, \forall i\})\end{aligned} \tag{32}$$

For PCD-SCT with multichannel data different from a single-energy CT, the correlated information between energy bins are often explored to reduce noise by a synergistic fashion, i.e., other data channels can participate in the processing of the targeted one. A representative work is in [Chang 2022] where VMI at 70 keV is used as a prior for its low-noise character to help denoising VMI at 50 keV.

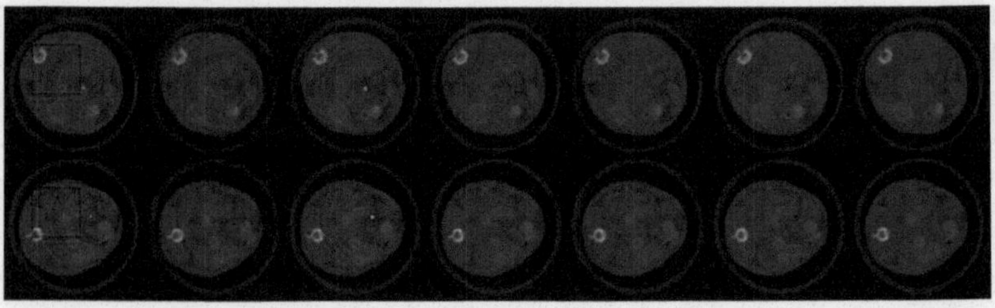

Fig. 3 Two cases of ring artifact reduction for PCD-SCT, courtesy of Wei Fang. From left to right: original images with artifact, artifact reduced by wavelet-Fourier filtering (WF) method in projection domain, WF method in polar coordinates, U-Net in image domain, U-Net in projection domain, U-Net in polar coordinates, and U-Net in both projection and image domains (comprehensive). Display window: [0, 0.46] [9]

Secondly, inhomogeneity for a PCD system is often more complex than for energy integrating detectors. It leads to streaks in sinograms and rings in reconstructions. Inhomogeneity among PCD detector pixels can be caused by variation in the energy bin thresholds or temperature changes during imaging acquisition time. We may follow the roadmap similar to denoising to restrain ring artifacts from detector inhomogeneity. Fang proposed a comprehensive model for deep learning that combined artifact learning in both sinogram domain and image domain [9]. The resulted artifacts from sinogram domain learning and image domain learning are averaged to give final output. It effectively removed ring artifacts in PCD-SCT images as shown in Fig. 3. Moreover, the performance of comprehensive model is much better than single-domain learning and traditional methods.

3.2 *Loss Functions for Deep Learning Methods in Spectral CT*

Loss function setting is one of the key factors in all deep learning methods. Various loss functions have been proposed and applied in CT reconstruction. Analogy to objective function in traditional iterative reconstructions, all kinds of terms can be included in loss functions (aka cost functions), such as fidelities, penalties, and regularizations that contribute to the objective of neural network training. Here, we firstly overview some basic and commonly used loss functions for supervised learning and then elaborate on feature space losses and cross-domain losses afterward.

A. Classic Loss Functions

Currently, most commonly used loss functions are l_p-norms (sometimes normalized, e.g., relative root mean square error), SSIM, PSNR, etc. in supervised learning that requires the accessibility of ground truths or labels:

$$\mathcal{L}^{p-\text{norm}}\left(\hat{\mathbf{x}}, \mathbf{x}^*\right) = \left\|\hat{\mathbf{x}} - \mathbf{x}^*\right\|_p \tag{33}$$

$$\mathcal{L}^{\mathrm{SSIM}}\left(\hat{\mathbf{x}},\mathbf{x}^*\right)=\frac{\left(2\overline{\hat{\mathbf{x}}}\overline{\mathbf{x}^*}+C_1\right)\left(2\sigma_{\hat{\mathbf{x}}\mathbf{x}^*}+C_2\right)}{\left(\overline{\hat{\mathbf{x}}}^2+\overline{\mathbf{x}^*}^2+C_1\right)\left(\sigma_{\hat{\mathbf{x}}}^2+\sigma_{\mathbf{x}^*}^2+C_2\right)} \tag{34}$$

where $\hat{\mathbf{x}}$ and $\mathbf{x}^*$ are the dummy vectors representing the output of a neural network and corresponding ground truth. For SSIM loss, $\overline{\hat{\mathbf{x}}}$ and $\overline{\mathbf{x}^*}$ are means of $\hat{\mathbf{x}}$ and $\mathbf{x}^*$, respectively, $\sigma_{\hat{\mathbf{x}}}$ and $\sigma_{\mathbf{x}^*}$ are standard deviations of $\hat{\mathbf{x}}$ and $\mathbf{x}^*$, and $\sigma_{\hat{\mathbf{x}}\mathbf{x}^*}$ is covariance. $C_1=(k_1R)^2$ and $C_2=(k_2R)^2$ with R denoting the dynamic range of pixel intensity and $k_1=0.01$ and $k_2=0.03$ in general. These losses can be straightforwardly adopted in PCD-SCT reconstruction neural networks.

B. Feature Space Loss

Here, we discuss some indirect ways for loss definition. Handcrafted features such as sharpness metric based on local binary patterns [10] have been adopted to explore common features between images. By using some neural network feature extractor, we can obtain a loss written as:

$$\mathcal{L}\left(\hat{\mathbf{x}},\mathbf{x}^*\right)=\left\|\varphi_{\boldsymbol{\theta}}\left(\hat{\mathbf{x}}\right)-\varphi_{\boldsymbol{\theta}}\left(\mathbf{x}^*\right)\right\|_2^2 \tag{35}$$

with $\varphi_{\boldsymbol{\theta}}$ being a neural network feature extractor parameterized by $\boldsymbol{\theta}$. Perceptual loss [11] is one type of feature space losses with the feature extractor defined by a VGG neural network for classification task on ImageNet dataset. It is well known in the computer vision field and been introduced to CT imaging [12]. This idea can be extended. In 2022, we proposed a feature space loss tailored for CT (CTIS):

$$\mathcal{L}^{\mathrm{CTIS}}\left(\hat{\mathbf{x}},\mathbf{x}^*\right)=\left\|\varphi_{\boldsymbol{\theta}}^{\mathrm{CT}}\left(\hat{\mathbf{x}}\right)-\varphi_{\boldsymbol{\theta}}^{\mathrm{CT}}\left(\mathbf{x}^*\right)\right\|_2^2 \tag{36}$$

with the feature extractor $\varphi_{\boldsymbol{\theta}}^{\mathrm{CT}}$ defined by training an encoder-decoder using high-quality CT dataset:

$$\mathcal{E}^{\mathrm{CT}},\mathcal{D}^{\mathrm{CT}}=\underset{\mathcal{E},\mathcal{D}}{\operatorname{aragmin}}\sum\nolimits_i\left\|\boldsymbol{\mu}_i-\mathcal{D}\left(\mathcal{E}\left(\boldsymbol{\mu}_i\right)\right)\right\|_2^2 \tag{37}$$

here, $\mathcal{E}$ and $\mathcal{D}$ represent an encoder and decoder, respectively, and $\boldsymbol{\mu}_i$ are high-quality CT images in a dataset. The neural network architecture of $\varphi_{\boldsymbol{\theta}}$ can be users' choice. An easy setting is the left part (encoder) of auto-encoder as shown in Fig. 4. $\varphi_{\boldsymbol{\theta}}^{\mathrm{CT}}\left(\hat{\mathbf{x}}\right)$ is basically composed of outputs of multiple layers in the encoder when inputting $\hat{\mathbf{x}}$. Of course, one may choose whether including some layer outputs or not according to tasks. A natural choice of the feature would be the output of encoder (shown by the red dash arrow in Fig. 4) that contains rather complete information of an image. If using same neural network architecture, the difference between feature extractor $\varphi_{\boldsymbol{\theta}}^{\mathrm{CT}}\left(\hat{\mathbf{x}}\right)$ and $\varphi_{\boldsymbol{\theta}}^{\mathrm{percept}}\left(\hat{\mathbf{x}}\right)$ is only in the neural network parameters, so that $\varphi_{\boldsymbol{\theta}}^{\mathrm{CT}}\left(\hat{\mathbf{x}}\right)$ is tailored for CT images, but $\varphi_{\boldsymbol{\theta}}^{\mathrm{percept}}\left(\hat{\mathbf{x}}\right)$ is for general visible light images.

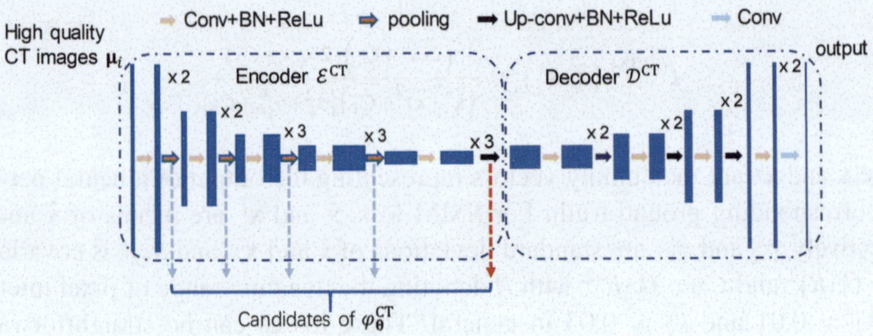

Fig. 4 The features defined by an encoder trained for CT images

An intuitive understanding of these losses is that l_p-norm and SSIM can be thinking of as low-level feature space losses that are of no difference to different domain or tasks, but $\varphi_{\boldsymbol{\theta}}^{\mathrm{CT}}(\hat{\mathbf{x}})$ and $\varphi_{\boldsymbol{\theta}}^{\mathrm{percept}}(\hat{\mathbf{x}})$ extract high-level features that are more of broad and variational sematic information. Semantically meaningful features could be different for different tasks. Because perceptual loss is from classification task training, it focuses on higher-level features not fully accord to our intended task. CTIS attends to image restoration and catch the characteristics of high-quality CT images than the perceptual loss better.

[13] also studied the effect of feature loss with $\varphi_{\boldsymbol{\theta}}()$ parameterized by random values (denoted as $\varphi_{\boldsymbol{\theta}}^{\mathrm{Rnd}}(\hat{\mathbf{x}})$), i.e., not trained by any task. Interestingly, it is found that the performance of $\varphi_{\boldsymbol{\theta}}^{\mathrm{Rnd}}(\hat{\mathbf{x}})$ is worse than $\varphi_{\boldsymbol{\theta}}^{\mathrm{CTIS}}(\hat{\mathbf{x}})$ but better than $\varphi_{\boldsymbol{\theta}}^{\mathrm{percept}}(\hat{\mathbf{x}})$. This tells us that feature space is useful strategy for deep learning supervision even with a random feature extractor.

These high-level feature space losses could be very valuable as a complementary knowledge to help deep learning neural network training. They can also be used by computing distance between results from random samples and alleviate the need of labels. However, we do need to carefully adjust their contributions.

C. Loss from Regularization Functions of Priors

A statistical interpretation of learning from training data is that learned solution represents a Bayes estimator that minimizes sampled mean error from a real distribution. Data distribution is expressed in training samples. Priors commonly used in MBIR promote desired properties including smoothness and edge-preserving by Huber prior, sparsity such as TV [14, 15] and dictionary-learning [16], etc. They are another source of information about probability distribution of images from common sense. Therefore, handcrafted priors can serve as additional knowledge of data/image distribution in the format of potential energy functions or regularization functions. Obviously, they can be easily included as a component of the loss function in training neural networks as:

$$\mathcal{L}^{\mathrm{prior}}(\hat{\mathbf{x}}) = \mathcal{R}(\hat{\mathbf{x}}) \tag{38}$$

Another valuable type of priors we would like to mention is learned priors relied on a neural network defined "manifold" for plausible images [17]. The corresponding potential energy function can be written as the distance from a plausible version after a neural network process:

$$\mathcal{L}^{\mathrm{manifold}}\left(\hat{\mathbf{x}}\right) = \left\|\hat{\mathbf{x}} - \mathcal{F}_{\boldsymbol{\theta}}\left(\hat{\mathbf{x}}\right)\right\|_p \tag{39}$$

with $\mathcal{F}_{\boldsymbol{\theta}}()$ being a neural network normally pretrained that can be an encode-decoder, a denoising neural network, and other generative machine learning models [18]. This $\mathcal{L}^{\mathrm{manifold}}\left(\hat{\mathbf{x}}\right)$ gives us a mechanism to incorporate information by learning from a separate valuable dataset or a related task.

Nowadays, generative adversarial network (GAN) has been extensively studied and adopted in wide imaging applications. It is a combination of a generative model and a discriminator. We can also treat an adversarial loss with a discriminator as a high-level feature loss, though the discriminator is not predetermined but trained along with the mainstream network. Since it is a very broad topic and out of the scope of this chapter, we would like refer readers to [18, 19].

Both the commonly used *lp*-norm, SSIM losses and feature space losses require paired labels (except adversarial loss), i.e., ground truth. Ground truth is normally only available in studies of numerical or Monte Carlo simulations. It is expensive or even unavailable in practical situations. Losses of potential energy functions from prior knowledge do not require labels, which gives it a big advantage for practical usage.

D. Transfer Loss Definition to Other Domains

Another way to alleviate the heavy demanding on labels in supervised learning is to shift loss definition to other domains that require no label or have easily accessible labels. That's also the key for self-supervised learning or unsupervised learning.

For CT reconstruction, data fidelity has served as the major cost function in model-based iterative reconstruction (MBIR) [20]. For a PCD-SCT, the data fidelity term with noise statistics modeled by Poisson is written as:

$$\mathcal{L}_i^{\mathrm{Fid}}\left(\boldsymbol{\lambda}_i, \hat{\boldsymbol{\lambda}}_i\right) = \left(\boldsymbol{\lambda}_i \bullet \log \hat{\boldsymbol{\lambda}}_i - \hat{\boldsymbol{\lambda}}_i\right)_\Sigma \tag{40}$$

with $\hat{\boldsymbol{\lambda}}_i$ denoting the estimation (expected measurement) of count detection in a PCD-CT for each energy bin, "$\bullet$" the Hadamard product, and $()_\Sigma$ the element summation. When used in training neural networks of deep learning PCD-SCT reconstruction, the fidelity loss becomes:

$$\mathcal{L}_i^{\mathrm{Fid}}\left(\boldsymbol{\lambda}_i, \hat{\boldsymbol{\lambda}}_i\left(\boldsymbol{\theta}\right)\right) = \left(\boldsymbol{\lambda}_i \bullet \log \hat{\boldsymbol{\lambda}}_i\left(\boldsymbol{\theta}\right) - \hat{\boldsymbol{\lambda}}_i\left(\boldsymbol{\theta}\right)\right)_\Sigma \tag{41}$$

where $\hat{\boldsymbol{\lambda}}_i\left(\boldsymbol{\theta}\right) = \sum_\tau s_{i,\tau} \exp\left(-\mathbf{H}\sum_b \mu_b\left(E_\tau\right) \Phi_{\boldsymbol{\theta}}^{\mathrm{NNdcmp}-\mu}\left(\left\{\boldsymbol{\mu}_i\right\}\right)\right)$ if $\Phi_{\boldsymbol{\theta}}^{\mathrm{NNdcmp}-\mu}$ is the neural network to be trained to output decomposition coefficients of basis materials. This is simply done by including the forward physical model in Eq. (3). If we want to apply another loss term from prior in another domain, e.g., $\mu(E_\tau)$, the additional loss would be:

$$\mathcal{L}_i^{\text{Prior}}(\boldsymbol{\theta}) = \mathcal{R}\left(\sum_b \mu_b(E_\tau)\,\Phi_{\boldsymbol{\theta}}^{\text{NNdcmp}-\mu}(\{\boldsymbol{\mu}_i\})\right) \tag{42}$$

here, $\mathcal{R}$ can be any regularization function.

Transferring loss definition to domains other than the domain of network output gives big convenience and flexibility to deep learning reconstruction especially in the problem of long track of physics like PCD-CT. As we talked in Section (b), there might be no ground truth available in the targeted output domain, so that computing loss in image domain where we have specific knowledge could be easily done. As long as we have forward physical models, we are able to include loss term in all data domain, so that we are able to exhaustively exploit all available information in network training. An extreme case of combined overall loss can be:

$$\mathcal{L}^{\text{all}} = \sum_i \beta_i \mathcal{L}_i^{\text{Fid}} + \beta^{\text{Prior}}\mathcal{L}^{\text{Prior}} + \beta^{\text{p}}\mathcal{L}^{p-\text{norm}} + \beta^{\text{SSIM}}\mathcal{L}^{\text{SSIM}} + \beta^{\text{CTIS}}\mathcal{L}^{\text{CTIS}} \tag{43}$$

with hyperparameters β_i, β^{Prior}, β^{p}, $\cdots$ balance the contribution of each term to optimize performance. The losses $\mathcal{L}^{\text{Prior}}$, $\mathcal{L}^{p-\text{norm}}$, $\mathcal{L}^{\text{SSIM}}$, and $\mathcal{L}^{\text{CTIS}}$ all can be in attenuation image domain or component coefficient image domain. As we can see, using the first two terms only would form a self-supervised learning, because no ground truth of output is required. The learning with $\mathcal{L}^{\text{all}}$ in Eq. (43) forms a semisupervised learning if some pixel values of some objects but not all are available in the training dataset.

Besides these types of losses, much more losses have been proposed such as attention loss to inject visual attention knowledge [21], identity, or consistent loss [22] for the cyclic relationship if a cycle GAN is adopted. In most deep learning works, multiple losses are combined to form an overall hybrid loss with some hyperparameter weighting.

3.3 Network Architecture Design for PCD-CT

Network architecture is another key factor affecting the performance of deep learning methods. In these days, more and more complicate neural network has been proposed and studied. Popular backbone networks we have seen in the literature are U-Net [23], ResNet [24], and DenseNet [25, 26]. Among them, U-Net-type architecture is still the most popularly used backbone in CT reconstruction field, though transformer has caught a lot of attention because of its big success in natural language processing (NLP). Some published works combine different network blocks to fit their tasks [27, 28]. In this chapter, we are not to address the state-of-art network architectures but present a customized explainable design idea for PCD-SCT based on basic convolutional layers.

For a PCD-SCT, the critical step that is beyond the spatial reconstruction of a traditional CT is spectral information decomposition or decoupling. In this section, we focus on the subnetwork design for processing input photons from multiple energy bins to output projections of monochromatic attenuation map. The design is based on a mathematical analysis of the physical model. For notation convenience, here we consider one ray only.

Considering the photons arrived at the surface of a detector, the photon counts within a discretized energy window could be written as:

$$s_j^{\text{ideal}} = \int_{E_j}^{E_{j+1}} S_0(E) \exp\left(-p(E)\right) \mathrm{d}E \tag{44}$$

here, we use j to index the discretized energy widow. The relationship of this ideal quantity and the photon counts detected by a realistic detector with response function $D_i(E)$ can be reorganized from Eq. (3) to summation of segments as:

$$\lambda_i = \sum_j \int_{E_j}^{E_{j+1}} S_0(E) \exp\left(-p(E)\right) D_i(E) \mathrm{d}E$$
$$= \sum_j \underbrace{\int_{E_j}^{E_{j+1}} S_0(E) \exp\left(-p(E)\right) \mathrm{d}E}_{s_j^{\text{ideal}}} \underbrace{\left(\int_{E_j}^{E_{j+1}} \frac{S_0(E) \exp\left(-p(E)\right)}{\int_{E_j}^{E_{j+1}} S_0(E) \exp\left(-p(E)\right) \mathrm{d}E} D_i(E) \mathrm{d}E \right)}_{D_{ij}} \tag{45}$$

The second equation of Eq. (45) is just a manipulation of the expression to make the spectrum in a normalized format of $\frac{S_0(E)\exp(-p(E))}{\int_{E_j}^{E_{j+1}} S_0(E)\exp(-p(E))\mathrm{d}E}$. We denote the second part as D_{ij} denoting the discrete detector response corresponding to an energy window setting. Obviously, it is a nonlinear function dependent on $D_i(E)$ and the local spectrum after passing a scanned object. We can extract the arithmetic mean of the windowed detector response function computed by $[\underline{\mathbf{D}}_0]_{i,j} \stackrel{\text{def}}{=} \frac{1}{E_{j+1}-E_j} \int_{E_j}^{E_{j+1}} D_i(E) \mathrm{d}E$, and Eq. (2.45) can be further transform in matrix format as:

$$\lambda_i = \sum_j s_j^{\text{ideal}} \left([\underline{\mathbf{D}}_0]_{i,j} + \int_{E_j}^{E_{j+1}} \left(\frac{S_0(E)\exp(-p(E))}{\int_{E_j}^{E_{j+1}} S_0(E)\exp(-p(E))\mathrm{d}E} - \frac{1}{E_{j+1}-E_j} \right) D_i(E) \mathrm{d}E \right) \tag{46}$$

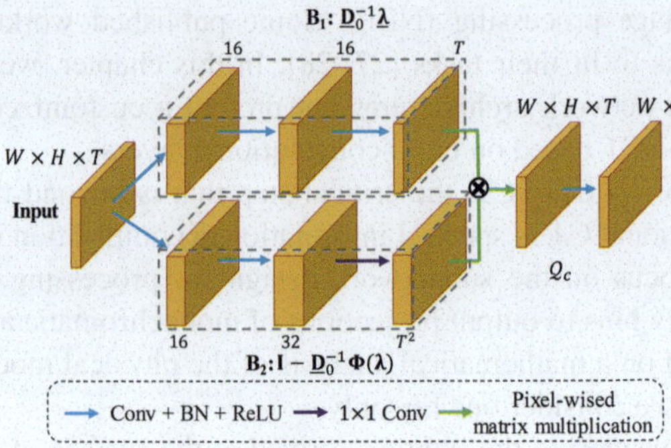

Fig. 5 Flow chart of spectral information decomposition network WeaveNet. The blocks $\mathcal{B}_1$ and $\mathcal{B}_2$ are multiplied together to follow the mathematical interpretation in Eq. (48) and $\mathcal{Q}_c$ is for a cascaded step to obtain monochromatic projections. T represents the number of energy bins

Locally, the normalized spectrum shall be close to its mean, so that the second term in the bracket is a small quantity of residual photons after mean extraction. For convenience, we further express it in a matrix format as:

$$\underline{\boldsymbol{\lambda}} = \left(\underline{\mathbf{D}}_0 + \varphi\left(\Delta\underline{\boldsymbol{\lambda}}\right)\right)\underline{\mathbf{S}}^{\text{ideal}} \tag{47}$$

Here, the second term $\varphi\left(\Delta\underline{\boldsymbol{\lambda}}\right)$ models the residual after mean extracted that is related to the detected spectrum. The vector is in the dimension of energy bins. The underline is to distinguish it from the spatial dimensional vectors. Hence, a simple solution for $\underline{\mathbf{S}}^{\text{ideal}}$ can be:

$$\begin{aligned}\underline{\mathbf{S}}^{\text{ideal}} &= \left(\left(\underline{\mathbf{D}}_0 + \varphi\left(\Delta\underline{\boldsymbol{\lambda}}\right)\right)^T \boldsymbol{\Sigma}^{-1}\left(\underline{\mathbf{D}}_0 + \varphi\left(\Delta\underline{\boldsymbol{\lambda}}\right)\right)\right)^{-1} \cdot \left(\underline{\mathbf{D}}_0 + \varphi\left(\Delta\underline{\boldsymbol{\lambda}}\right)\right)^T \boldsymbol{\Sigma}^{-1} \cdot \underline{\boldsymbol{\lambda}} \\ &\approx \underbrace{\left(\mathbf{I} - \underline{\mathbf{D}}_0^{-1}\boldsymbol{\Phi}\left(\underline{\boldsymbol{\lambda}}\right)\right)}_{\mathcal{B}_2}\underbrace{\underline{\mathbf{D}}_0^{-1}\underline{\boldsymbol{\lambda}}}_{\mathcal{B}_1}\end{aligned} \tag{48}$$

where $\boldsymbol{\Phi}\left(\underline{\boldsymbol{\lambda}}\right)$ is a matrix of local variation of φ respecting to $\underline{\boldsymbol{\lambda}}$, and the approximation in the second line is according to Taylor expansion. After $\underline{\mathbf{S}}^{\text{ideal}}$ obtained, it is easy to compute monochromatic projections. We denote the operation of computing projection at energy E from as $\underline{\mathbf{S}}^{\text{ideal}}$ as $\mathcal{Q}_c$:

$$p(E) = \mathcal{Q}_c \circ \underline{\mathbf{S}}^{\text{ideal}} = \mathcal{Q}_c \circ \left(\mathbf{I} - \underline{\mathbf{D}}_0^{-1}\boldsymbol{\Phi}\left(\underline{\boldsymbol{\lambda}}\right)\right)\underline{\mathbf{D}}_0^{-1}\underline{\boldsymbol{\lambda}} \tag{49}$$

for a specific energy E.

This gives a clue to construct a neural network for solving the problem. According to Eq. (48), we can set up a neural network as shown in Fig. 5. We simply chose convolutional layers for blocks $\mathcal{B}_1$, $\mathcal{B}_2$, and $\mathcal{Q}_c$ and call this architecture as WeaveNet.

This WeaveNet of interpretable architecture performs much better than a plain CNN conditioned on similar number of parameters as shown in the results in Fig. 6

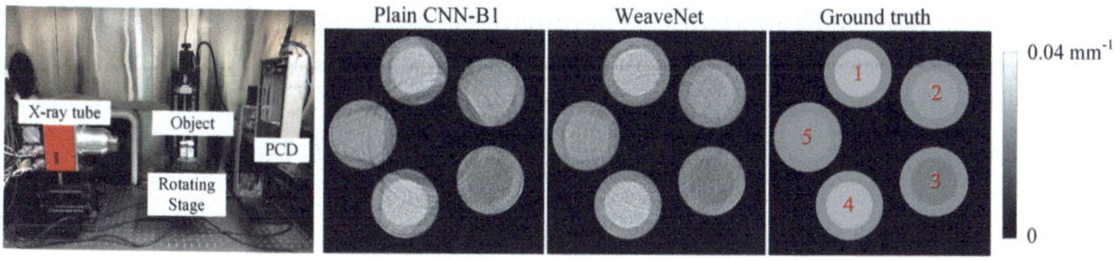

Fig. 6 [29]. Left, a laboratory PCD-SCT system with a micro-focus X-ray source and a photon counting detector; right, reconstructions of experimental data from the laboratory PCD-SCT by WeaveNet and Plain CNN-B1. ROI 1-5 were 25% NaCl (aq), 15% NaCl (aq), H_2O, 10% $CuSO_4$ (aq), and 5% $CuSO_4$ (aq)

on an experimental study on a laboratory PCD-SCT with micro-focus source and an XCounter PCD. This design of WeaveNet tells us that it is possible to incorporate physical models in the designing of network architecture to improve the learning potential of deep learning methods.

3.4 Datasets

As we know that deep learning is data-driven and heavily sometimes overly dependent on the quantity and quality of training data. Network architecture, loss function, and dataset together form a triangular support to obtain reliable performance for a deep learning method. In PCD-SCT, it is generally not easy to acquire big-size dataset for network training. Collecting PCD-SCT data can be costly and time-consuming. Simulation is commonly used to generate synthetic spectral images, mono-energetic images and iodine maps, etc. Data augmentation is often necessary to add more training data and improve the data distribution or representability of data, so that the generalization ability of trained neural networks can be increased. Some basic data augmentation methods extensively used in all kinds of deep learning tasks are easy-to-implement geometric transformations such as flip, rotation, translation, rescaling, gray-scaling, etc. Noise injection and gray-scaling are commonly used in CT imaging. They can help solving the challenge caused by the variation in X-ray source spectra and mAs. Different voltage/spectrum from X-ray source results in different magnitude in attenuation. Noise level changes a lot in case of different attenuation length and tube current (or photon flux). The convenience of these data transformations is that it is easy to preserve labels after the transformations.

In these days, there are more techniques available for data augmentation. For example, generative adversarial networks (GANs) can produce new samples that resemble existing data [30]. Neural style transfer can produce new data blending targeted content and style, respectively, defined by different dataset. Recently, artificial intelligence-generated content (AIGC) gained a lot of attention. Large-scale models and multimodality generative AI such as DALL-E-2 [31] have achieved big success. This development makes the creation of large amounts of content efficiently with high quality. Surely, AIGC can be a very power tool for data augmentation. A more detailed review on data augmentation can be found by [32].

3.5 Other Deep Learning Methods for PCD-SCT Reconstruction

Deep learning is booming in all kinds of fields. Many deep learning methods for CT and PCD-SCT reconstruction have been proposed. Roughly, we can categorize deep learning methods into three categories: end-to-end learning methods, unrolling algorithms, and diffusion model-based iterative methods. Of course, hybrid approaches can also be good choices [33]. Here, we limit our topic and focus on end-to-end methods following the roadmap of basic operations to explicitly exploit the knowledge of fundamental theories and mathematical models of PCD-SCT. Unrolling iterative reconstruction method is another important branch of deep learning base reconstruction attracted great attention. The basic idea is replacing the update in an iterative algorithm with a neural network operator. The neural network operator is usually of small size (small number of layers and neurons), so that is low demanding in the size of training datasets. The advantage of such an unrolling algorithm lies in its developing efficiency, interpretable, and high performance. However, the overall reconstruction is iterative and could be time-consuming.

Diffusion model-based methods, including denoising diffusion probabilistic models (DDPMs), score-based generative models (SGMs), and stochastic differential equations (Score SDEs), have been shown to be a powerful generative model to solve inverse problem [34]. They are the family of methods that progressively degrade data by noise injection and learn to reverse the degrading process, so that high-quality samples can be generated. In many fields, diffusion model-based methods become the state-of-the-art generative model. It also has been explored using a diffusion model to solve ill-posed problems in X-ray CT imaging and great results are achieved [35, 36].

4 Discussion

Deep learning-based reconstruction for tomographic imaging has gained great success in past years. Extensive works have been published to tackle all kinds of problems in this field including X-ray CT image reconstruction, denoising, beam hardening correction, scatter correction, etc. Their application tailored to spectral CT remains an active area of research. We only give a very limited description and explanation on deep learning for PCD-CT. PCD-CT is the biggest revolution of the X-ray CT imaging field in this century. Only a few commercial PCD-CT have been deployed in hospitals currently. Designing creative deep learning strategies and applying state-of-art deep learning methods to PCD-CT reconstruction and related tasks could be a very powerful engine to drive its development and smooth its path to clinic.

Interpretability and generalization ability have been the big concern since the flourishing of deep learning methods. It is crucial to clarify the suitable scope of deep learning methods and identify the limitations and the possible failure conditions, especially in medical applications. Interpretability can play a fundamental role there. In this chapter, we presented a roadmap of deep learning following the mathematical models of PCD-CT. Incorporating models of imaging physics and domain knowledge could be of great help to interpretable deep learning. Standing on known reliable points, we can further improve with rich and fast developing edgy technologies in deep learning, including data augmentation, new network architecture, contrastive learning, teacher student learning, diffusion, and reinforcement learning. Scaling deep learning models further to mine out the hidden capabilities of models is another fast-growing research direction to a bright future.

References

1. K. Hornik, M. Stinchcombe, H. White, Multilayer feedforward networks are universal approximators. Neural Netw. **2**(5), 359–366 (1989)
2. G.V.J. Cybenko, Approximation by superpositions of a sigmoidal function. Math. Control Signals Syst. **2**, 303–314 (1989)
3. R.C. Murty, Effective atomic numbers of heterogeneous materials. Nature **207**(4995), 398–399 (1965)
4. A.K. Maier et al., Learning with known operators reduces maximum error bounds. Nat. Mach. Intell. **1**(8), 373–380 (2019)
5. C. Feng, K. Kang, Y. Xing, Fully connected neural network for virtual monochromatic imaging in spectral computed tomography. J. Med. Imaging (Bellingham) **6**(1), 011006 (2019)
6. M. Geng et al., PMS-GAN: Parallel multi-stream generative adversarial network for multi-material decomposition in spectral computed tomography. IEEE Trans. Med. Imaging **40**(2), 571–584 (2021)
7. C. Feng et al., A multi-energy material decomposition method for spectral CT using neural network, in *SPIE Medical Imaging : Physics of Medical Imaging*, (2018)
8. T. Su et al., DIRECT-Net: A unified mutual-domain material decomposition network for quantitative dual-energy CT imaging. Med. Phys. **49**(2), 917–934 (2022)
9. W. Fang, L. Li, Z. Chen, Removing ring artefacts for photon-counting detectors using neural networks in different domains. IEEE Access **8**(99), 42447–42457 (2020)
10. X. Yi, M. Eramian, LBP-based segmentation of defocus blur. IEEE Trans. Image Process. **25**(4), 1626–1638 (2016)
11. J. Johnson, A. Alahi, L. Fei-Fei, Perceptual losses for real-time style transfer and super-resolution, in *Computer Vision – ECCV 2016, 2016*, (Springer, Cham)
12. Q. Yang et al., Low-dose CT image denoising using a generative adversarial network with wasserstein distance and perceptual loss. IEEE Trans. Med. Imaging **37**(6), 1348–1357 (2018)
13. A. Zheng et al., A CT image feature space (CTIS) loss for restoration with deep learning-based methods. Phys. Med. Biol. **67**(5), 055010 (2022)

14. E.Y. Sidky, X. Pan, Image reconstruction in circular cone-beam computed tomography by constrained, total-variation minimization. Phys. Med. Biol. **53**(17), 4777–4807 (2008)
15. L. Shen, Y. Xing, L. Zhang, Joint reconstruction and spectrum refinement for photon-counting-detector spectral CT. IEEE Trans. Med. Imaging **42**(9), 2653–2665 (2023)
16. Y. Zhang et al., Tensor-based dictionary learning for spectral CT reconstruction. IEEE Trans. Med. Imaging **36**(1), 142–154 (2016)
17. D. Wu et al., Iterative low-dose CT reconstruction with priors trained by artificial neural network. IEEE Trans. Med. Imaging **PP**(12), 1–1 (2017)
18. M. Duff, N.D.F. Campbell, M.J. Ehrhardt, Regularising Inverse Problems with Generative Machine Learning Models. arXiv preprint. (arXiv:2107.11191, 2021) (2021)
19. A. Dash, J. Ye, G. Wang, A review of Generative Adversarial Networks (GANs) and its applications in a wide variety of disciplines – From medical to remote sensing (2021)
20. Elbakri et al., Statistical image reconstruction for polyenergetic X-Ray computed tomography. IEEE Trans. Med. Imaging **21**(2), 89–99 (2002)
21. W. Du et al., Visual attention network for low-dose CT. IEEE Signal Process. Lett. **26**(8), 1152–1156 (2019)
22. J.Y. Zhu et al. *Unpaired Image-to-Image Translation Using Cycle-Consistent Adversarial Networks*. in 2017 IEEE International Conference on Computer Vision (ICCV) (2017)
23. O. Ronneberger, P. Fischer, T. Brox, *U-Net: Convolutional Networks for Biomedical Image Segmentation* (Springer, Cham, 2015)
24. K. He et al., *Deep Residual Learning for Image Recognition*. in 2016 IEEE Conference on Computer Vision and Pattern Recognition (CVPR) (2016)
25. G. Huang et al., *Densely Connected Convolutional Networks*. in Proc IEEE Conf Comput Vis Pattern Recogn (2017)
26. A. Vaswani et al., Attention Is All You Need. arXiv e-prints (2017) (arXiv:1706.03762)
27. W. Wu et al., Deep learning based spectral CT imaging. Neural Netw. **144**, 342–358 (2021)
28. X. Chen, et al., SOUL-Net: A Sparse and low-rank UnroLling Network for spectral CT image reconstruction. arXiv e-prints, (2022) (arXiv:2207.12639)
29. A. Zheng et al., Interweaving network: A novel monochromatic image synthesis method for a photon-counting detector CT system. IEEE Access **8**, 217701–217710 (2020)
30. M. Moradi et al., Chest x-ray generation and data augmentation for cardiovascular abnormality classification, in *SPIE Medical Imaging (Image Processing)*, (2018)
31. A. Ramesh et al., Hierarchical Text-Conditional Image Generation with CLIP Latents. arXiv e-prints (2022) (arXiv:2204.06125)
32. C. Shorten, T.M. Khoshgoftaar, A survey on image data augmentation for deep learning. J. Big Data **6**(1), 60 (2019)
33. E.Y. Sidky, X. Pan, Report on the AAPM deep-learning spectral CT grand challenge. Med. Phys., 1–14 (2023)
34. L. Yang et al., Diffusion Models: A Comprehensive Survey of Methods and Applications. arXiv e-prints (2023) (arXiv:2209.00796)
35. Y. Song et al., Solving Inverse Problems in Medical Imaging with Score-Based Generative Models. arXiv e-prints (2022) (arXiv:2111.08005)
36. H. Chung et al., Solving 3D Inverse Problems using Pre-trained 2D Diffusion Models (2022) (arXiv:2111.08005)

Deep Learning Methods in Dual Energy CT Imaging

Tianling Lyu, Wentao Zhu, Yikun Zhang, Wei Zhao, Jian Yang, and Guisheng Wang

1 Background

1.1 From CT to Dual-Energy CT

Visual perception is the primary means by which people see the world. Nevertheless, the ability of human vision is greatly restricted by the lack of transparency in both natural and artificial things. The concept of attaining "an inner vision" has been contemplated in numerous contexts throughout history [1]. X-ray computed tomography (CT) is the very first modality that enables precise and noninvasive interior imaging of an object from sufficient X-ray projections. Since its first introduction by Hounsfield [2], X-ray CT has completely revolutionized medical diagnostics and treatments by enabling noninvasive imaging of internal structures with unprecedented detail and has become an essential component of every contemporary radiology department.

Shortly after the invention of X-ray CT, researchers realized that the physical properties of the material being scanned, such as density and atomic number, could be obtained by combining scanning at multiple energies. This discovery directly

T. Lyu · W. Zhu (✉)
Zhejiang Lab, Hangzhou, China

Y. Zhang
Southeast University, Nanjing, China

W. Zhao
Beihang University, Beijing, China

J. Yang
Beijing Institute of Technology, Beijing, China

G. Wang
Chinese PLA General Hospital, Beijing, China

K. (K.) Iniewski, L. (K.) Cai (eds.), *Deep Learning for Advanced X-ray Detection and Imaging Applications*, https://doi.org/10.1007/978-3-031-75653-5_3

promoted the development of multi-energy CT [3, 4]. Conventional single-energy computed tomography (SECT) technology uses the different absorption coefficients of substances to X-rays, quantified in Hounsfield units (HU), to distinguish different substances. However, different substances may have similar absorption coefficients under a single-energy spectrum, and the existence of image noise enhances the challenge in differentiating compounds with similar X-ray attenuation coefficients, such as iodine and calcium. The advent of dual-energy CT (DECT) was aimed at resolving this challenge. In recent years, DECT (that is, multi-energy CT using two energy levels of X-rays) technology has developed rapidly and has gradually become one of the commonly used examination methods in clinical practice [5]. DECT, in contrast to typical SECT, employs X-rays of dual-energy spectra to capture and visualize the patient's anatomy. This technique generates two distinct tomographic images (a low-energy image and a high-energy image) of the same area. By employing material decomposition technology (either in the projection domain or image domain), it is possible to convert these two images into atomic numbers and electron densities for every pixel point. Alternatively, the images can also be transformed into a representation of the relative abundance of various basis materials, facilitating quantitative analysis [6].

The primary objective of DECT was to separate linear attenuation coefficients into the major interactions present within the clinical CT energy range (i.e., 30–200 keV). In the energy range mentioned, the primary interactions of X-ray photons are Compton scattering and the photoelectric effect. The linear attenuation coefficient can be represented by the following equation:

$$\mu(E) = a_1 \frac{1}{E^3} + a_2 f_{KN}(E). \tag{1}$$

Among them, $a_1 \approx K_1 \frac{\rho}{A} Z^n, n \approx 4$ is the coefficient related to the photoelectric effect; $a_2 \approx K_2 \frac{\rho}{A} Z$ is the coefficient for Compton scattering; $\mu(E)$ is the absorption coefficient of X-ray photons with energy E; $f_{KN}(E)$ is the Kleinn-Nishina function used to characterize Compton scattering; K_1 and K_2 are constants; ρ is the mass density of the corresponding material; A and Z are the average atomic mass and the average atomic number of the substance, respectively; Given the assumption that the linear attenuation coefficient of matter in this energy range is solely influenced by these two interactions, and considering that two sets of data have been acquired through DECT scanning, we can employ Eq. (1) to formulate two equations for the mass density ρ and the average atomic number Z of the material. Empirical evidence has demonstrated that the margin of error in calculations is within 1% for elemental substances with atomic numbers Z below 25 in the human body [7].

More widely used techniques decompose the original data into different basis materials [8]. The ways to perform material decomposition vary. Most of the existing approaches can be divided into three categories: projection-domain-based [7], image-domain-based [9–11], and model-based [12–15]. These methods are based on the equations listed according to Beer-Lambert Law under polyenergetic spectrum and multiple attenuation materials as follows:

$$\begin{cases} P_L = -log(\int S_L(E)e^{-\Sigma_{i=1}^{N}\mu_i(E)L_i}dE) \\ P_H = -log(\int S_H(E)e^{-\Sigma_{i=1}^{N}\mu_i(E)L_i}dE) \end{cases}, \tag{2}$$

where P_L and P_H stand for the low- and high-energy projection data, respectively, $S_L(E)$ and $S_H(E)$ are the low- and high-energy spectra, respectively, $\mu_i(E)$ is the attenuation coefficient of the ith material at energy E, and L_i is the equivalent material thickness along the projection ray-path, which directly corresponds to the material map.

Material decomposition and its derivative technologies are extensively employed in clinical settings to offer clinicians supplementary diagnostic information. Conventionally, standard CT requires two sets of data for contrast-enhanced imaging: one with the contrast agent injected and one non-enhanced scan without the contrast agent. The virtual non-contrast (VNC) imaging technique acquires adequate data with the contrast agent in a single scan and employs algorithms to reconstruct the corresponding non-enhanced image. As a result, it replaces conventional CT in this particular application and reduces the dose by 20–48.9% [16]. The virtual monochromatic imaging (VMI) technique reconstructs CT images under monoenergetic spectra. As materials show different attenuation properties under different photon energies, radiologists have the flexibility to select appropriate images based on specific clinical applications [17, 18].

1.2 Implementations of DECT

As early as 1973, Hounsfield [2] proposed the idea of DECT imaging, and this idea was further expanded by Alvarez and Marcovski [7] in 1976 and became the basis of modern DECT imaging. However, the clinical applications of DECT lag far behind the initial recognition of its clinical feasibility. It was not until Siemens released the SOMATOM Definition system in 2006 that DECT truly became commercially available. The primary focus of these 30 years was dedicated to research on DECT implementation solutions.

A straightforward implementation is to change the tube potential once the first energy level scan is finished and then proceed to scan the second energy level (Fig. 1a) [19]. However, it is unsuitable for scanning objects that display temporal motion. The motion leads to inconsistent readings, causing significant deterioration of the decomposed material-specific images [20]. Thus, this particular technique of potential sequential scanning is mostly employed in laboratory simulations of DECT data and industrial DECT. An alternative approach to implementing DECT involves alternating the potential of the X-ray tube during the scanning process. By ensuring that the time interval between high and low energy is always limited to one frame, it is possible to maintain the dual-energy image consistency to the greatest extent possible, even when the target object is in motion. The issue with this solution is the challenge of achieving swift and precise transitions between X-ray energy levels

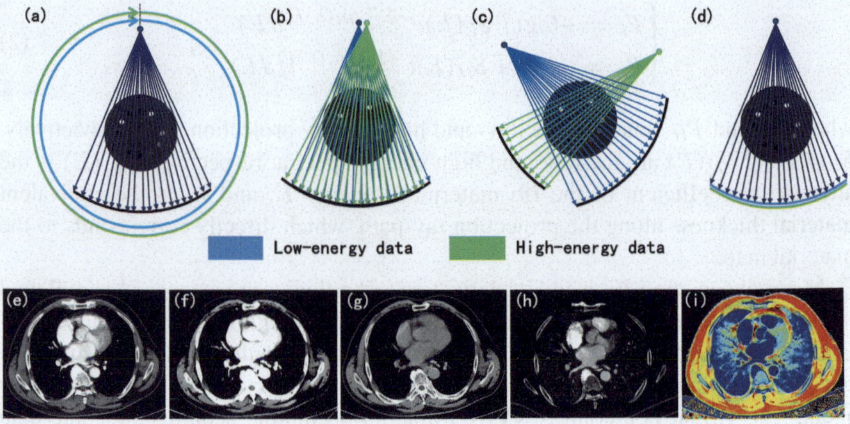

Fig. 1 DECT implementations and imaging results on a sample slice. (**a**) Dual-scan; (**b**) Rapid kVp switching; (**c**) Dual-source system; (**d**) Dual-layer detector; (**e**) Polychromatic image under 120 kVp; (**f**) Monochromatic image under 40 keV; (**g**) Monochromatic image under 200 keV; (**h**) Iodine map; (**i**) Effective atomic number map

using only one X-ray source. Commercial DECT systems used for medical purposes can be categorized into three groups based on their implementation.

- **Rapid kVp switching:** Rapid kVp switching enables continuous projection measurements at alternating tube voltages in less than a millisecond. This method permits the acquisition of both low- and high-energy projection data in a single complete scan (Fig. 1b). Projection acquired in this manner results in an interleaved sinogram [21, 22]. The rapid kVp switching approach demonstrates a strong temporal consistency between DECT images. Nevertheless, the spectra produced by this kind of system exhibit a strong correlation due to the limited applicability of filtration, which exacerbates challenges in material decomposition.
- **Dual-source systems:** Dual-source systems employ two sets of X-ray measurement systems that are installed on the rotating gantry and are oriented perpendicular to each other (Fig. 1c) [23]. They circumvent the technical issues encountered in rapid kVp switching and filtration can be easily applied for better spectra separation. Nevertheless, these systems are typically more expensive and may encounter temporal consistency issues when objects are in motion.
- **Dual-layer detector:** The two layers of detectors are specifically constructed to collect X-ray photon data within two distinct energy ranges (Fig. 1d). The initial layer of a dual-layer detector often employs detecting materials with low effective atomic numbers, which exhibit limited absorption capabilities for high-energy photons and enable the passage of such photons through the detector. The second layer of detectors beneath is composed of either high atomic number detection material or has sufficient thickness to capture the remaining photons [20, 24, 25]. These systems exhibit perfect temporal consistency between DECT images, but the crosstalk between two detector layers is challenging to eliminate.

Additionally, DECT systems can be built using alternative techniques such as beam filtration [26, 27] or dual-focus X-ray tube [28]. These systems adhere to the same physical principles of dual-energy imaging as other systems, although currently, they are still relatively less common.

1.3 Current Status

DECT is suitable for any clinical scenario that requires noninvasive tomographic imaging. In clinical settings, DECT can be used to scan patients in three dimensions, yielding detailed tomographic data that includes material composition. This information is particularly useful for the accurate quantification of diseased tissue. Therefore, DECT imaging is adopted in the clinical diagnosis of renal masses [29], differentiating intracerebral hemorrhage from iodinated contrast [30], automated bone removal in CT angiography [31–34], urinary stone characterization [35–38], etc.

Although DECT has achieved success in certain clinical scenarios, it still faces many issues hindering its further utilization.

As stated in Sect. 1.2, the implementation of DECT poses a formidable challenge. While there are multiple established commercial systems available, the hardware-based solutions considerably amplify the intricacy and expense of implementing the imaging system, therefore impeding the extensive adoption of DECT in clinical settings. Typically, the majority of DECT scanners have a price range of $1.6 to $2.5 million, while conventional multi-slice CT scanners are priced at approximately $0.55 million [39]. Furthermore, each solution exhibits inherent restrictions. For instance, the rapid kVp switching approach is hindered by inadequate energy resolution, whereas dual-source systems provide challenges in maintaining temporal consistency.

The DECT system is an extension of conventional CT systems and has similar issues. To begin with, there is the issue of radiation dose. The use of CT imaging while adhering to the as low as reasonably possible (ALARA) concept has become widely recognized as routine practice. There has been significant research conducted over the past two decades on the clinical benefits of further reducing the radiation dose from CT scanning. While research indicates that DECT achieves lower doses compared to SECT [40–42], the quest for additional reductions in radiation doses is ongoing and perpetual. DECT implementation also encounters image geometry issues such as limited-angle, sparse-view, and data truncation. Researchers are actively working on resolving these challenges.

Additionally, there are issues about material decomposition. Projection-domain decomposition is not a viable solution for DECT systems with inconsistent dual-energy projections. Image-domain decomposition suffers from accuracy limitations, whereas model-based iterative approaches exhibit impractical time requirements for clinical applications. Furthermore, the process of decomposition typically intensifies the disparities between low- and high-energy data, hence potentially magnifying noise and artifacts present in the resulting images.

Deep learning (DL) has lately demonstrated its efficacy as a potent tool for mapping complex relationships and integrating prior information into an inference model. It has already been utilized for low-dose CT imaging and CT artifact reduction. To address the previously described issues, researchers implemented deep learning methods for DECT imaging and produced significant advancements. In the rest of the chapter, we go over the applications of deep learning in DECT imaging, specifically focusing on data synthesis, image reconstruction, and material decomposition.

2 Deep Learning in DECT Synthesis

Currently, limited by the acquisition modes of existing DECT systems, the obtained DECT images often require compromising between equipment cost, radiation dose, image noise, time resolution, and energy resolution. Due to the similarity in anatomical features between the low- and high-energy CT images obtained from DECT scanners, there exists a significant amount of redundant information in the DECT images [43]. Researchers have recognized that deep learning, which has been successful in tasks such as image generation and image-to-image translation [44, 45], can also be applied to synthesize DECT images or derivative images from SECT images. This allows for DECT imaging to be achieved with only SECT systems, thereby reducing costs. Numerous researchers have investigated this concept and obtained specific outcomes.

2.1 CD-ConvNet

To the best of our knowledge, Li's preprint paper [46] represents the pioneering work on this concept. This study introduces a cascaded deep ConvNet (CD-ConvNet) to generate pseudo-high-energy images I_{pseudo} from SECT data. As depicted in Fig. 2a, this network employs a cascade of multiple basic convolutional neural networks (ConvNet) to tackle the trade-off between the subpar synthesis quality achieved by small networks with limited receptive fields and the substantial amount of training samples needed for training larger networks. The initial ConvNet is trained to establish the nonlinear relationship between the low-energy image I_L and the corresponding high-energy image I_H. This generates an imperfect intermediate synthesis image I^1_{pseudo}. The following ConvNets are designed to learn the transformation from the output of the previous ConvNet, denoted as I^{K-1}_{pseudo}, to the target I_H to enhance performance. In their subsequent study, the same network is also utilized for material map synthesis [47]. The training of their network involved a dataset consisting of 7 patients and 2121 slices. The model was evaluated by testing it on 334 slices from an extra patient. The predicted high-energy CT images

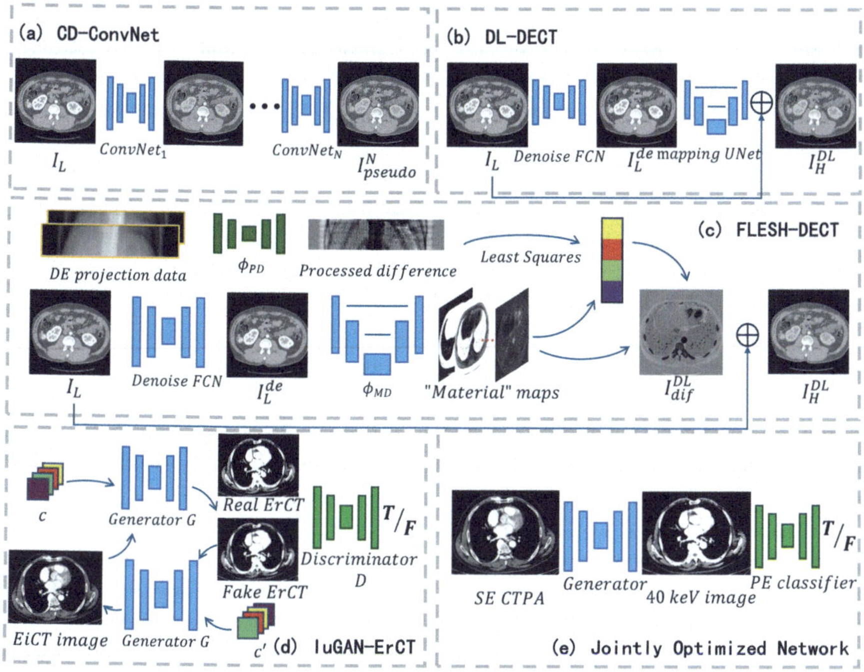

Fig. 2 Deep-learning-based DECT synthesis methods. (**a**) CD-ConvNet; (**b**) DL-DECT; (**c**) FLESH-DECT; (**d**) IuGAN-ErCT; (**e**) jointly optimized deep neural network

closely resemble the reference in terms of visual characteristics. The qualitative assessments with line profiles demonstrate a close agreement between the pseudo- and reference high-energy CT images.

2.2 *DL-DECT*

Zhao et al. further extend the idea of DECT synthesis to achieve diagnostic accuracy by incorporating pre-denoising and residual learning [48, 49]. This approach has been subsequently referred to as DL-DECT in a separate publication by the same research group [43]. As depicted in Fig. 2b, this method can be expressed with the following equation:

$$I_H^{DL} = I_L + \phi_{map}(\phi_{FCN}(I_L)), \tag{3}$$

where I_H^{DL} is the estimated high-energy image by the DL-DECT method, and I_L is the input low-energy image.

Prior to training and testing, the DECT images undergo denoising to reduce the negative impact of noise. To achieve this goal, they developed a fully convolutional network ($\phi_{FCN}(\cdot)$) that uses the raw CT image as input and generates the difference image between the raw CT image and a low-noise (high-dose) CT image as output. The denoising FCN model is trained using paired low- and high-dose CT images from the American Association of Physicists in Medicine (AAPM) Low-Dose CT Grand Challenge. After the model has been trained, the predicted image from inference is subtracted from the raw CT image. This results in images with a substantial noise reduction. The model is subsequently deployed on all DECT images to suppress image noise.

Following the denoising process, a convolutional neural network $\phi_{map}(\cdot)$ is trained to learn the mapping from the denoised low-energy image I_L^{de} to the difference between the denoised DECT images ($I_{dif} = I_H^{de} - I_L^{de}$), rather than adopting the strategy used in CD-ConvNet to learn the mapping from low-energy image to high-energy image. The rationale of this design is two-fold. First, the inherent rich details in DECT images have been greatly reduced in I_{dif}, and the network can be trained to focus more on the dual-energy difference (such as the iodine contrast) between I_L and I_H, resulting in a less complicated regression model and more robust and accurate performance for dual-energy mapping. Furthermore, convolutional neural network (CNN) models have a tendency to generate excessively smoothed images, particularly when trained using L2 loss. Radiologists do not prefer these excessively smoothed images as they likely lead to a loss of image information. Learning the residual map perfectly avoids this problem, as the noise characteristics and image details can be restored by adding I_L.

The DL-DECT model demonstrates exceptional synthesis capabilities on the testing dataset (Fig. 3). The analysis [49] reveals that the discrepancy in Hounsfield Units (HU) between the predicted and target high-energy CT images is 1.94 HU, 3.32 HU, 1.83 HU, and 1.10 HU for the regions of interest (ROIs) located on the spine, aorta, liver, and stomach, respectively. The subsequent analysis on VNC images and iodine maps demonstrates the clinical promise of DL-DECT. In addition to the excellent consistency, the material maps produced from DL-DECT exhibit significantly lower noise compared to those obtained from real DECT images. A noise reduction of up to 68% and iodine contrast-to-noise improvement of up to 55% have been achieved [50]. This improvement arises from the noise correlation between the synthetic DECT images. As stated in Sect. 1.3, material decomposition based on matrix inversion increases the disparities between low- and high-energy data and yields magnified image noise. The residual learning technique correlates the noise of the synthetic images with that of the original low-energy images and therefore mitigates the noise amplification problem.

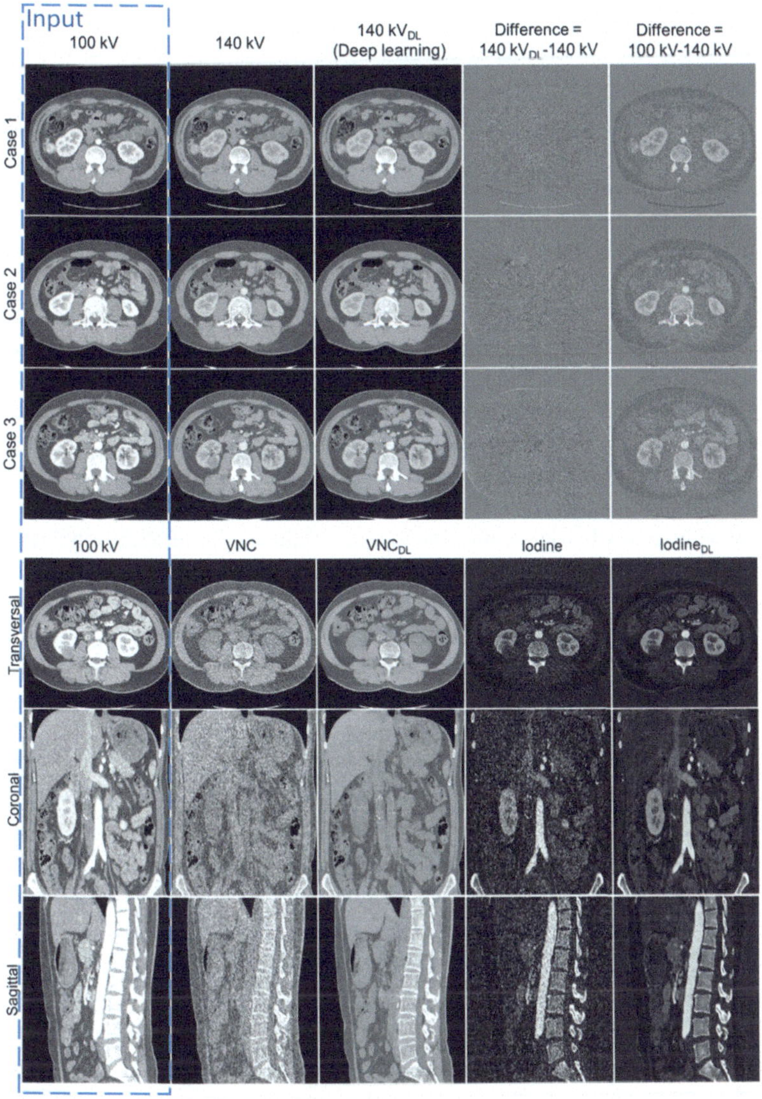

Fig. 3 Results from DL-DECT and comparison to real DECT images

2.3 FLESH-DECT

CD-ConvNet and DL-DECT directly infer the high-energy images from the low-energy ones. The rationale behind these two strategies is plain. Nevertheless, researchers have expressed concerns over the potential decrease in reliability when depending exclusively on deep learning without incorporating any other auxiliary data.

The study conducted by Lyu et al. [43] presents an approach for incorporating high-energy spectrum knowledge into the image translation network. The researchers used additional data from a high-energy projection view to enhance the synthesis of DECT and introduced a deep learning technique known as fully low-energy and single high-energy DECT imaging (FLESH-DECT). This supplementary high-energy projection view can be acquired soon before or after the usual SECT data acquisition with kVp switching. Due to the need for only one more projection view, there is unlikely to be a significant misalignment between the low- and high-energy data. Existing SECT systems can execute the acquisition protocol without any hardware modifications, and the negligible additional radiation dose can be disregarded.

This method aligns with the concept of image-domain dual-energy material decomposition. The attenuation coefficient at each location can be estimated by combining the attenuation coefficients of the basis materials in a linear manner [7]:

$$\mu = \alpha_1\mu_1 + \alpha_2\mu_2 + \cdots + \alpha_m\mu_m, \tag{4}$$

where m is the number of basis materials, α_i is the percentage of the ith basis material, and μ_i is the attenuation coefficient of the ith basis material. Considering the whole image slice under dual-energy scan, this equation can be rewritten in the following form:

$$\begin{cases} I_L = M_\alpha \cdot \boldsymbol{\mu}_L^T \\ I_H = M_\alpha \cdot \boldsymbol{\mu}_H^T \end{cases}, \tag{5}$$

where $M_\alpha = [\alpha_{ij}]_{n_{pix}\times m}$ is the basis material percentage matrix, n_{pix} is the number of pixels in the image, and $\boldsymbol{\mu}_L$ and $\boldsymbol{\mu}_H$ are the equivalent attenuation coefficients for each basis material corresponding to the low- and high-energy spectra, respectively. It is evident from the given formula that M_α is the shared component between high-energy and low-energy images, and the authors developed a network $\phi_{MD}(\cdot)$ to estimate M_α from I_L:

$$\begin{cases} I_L = \phi_{MD}(I_L) \cdot \boldsymbol{\mu}_L^T \\ I_H = \phi_{MD}(I_L) \cdot \boldsymbol{\mu}_H^T \end{cases}. \tag{6}$$

The method prioritizes the high-energy image synthesis as the target rather than obtaining material component labels. It is not required to explicitly indicate each channel in $\phi_{MD}(\cdot)$ to represent actual material or a combination of multiple materials. The comparable attenuation coefficients $\boldsymbol{\mu}_H$ are thus determined based on the high-energy projection view.

The forward projection in CT imaging can be modeled with a linear operation with the system matrix A_{proj} on the image. Taking the additional high-energy projection view P_H into consideration, one can get the following equation:

$$P_H = A_{proj} \cdot \phi_{MD}(I_L) \cdot \boldsymbol{\mu}_H^T. \tag{7}$$

With M_{proj} obtained according to the geometry and $\phi_{MD}(I_L)$ obtained from the network, $\boldsymbol{\mu}_H$ can be optimized by solving the least squares problem. The entire synthesis framework can be expressed with the following equations:

$$\begin{cases} I_H = \phi_{MD}(I_L) \cdot \boldsymbol{\mu}_H^T \\ \boldsymbol{\mu}_H = \underset{\boldsymbol{\mu}}{\operatorname{argmin}} \left\| A_{proj} \cdot \phi_{MD}(I_L) \cdot \boldsymbol{\mu}^T - P_H \right\|_2^2 \end{cases}. \tag{8}$$

The authors further optimized the framework by incorporating the pre-denoising and residual learning in DL-DECT into this new algorithm, as well as applying a projection-domain network $\phi_{PD}(\cdot)$ to mitigate the inconsistency caused by noises in the projection domain. The final framework can be mathematically expressed with the following equations:

$$\begin{cases} I_H = I_L + \phi_{MD}(\phi_{FCN}(I_L)) \cdot \boldsymbol{\mu}_{dif}^T \\ \boldsymbol{\mu}_{dif} = \underset{\boldsymbol{\mu}}{\operatorname{argmin}} \left\| A_{proj} \cdot \phi_{MD}(\phi_{FCN}(I_L)) \cdot \boldsymbol{\mu}^T - \phi_{PD}(P_H - P_L) \right\|_2^2 \end{cases}, \tag{9}$$

where P_L is the low-energy projection view with the same geometry as P_H. The general framework of FLESH-DECT is depicted in Fig. 2c.

By incorporating information from the additional projection view, FLESH-DECT achieves improved synthesis quality with reduced parameter numbers. Compared to DL-DECT, MSE values between synthetic and real 140 kV images were reduced by nearly 50% with FLESH-DECT, while Peak Signal-to-Noise Ratio (PSNR) values were improved by about 10%. $\phi_{MD}(\cdot)$ in FLESH-DECT also adopts U-Net architecture, but the parameter number has been reduced from 91M to 18M. In FLESH-DECT, the high-energy projection adds a constraint to ensure that the projection made by the predicted high-energy images aligns with the measurement. This enhances the precision of the predicted images.

One additional benefit of FLESH-DECT is its potential to synthesize images under different spectra with the same image-domain network. With the X-ray energy levels varying, the fundamental composition remains constant. This indicates that accurate material decomposition maps can give CT images under any spectrum if they are combined with the projection view obtained using that spectrum (for example, a 120 kV image from a 120 kV projection or a 150 kV image from a 150 kV projection). The authors trained and tested their model under the 100 kV/Sn 140 kV protocol, and the ϕ_{MD} obtained is likely to be optimized for this specific protocol. Nevertheless, if the model underwent training using images obtained from various spectra, the "material decomposition" maps would closely resemble the actual ones, enabling the model to generate images under varied spectra without the need for retraining ϕ_{MD}.

2.4 GAN-Based Methods

The utilization of generative adversarial nets (GAN) in image generation challenges has been extensive ever since its initial proposition by Goodfellow et al. [51]. GAN models employ a dual-network architecture to facilitate generative adversarial learning on data. One network is responsible for generating data based on input, while the other network discriminates between the produced data and real data. The networks employ a zero-sum game framework to accurately model the distribution of the data and generate new data. These models outperformed traditional CNNs in image-style transfer tasks. Researchers studying the DECT synthesis problem quickly discover that the conversion of images with varied kVp settings may also be considered as style transfer. Consequently, GAN models can be used for this purpose.

Yao et al. are the first to use GAN models in DECT synthesis [52, 53]. The authors initially introduced a unified generative adversarial network (uGAN) as a means to generate monochromatic CT pictures at various energy levels (referred to as EiCT in the study) using the input polychromatic CT images (referred to as ErCT in the chapter). The uGAN model differs from the original GAN model [51]. Instead, it is more similar to the conditional GAN model [54] as it incorporates additional spectral information as a condition. To achieve multiple image generations with one network, an additional target label c is introduced as the input to the generator. The label c serves as a domain classification condition for the output energy level, which should also be discriminated by the discriminator network. Two additional losses, $\mathcal{L}_{cls}^{f}$ and $\mathcal{L}_{cls}^{b}$, were introduced on c for the forward and backward image generation, respectively, as follows:

$$\begin{cases} \mathcal{L}_{cls}^{f} = E_{x,c}[-\log D_{cls}(c|G(x,c))] \\ \mathcal{L}_{cls}^{b} = E_{x,c'}[-\log D_{cls}(c'|x)] \end{cases}, \tag{10}$$

where $D_{cls}(\cdot)$ is the discriminator network and $G(\cdot)$ is the generator network; x is the input polychromatic image, and c' is the label corresponding to x. The uGAN model underwent training using 8304 image slices, which were enhanced from a dataset consisting of 1384 slices through rotation and flip techniques. Subsequently, the model was tested on a separate set of 130 slices. The monochromatic image generation outcomes at 70, 90, 110, and 130 keV were reasonably attained, particularly in soft-tissue regions. Nevertheless, it was unable to accurately restore intricate features in regions of high intensity, such as cortical bone. To overcome this problem, the authors introduced an improved uGAN model called "IuGAN-ErCT" [53]. The IuGAN-ErCT shared an identical network structure with uGAN (depicted in Fig. 5d). The only difference lies in that groundtruth monochromatic images are used to supervise the generator training, therefore introducing a fidelity term to the loss function. This fidelity loss can be expressed as

$$\mathcal{L}_{fid} = E_{x,y,c}[||y - G(x, c)||_1], \tag{11}$$

where y is the corresponding monochromatic images from real DECT data. Not surprisingly, the introduction of supervised learning has a positive effect on image synthesis, and the network performs slightly better on cortical bone areas.

Kawahara et al. conducted a thorough comparison between GAN-generated monochromatic CT images and DECT-calculated ones [55]. The researchers utilized the original GAN paradigm, except for substituting the generator network with a U-Net design. The generation model was trained and tested using a significantly larger dataset consisting of over 18,000 photos obtained from 28 patients. The authors also conducted a threefold cross-validation to demonstrate the robustness of the method. According to the findings, the authors noted that the 70 keV monochromatic images produced by their model exhibited more conformity with the groundtruth in comparison to the 40 keV and 140 keV images. However, the difference is not immediately apparent. The cause of this phenomenon can be ascribed to the proximity of 70 keV to the effective X-ray energy of the spectrum employed for polychromatic scanning. Jeong et al. [56] compared two state-of-the-art GAN models for DECT image synthesis, namely the Pix2PixHD model [57] and the CycleGAN model [58]. The authors used 70 keV images as inputs, 50 keV, iodine, and VNC images as outputs. The results show that DL-generated images are qualitatively indistinguishable from real ones by a radiologist, and Pix2PixHD performs slightly better than CycleGAN quantitatively.

With all the abovementioned DECT synthesis algorithms, researchers have begun to question the specific scenarios in which these synthesized images are indeed necessary. The research by Fink et al. offers a possibility in this regard [59]. The authors concentrated on the specific objective of diagnosing pulmonary embolism (PE) and proposed jointly optimized deep neural networks to synthesize monoenergetic images from SECT. The networks employed in this work should not be categorized into GAN-style, but, similar to all GAN-style networks, this framework includes both a generator and a classifier (Fig. 2e). The generator used was a fully convolutional 9-block ResNet, which can predict 40 keV monochromatic images from polychromatic Computed Tomography Angiography (CTA) inputs. The resulting monochromatic images were then fed into a ResNet50 network to discriminate the presence or absence of PE. The two networks were optimized in an end-to-end manner, with a joint loss expressed as

$$\mathcal{L}_{joint} = (1 - m) \cdot \mathcal{L}_{gen} + m \cdot \mathcal{L}_{cls}, \tag{12}$$

where

$$\begin{cases} \mathcal{L}_{gen} = E_{x,y}[||G(x) - y||_1] \\ \mathcal{L}_{cls} = E_{x,z}[-z \log \sigma(D(G(x))) - (1 - z) \log(1 - \sigma(D(G(x))))] \end{cases} \tag{13}$$

are the generation loss and classification loss, respectively, and x, y, and z are the input image, the target monochromatic image, and the classification label, respectively. By incorporating the auxiliary task of classification, the generator network resulted in close or slightly improved synthesis quality compared to other GAN-based networks. More importantly, by incorporating the image synthesis network, the discriminator network achieved significantly improved classification performance compared to using SECT images directly. The area under the curve (AUC) value showed an increase from 0.8142 to 0.8420. This enhancement suggests that DECT synthesis networks include supplementary prior knowledge into the resulting images, hence enhancing the ability to improve the precision of clinical diagnosis.

The study by Jiang et al. [60] presents more support for the clinical viability of deep-learning-generated DECT images. The researchers developed a SECT-to-DECT generative adversarial network (S2DGAN) capable of simultaneously generating a VNC image and an iodine map from a single SECT (120 kV) image. This enables the real-time diagnosis of post-interventional cerebral hyper-density (PCHD). The Transformer architecture [61], recently applied to computer vision tasks, is adopted as the backbone under an encoder-decoder style [62]. The encoder network first encodes the input SECT image before decoding it with a VNC decoder and an IOM decoder for the VNC image and the iodine map, respectively. The authors additionally suggested a shared attention technique to facilitate communication between the two decoders. This involved calculating attention levels from the VNC decoder and utilizing them in the IOM decoder. The authors evaluated the generated images in subsequent tasks aimed at differentiating hemorrhaging from contrast extravasation. Distinguishing between these two phenomena on single-energy CT scans is challenging due to their similar, slightly elevated HU values in comparison to the surrounding tissues. The findings demonstrate a substantial enhancement in classification accuracy when clinical physicians utilize the synthesized VNC images and iodine maps. The presented evidence demonstrates the dependable usability and significant potential of the DECT images produced by DL techniques for subsequent tasks.

3 Deep Learning in DECT Reconstruction

In the previous section, we introduced a series of methods that employ deep learning to approximate DECT from conventional SECT. These approaches primarily acquire knowledge regarding the relationship between dual-energy images

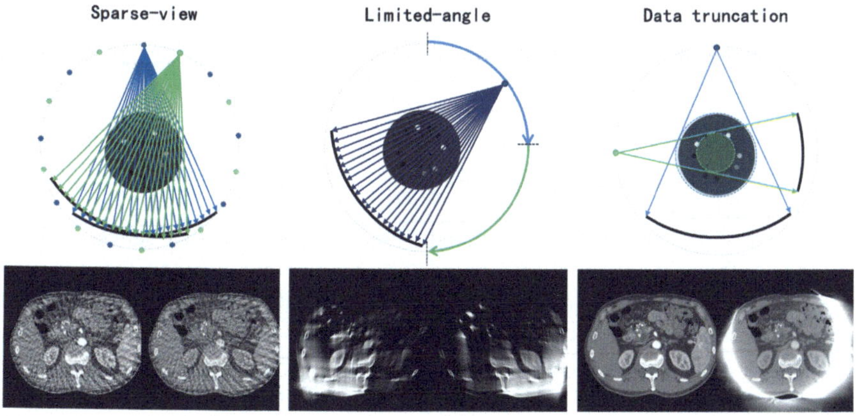

Fig. 4 Illustration for the geometrical artifacts in DECT imaging and the corresponding reconstruction results on an example slice. The images were reconstructed with the standard filtered backprojection algorithm

by training with extensive data. They subsequently utilize this knowledge to conject the associated DECT image from the available SECT data. The conjecture's precision is deficient in a theoretical foundation, so it remains distant from practical implementation in a clinical setting. On the other hand, even if we collect raw data at two different energy levels, there are still numerous avenues to be explored in the field of dual-energy image reconstruction. As illustrated in Fig. 4, sparse-view artifacts, limited-angle artifacts, and data truncation artifacts are a few of the most significant issues. Deep-learning-based reconstruction techniques have been extensively employed to reduce artifacts in conventional SECT. Consequently, they hold promise for addressing analogous issues in DECT data. However, in contrast to traditional SECT, there exists an inherent link between the DECT projection data obtained from distinct energy levels. By employing deep learning techniques to thoroughly investigate and exploit this inherent association, it is anticipated that superior reconstruction outcomes can be attained compared to conventional SECT data. Currently, researchers have undertaken numerous studies in this field.

3.1 Sparse-View Reconstruction

Sparse-view reconstruction typically pertains to the issue where the angular separation between neighboring projections is excessively wide. In this scenario, the acquired data fails to satisfy the criteria for precise reconstruction, resulting in the introduction of a series of streak artifacts into the image. This issue commonly arises when there is a need for quicker scanning durations. State-of-the-art commercial spiral CT scanners can typically do a scan with over 1,000 projection angles in just one second. Nevertheless, the scanning velocity of certain other devices, like as CBCT, is considerably more sluggish. Certain equipment requires over a minute to perform a scan consisting of 600 projections. The primary factor constraining the rotation speed is the stability of the hardware. In order to minimize motion artifacts during scanning, it would be necessary to decrease the number of projections acquired per cycle. However, this would result in the issue of sparse-view reconstruction.

When it comes to DECT, specifically DECT equipment that relies on kVp switching, effectively addressing the issue of sparse view reconstruction is of utmost significance. The fundamental reason for this is that the implementation of kVp switching back and forth will result in a reduction of the number of projections collected at each energy level by half. Simultaneously, as the process of kVp switching also requires time, this mode has the potential to significantly decrease the number of projections acquired in each scanning cycle. Hence, investigating deep-learning-based dual-energy sparse angle reconstruction techniques can enhance the image reconstruction quality in this mode and substantially alleviate the hardware burden.

Comprehensive domain network (CD-Net) [63] was proposed to address the DECT sparse-view reconstruction problem. As many dual-domain CT reconstruction networks [64, 65] do, CD-Net comprises a projection-domain network (ϕ_{PD}), an image-domain network (ϕ_{ID}), and an analytical reconstruction operator ($\mathcal{R}^{-1}$) performing domain transformation (Fig. 5a). CD-Net adopts the residual learning concept and takes interpolated dual-energy projections $\binom{P_L^{LI}}{P_H^{LI}}$ as inputs. The reconstruction algorithm can be expressed with the following equation:

$$\begin{cases} \binom{P_L^{DL}}{P_H^{DL}} = \binom{P_L^{LI}}{P_H^{LI}} + \phi_{PD}\binom{P_L^{LI}}{P_H^{LI}} \\ \binom{I_L^{DL}}{I_H^{DL}} = \mathcal{R}^{-1}\binom{P_L^{DL}}{P_H^{DL}} + \phi_{ID}(\mathcal{R}^{-1}\binom{P_L^{DL}}{P_H^{DL}}) \end{cases} . \tag{14}$$

where I_L^{DL} and I_H^{DL} are the predicted DECT images. To fully utilize the intrinsic correlation between dual-energy data, a dual-stream ResUNet was developed for both ϕ_{PD} and ϕ_{ID}. The U-Net architecture was adopted, with the decoder replaced by a dual-stream structure. The mainstream component receives input DE data and encodes it into shared features. Then these features simultaneously flow into the low- and high-energy stream components, which perform differential learning to produce the corresponding low- and high-energy projections or images. An additional loss $\mathcal{L}^{diff}$ was introduced to supervise the difference between low- and high-energy data during network training, therefore preserving the accuracy of material decomposition:

$$\mathcal{L}^{diff} = ||(D_L - D_H) - \left(D_L^{DL} - D_L^{DL}\right)||_1 \tag{15}$$

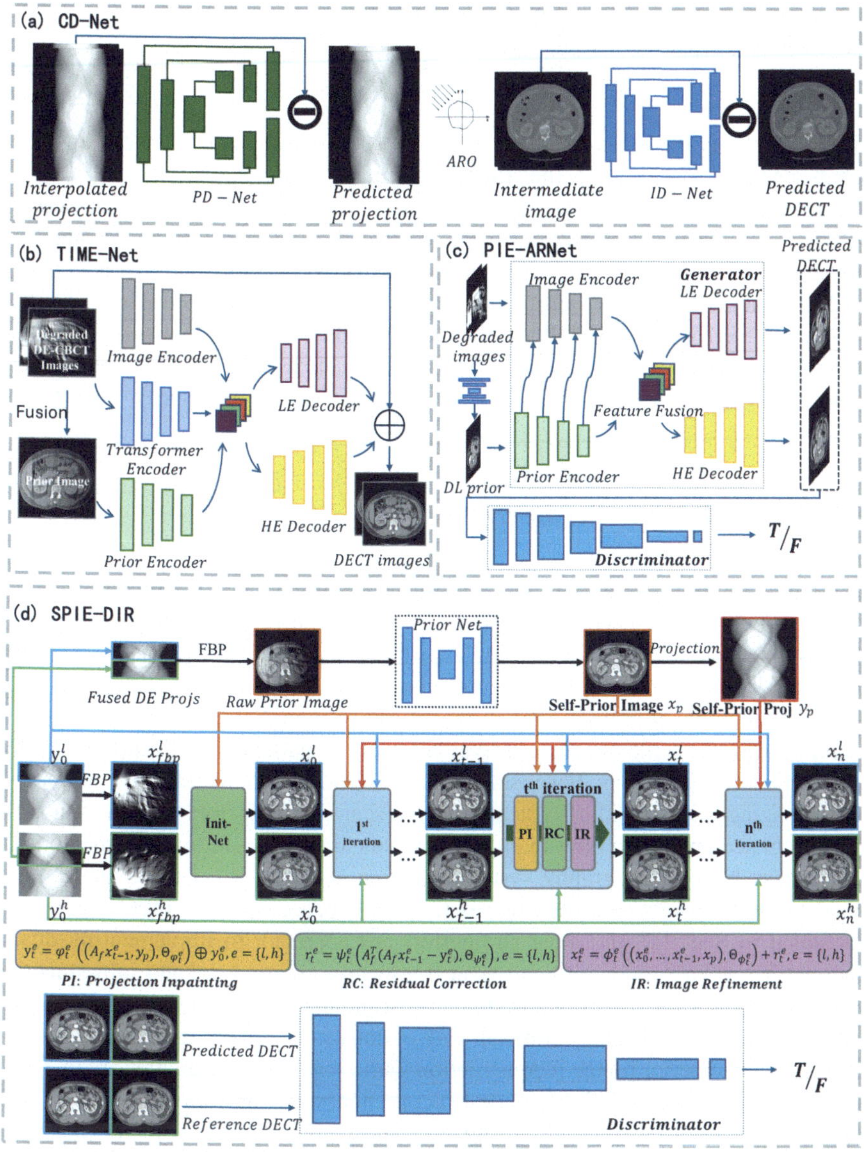

Fig. 5 Deep-learning-based DECT reconstruction methods. (**a**) CD-Net for sparse-view reconstruction; (**b**) TIME-Net for limited-angle reconstruction; (**c**) PIE-ARNet for limited-angle reconstruction; (**d**) SPIE-DIR for limited-angle reconstruction

where D_L and D_H are the groundtruth data (either projection domain or image domain) under low- and high-energy, respectively, and D_L^{DL} and D_H^{DL} are the corresponding results produced by CD-Net. The composite loss function for both

PD-Net and ID-Net was set to the following form:

$$\mathcal{L} = \mathcal{L}^{MAE} + \alpha \cdot \mathcal{L}^{SSIM} + \beta \cdot \mathcal{L}^{diff}, \tag{16}$$

where $\mathcal{L}^{MAE}$ is the mean absolute difference between predicted images and the labels, and $\mathcal{L}^{SSIM}$ represents the structural similarity loss.

CD-Net has been tested on a simulated sparse-view DECT dataset, where 90-view and 72-view projections were extracted from the 720-view complete projection. CD-Net shows superior sparse-view reconstruction ability by achieving an MAE value of 8 HU, lower than other comparative methods. More importantly, CD-Net outperforms its SECT version, which adopts the same network structure except for having only one stream in the decoder. It indicates that better artifact suppression can be achieved by exploring the spectral complementarity.

3.2 Limited-Angle Reconstruction

Limited-angle reconstruction is another issue often discussed in CT reconstruction. In contrast to the sparse-view reconstruction problem, this issue mostly arises from the insufficient range of scanning angles, which hinders correct reconstruction. In medical CT, whether using fan-beam or cone-beam geometries, it is necessary to scan over an angle of at least $180° + 2\gamma_m$ to gather enough data for reconstruction. Here, γ_m represents the fan angle. Nevertheless, it is challenging to fulfill this angle range in certain situations. For instance, C-arm CBCT systems used for intraoperative imaging in orthopedics typically have a scanning trajectory of less than 180 degrees to accommodate physicians' movements and the positioning of the couch. In contrast, breast tomography systems have an even lower scanning angle. The issue of limited-angle reconstruction is frequently encountered in DECT imaging. As stated in Sect. 1.2, the most straightforward approach to implementing dual-energy CT involves conducting one scan at a specific energy level and subsequently switching to the second energy level to perform another scan. Nevertheless, the issue with this method lies in the significant lack of consistency in the time dimension. If we can attain superior image reconstruction while scanning at a smaller angle, such as 90°, we can diminish the temporal inconsistency to one fourth of its original value. Simultaneously, this will result in a substantial reduction in the radiation exposure to the patient.

CT-Net [66] is a recent contribution to the field of limited-angle DECT reconstruction. CT-Net applies CD-Net to the limited-angle reconstruction problem. An element that sets CT-Net apart from CD-Net is the preprocessing stage. The linear interpolation to limited-angle dual-energy projections is useless for assisting feature extraction. The absent perspectives are consequently replaced with input values of 0. The network is tested on limited-angle projection data simulated from fully scanned DECT images. The strategy to select limited-angle DE projections P^l from full-view projections P^f is defined by the following formulas:

$$\begin{cases} P_L^l = P_L^f(\theta_{start} : \theta_{start} + \alpha) \\ P_H^l = P_H^f(\theta_{start} + \alpha : \theta_{start} + 2\alpha) \end{cases}, \tag{17}$$

where θ_{start} was set to 0°, and α was set to 90°. The results from CT-Net show high consistency with the reference DECT images. The artifacts introduced by limited-angle scanning can hardly be observed.

CD-Net (or CT-Net) inputs the low- and high-energy data into the network and instructs the network to extract the correlation between them. Building upon this foundation, researchers contemplated methods for incorporating prior knowledge on the relationship between low- and high-energy data. As a result, they put forth the transformer-integrated multi-encoder network (TIME-Net) as a solution [67]. The underlying concept of this network is to combine low- and high-energy projection data to reconstruct an image (referred to as a fused image) that possesses the characteristics of both low- and high-energy, while minimizing limited-angle artifacts to a large extent. As a result, the fused image can serve as prior knowledge for the network, as it encompasses anatomical details of the intended object. TIME-Net comprises three different encoders: an image encoder like the encoder in CD-Net, a transformer encoder that adheres to the latest advancements in the visual transformer, and a prior encoder that extracts information from the prior image. A feature fusion module is utilized to integrate the outputs from three encoders before inputting them into the decoders. The overall structure of TIME-Net is illustrated in Fig. 5b. In addition to the simulated limited-angle DECT dataset, TIME-Net was also evaluated using a real rat dataset. The results from both datasets demonstrate that TIME-Net exhibits a significantly enhanced ability to reduce artifacts compared to SECT-based approaches.

The strategy of introducing the prior image is further extended by prior image enhanced artifact removal network (PIE-ARNet) [68] and self-prior information enhanced deep iterative reconstruction (SPIE-DIR) [69]. As illustrated in Fig. 5c, PIE-ARNet employed an additional prior network to improve the fused image, therefore providing more effective guidance. In the PIE-ARNet model, the features obtained from the shallow layers of the prior encoder are transferred to the image encoder to enhance the fusion of features. Moreover, PIE-ARNet adopts the adversarial learning mechanism to avoid producing over-smoothed results. TIME-Net or PIE-ARNet makes use of the prior information only in the image domain. SPIE-DIR further takes the prior information to enhance the projection inpainting. Inspired by Zhang et al. [70] and Wu et al. [71, 72], SPIE-DIR integrates projection inpainting, residual correction, and image reinforcement into an unrolled reconstruction framework, corresponding to the following three update steps:

$$\begin{cases} y_t^e = \varphi_t^e\left((Ax_{t-1}^e, y_p), \Theta_{\varphi_t^e}\right) \oplus y_0^e \\ r_t^e = \psi_t^e\left(A^T(Ax_{t-1}^e - y_t^e), \Theta_{\psi_t^e}\right) \\ x_t^e = x_{t-1}^e - \phi_t^e\left((x_0^e, \ldots, x_{t-1}^e, x_p), \Theta_{\phi_t^e}\right) - r_t^e \end{cases}, \quad e = \{l, h\}. \tag{18}$$

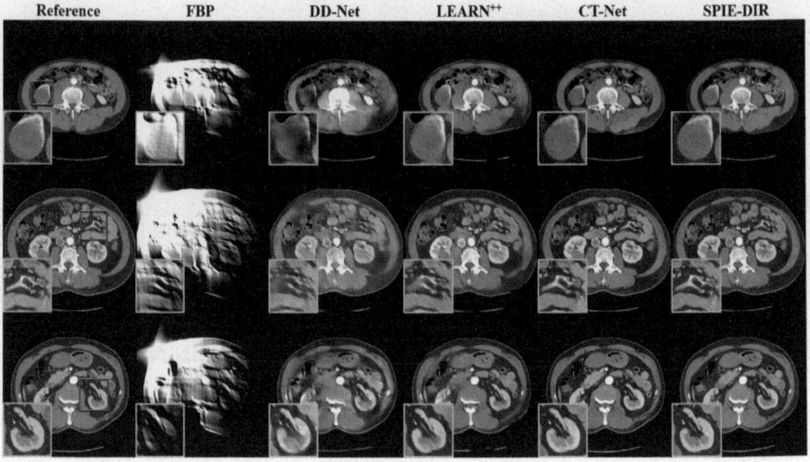

Fig. 6 Qualitative comparison between different limited-angle reconstruction methods on three slices. Two SECT-based methods (DD-Net [73] and LEARN++ [70]) and two DECT-based methods mentioned in this chapter (CT-Net [66] and SPIE-DIR [69]) are compared here

where x_t^e is the reconstructed image at iteration t under energy e, which can either be low- or high-energy; A stands for the full-sampling system matrix; y_0^e is the available projection data; x_p is the prior image; $y_p = Ax_p$ is the prior projection; $\varphi_t^e(\cdot)$, $\psi_t^e(\cdot)$, and $\phi_t^e(\cdot)$ are the corresponding deep neural networks in projection inpainting, residual correction, and image refinement, respectively. As a result of implementing the iterative reconstruction framework, SPIE-DIR will have a slower computation time compared to other approaches. However, this method offers clear benefits in terms of interpretability and imaging quality, as shown in Fig. 6.

3.3 Truncated Data Reconstruction

Data truncation often occurs when the width of the detector is insufficient to encompass the full object. In this scenario, the theoretical foundation of CT imaging cannot be fulfilled, and often, only an approximate image may be produced. In the presence of data truncation, the conventional filtered back-projection reconstruction method will result in significant truncation artifacts around the boundary of the scanning field of view (FoV). The artifact typically manifests as a prominent ring, with areas beyond the field of view being affected by the artifact caused by the constrained angle. Dual-energy imaging technology, which relies on dual-ray source systems, is more susceptible to data truncation compared to conventional single-energy spiral CT. As a result of limited rack space, one of the detectors, often utilized for high energy, had to decrease its width, making it prone to inadequate coverage of the human body.

To extend the DECT imaging FoV, Maier et al. proposed a deep learning method based on raw data consistency. This method falls in an iterative form which can be expressed as

$$\begin{cases} I_t^H = \phi\left(I_{t-1}^H, A^{-1}\left(A \cdot I_{t-1}^H - P_H\right)\right) \\ I_0^H = I_L \end{cases}, \tag{19}$$

where I_t^H is the predicted high-energy image at iteration t, I_L is the corresponding low-energy image, A is the system matrix, P_H is the high-energy projection data, and $\phi(\cdot)$ is a convolutional neural network. This method involves the introduction of prior information from the other energy level during initialization. A CNN is employed to estimate the values outside the FoV while also incorporating information on projection-domain consistency.

The method is evaluated using both simulated and real-scanned DECT datasets. The iterative approach described in the research produces images that exhibit greater proximity to the groundtruth when compared to direct image synthesis methods. Additionally, the proposed method is capable of addressing the issue of undifferentiable materials or failures in reconstructing low-energy images. To demonstrate this, the authors deliberately introduced several circular lesions into the low-energy images while ensuring that these lesions were not present in the high-energy projection data. The proposed approach effectively eliminated the lesions in the high-energy images as predicted. The presence of lesions persisted in the results obtained from other methods, suggesting that the images were inconsistent with the projection data.

4 Deep Learning in Material Decomposition

The primary advantage of dual-energy CT over single-energy CT is its capacity to perform material decomposition. Nevertheless, as stated in Sect. 1.3, current analytical techniques for material decomposition continue to experience artifacts and inaccuracies due to the utilization of a simplified model. Conversely, iterative approaches are frequently computationally demanding and may encounter problems with convergence. Direct inversion in the image domain is now the prevailing algorithm utilized in dual-energy material decomposition research. This approach condenses the physics underlying dual-energy CT images into the subsequent equation:

$$\begin{pmatrix} I_L \\ I_H \end{pmatrix} = \begin{pmatrix} \mu_{1L} & \mu_{2L} \\ \mu_{1H} & \mu_{2H} \end{pmatrix} \begin{pmatrix} I_1 \\ I_2 \end{pmatrix}, \tag{20}$$

where I_1 and I_2 are material maps of the two basis materials, and μ_{iL} and μ_{iH} are the equivalent attenuation coefficients of material i under the low- and high-energy spectra, respectively. The estimation of material maps I_1 and I_2 can be achieved using a straightforward matrix inversion procedure. In the presence of noise or artifacts in the DECT images, the direct inversion method is prone to amplifying the noise or artifacts, leading to material maps of low image quality. Researchers are continuously researching solutions to this issue, and material decomposition methods that rely on deep learning have also been extensively investigated.

The work by Zhang et al. [74] was one of the earliest attempts in this direction. The authors noticed that the direct inversion method can be written as the following equation:

$$\begin{pmatrix} I_1 \\ I_2 \end{pmatrix} = A^{-1}\begin{pmatrix} I_L \\ I_H \end{pmatrix} = \begin{pmatrix} a_{11} & a_{12} \\ a_{21} & a_{22} \end{pmatrix}\begin{pmatrix} I_L \\ I_H \end{pmatrix}, \tag{21}$$

and therefore having

$$\begin{cases} I_1 = a_{11}I_L + a_{12}I_H \\ I_2 = a_{21}I_L + a_{22}I_H \end{cases}. \tag{22}$$

Following the form in (22), they designed a network for material decomposition with the form of

$$\begin{cases} I_1 = \phi_{11}(I_L) + \phi_{12}(I_H) \\ I_2 = \phi_{21}(I_L) + \phi_{22}(I_H) \end{cases}, \tag{23}$$

where ϕ_{ij} are the four subnetworks. To enhance data processing efficiency, the authors incorporated supplementary convolution blocks as both input and output layers. The network was named the Butterfly network (Butterfly-Net) because of its dual-entry and dual-output crossover architecture (refer to Fig. 7a). In addition to the Butterfly-Net, Xu et al. [75] proposed a network for projection-domain material decomposition. Similarly, Wang et al. [76] developed a network to decompose DECT images into water and calcium images by incorporating advanced techniques in GAN models and transformer models. All of these approaches demonstrate remarkable performance on their respective testing datasets.

The dual material decomposition technique discussed above decomposes dual-energy images into two independent material maps, such as bone and water. Nevertheless, the material makeup within the human body is highly intricate and encompasses more than just two components. The dual-material model fails to accurately portray the real scenario. Additionally, the material maps generated from this model do not allow for precise measurement of the quantity of these substances. Researchers, therefore, implemented triple-material decomposition [77] by including an extra constraint that ensures the total of material fractions equals one:

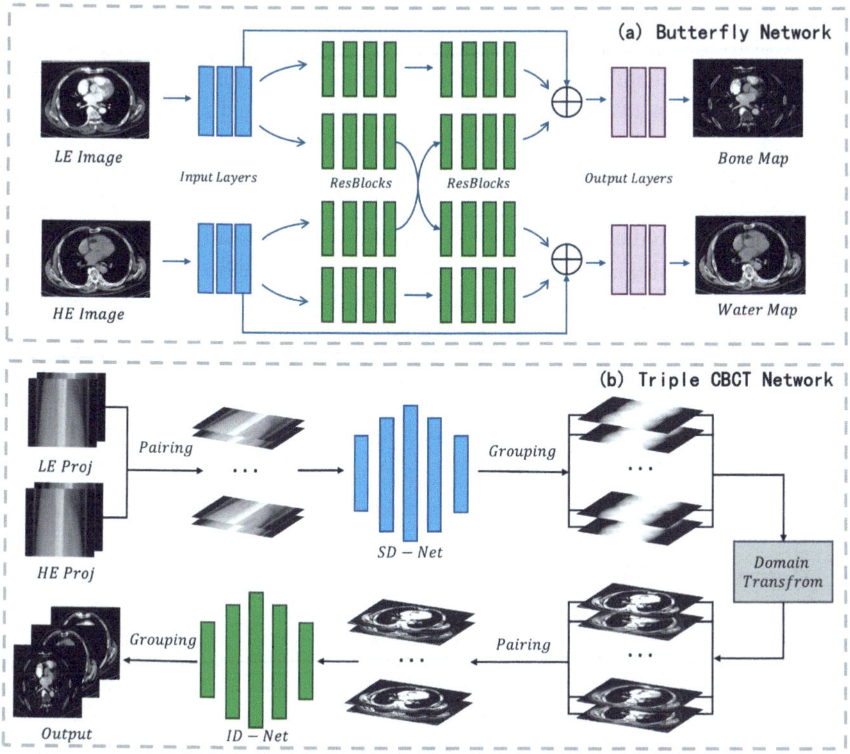

Fig. 7 Deep-learning-based material decomposition methods. (**a**) Butterfly-Net; (**b**) triple CBCT network by Zhu et al.

$$\begin{pmatrix} I_L \\ I_H \\ \mathbf{1} \end{pmatrix} = \begin{pmatrix} \mu_{1L} & \mu_{2L} & \mu_{3L} \\ \mu_{1H} & \mu_{2H} & \mu_{3H} \\ 1 & 1 & 1 \end{pmatrix} \cdot \begin{pmatrix} I_1 \\ I_2 \\ I_3 \end{pmatrix} \tag{24}$$

where $\mathbf{1}$ is all 1 array with same size of I_L and I_H. Deep-learning-based triple-material decomposition methods have also emerged in recent years.

Building upon a dual-material decomposition network developed by Su et al. [78], Zhu et al. [79] introduced a dual-domain triple-material decomposition network. This network is designed to generate iodine, bone, and water maps from the dual-energy raw data. The employed architecture bears resemblance to numerous dual-domain reconstruction networks [63, 64], including two networks: the sinogram-domain network (SD-Net) and the image-domain network (ID-Net), interconnected by a domain transform operator (refer to Fig. 7b). Instead of employing DECT images, the authors generated the training base images by synthesizing them with natural photos sourced from the ImageNet dataset [80]. The initial image I_R was obtained by choosing the red channel from an RGB image. The values in I_R were normalized to a range of $[0, 1]$ and subsequently blurred using down-sampling and up-sampling operations. The training material maps were subsequently computed using predetermined thresholds. The results obtained via their network demonstrate significantly reduced noise in comparison to the analytical multi-material decomposition method. The concept of utilizing natural image datasets for training purposes is valuable and can be considered in comparable applications.

Several recent studies (e.g., [81–84] etc.) have focused on utilizing deep learning techniques for multiple material decomposition. The methods get inputs from photon-counting CT, which is not directly pertinent to the topic of this section. Nevertheless, the majority of the innovations made in these studies can also be used for the DECT material decomposition. As an illustration, Gong et al. [81] incorporated a regularization term into the loss function to preserve edges in images. Fang et al. [82] integrated self-supervised Noise2Noise prior [85] into an iterative material decomposition framework.

5 Remaining Challenges

Despite the extensive research conducted on the application of deep learning technology in DECT imaging, particularly in the medical field, there are still numerous unresolved concerns about the development and training of deep models.

- **Model Robustness:** How to ensure the robustness of the model in a wider range of application scenarios is an issue that all deep learning researchers need to consider. This issue is most evident in the DECT synthesis task. If the low- or high-energy spectrum for testing data deviates from the training data, it will inevitably result in inaccuracies in the network's predictions. Lyu's method [43] is expected to work with different high-energy spectra by introducing additional high-energy spectrum information. However, this approach is insufficient to fully resolve the problem. Exploring the possibility of using Yao's idea [53] and using the two spectra as supplementary data in the network to develop a universal estimation method is a potential avenue for future research. Moreover, current network models are typically limited to addressing imaging tasks within a given anatomical region, such as the abdomen or head. Studying the implementation of a universal model for various anatomical regions is also worth considering.
- **Training Data:** As the number of parameters in new models continues to rise, the amount of data needed to train these models likewise grows. Acquiring training data for dual-energy CT-related research is more challenging compared to the field of natural image processing. DECT systems are not universally implemented in all hospitals, and the acquired DECT images may also exhibit temporal inconsistency issues, particularly in cardiac dual-energy CT images. There are mainly two ways to resolve this issue. Initially, the dataset can be expanded automatically. The approach proposed by Zhu [79] serves as a nice

illustration. One may also use newly developed foundation models to simulate DECT datasets. Additionally, a potential approach is to investigate the few-shot learning techniques for training models using a small amount of data.
- **Evaluation Metrics:** The majority of these contemporary works employ similarity evaluation measures that are extensively utilized in image processing, such as RMSE, PSNR, and SSIM, to assess the quality of images. Nevertheless, medical images lack the piecewise consistent qualities found in natural images, and these indicators fail to sufficiently reflect the way the human eye perceives image quality. Thus, it is imperative to study image quality assessment measures that align more closely with human visual perception to guide training deep learning models.

Undoubtedly, the ultimate objective of all the deep-learning-based dual-energy imaging models discussed in this chapter is toward clinical application. However, which specific deep learning model is currently accessible for clinical use? We contend that the subsequent three possible paths are worthy of investigation:

- **Diagnostic Values:** The primary goal of medical imaging is diagnosis and treatment. The mining of downstream tasks is crucial for dual-energy imaging based on deep learning, particularly for image synthesis tasks. Fink's [59] and Jiang's [60] research are commendable beginnings in this context, as well as Sheikhi's work [86] for kidney stones. These studies demonstrate that DECT synthesis can enhance the diagnostic precision of subsequent models or physicians in specific tasks. Furthermore, it is worthwhile to investigate the potential of utilizing deep learning in dual-energy imaging for the automated removal of bone structures and the computation of the stopping power ratio.
- **System Adaptation:** Prior to implementation, DL-based DECT imaging techniques must be tailored to a realistic CT imaging system. The significance of this matter is especially crucial for DECT reconstruction tasks, as the efficacy of these algorithms is intricately linked to the imaging geometry. Researchers must validate the algorithms in a scanning environment that closely resembles the real-world scenario or investigate models that can handle various geometries without being degraded.
- **Interpretability:** While clinical application may not necessarily require model interpretability, models that possess greater interpretability are more likely to persuade doctors of their potential. Furthermore, more comprehensible models could assist doctors and physics researchers in examining the physical characteristics of lesions and advancing the field of medical physics.

Generally, the current research on DL-based DECT imaging is at an early stage of development. In the future, more model methods will progressively arise to enhance the quality of dual-energy imaging. We also anticipate that future-related research will focus more on commercial systems and clinical practices, thereby truly promoting DL-based DECT imaging to clinical applications.

6 Summary

The use of DECT in clinical practice is highly significant since it can provide precise information about different materials. Despite the development and implementation of many DECT solutions, concurrent DECT systems still face challenges related to the trade-off between configuration cost, imaging quality, and radiation exposure. With the thriving of deep learning, deep neural networks have been integrated into several areas of DECT imaging, including data synthesis, image reconstruction, and material decomposition. These DL-based methods show a strong ability to improve DECT imaging quality. Looking forward, deep-learning-based DECT imaging holds significant promise to revolutionize the field of medical imaging.

Acknowledgments This work is supported in part by the Key Research and Development Program of Zhejiang Province under Grant 2021C03029, the Natural Science Foundation of Zhejiang Province under Grant LZ23A050002, and the National Natural Science Foundation of China (12175012) and the '111' center (B20065).

References

1. G. Wang, H. Yu, B.D. Man, An outlook on x-ray CT research and development. Med. Phys. **35**(3), 1051–1064 (2008)
2. G.N. Hounsfield, Computerized transverse axial scanning (tomography): Part 1. description of system. Br. J. Radiol. **46**(552), 1016–1022 (1973)
3. R.A. Brooks, A quantitative theory of the Hounsfield unit and its application to dual energy scanning. J. Comput. Assist. Tomogr. **1**(4), 487–493 (1977)
4. R.A. Rutherford, B.R. Pullan, I. Isherwood, Measurement of effective atomic number and electron density using an emi scanner. Neuroradiology **11**, 15–21 (1976)
5. K. Branch, S. Patel, Future in dual energy CT, in *Dual-Energy CT in Cardiovascular Imaging* (2015), pp. 259–267
6. P. Engler, W.D. Friedman, Review of dual-energy computed tomography techniques. Mater. Eval. **48**, 623–629 (1990)
7. R.E. Alvarez, A. Macovski, Energy-selective reconstructions in x-ray computerised tomography. Phys. Med. Biol. **21**(5), 733 (1976)
8. X. Liu, L. Yu, A.N. Primak, C.H. McCollough, Quantitative imaging of element composition and mass fraction using dual-energy CT: three-material decomposition. Med. Phys. **36**(5), 1602–1609 (2009)
9. T.P. Szczykutowicz, G.-H. Chen, Dual energy CT using slow kvp switching acquisition and prior image constrained compressed sensing. Phys. Med. Biol. **55**(21), 6411 (2010)
10. T. Niu, X. Dong, M. Petrongolo, L. Zhu, Iterative image-domain decomposition for dual-energy CT. Med. Phys. **41**(4), 041901 (2014)
11. W. Zhao, T. Niu, L. Xing, Y. Xie, G. Xiong, K. Elmore, J. Zhu, L. Wang, J.K. Min. Using edge-preserving algorithm with non-local mean for significantly improved image-domain material decomposition in dual-energy CT. Phys. Med. Biol. **61**(3), 1332–1351 (2016)
12. C. Maaß, E. Meyer, M. Kachelrieß, Exact dual energy material decomposition from inconsistent rays (MDIR). Med. Phys. **38**(2), 691–700 (2011)
13. R.F. Barber, E.Y. Sidky, T.G. Schmidt, X. Pan, An algorithm for constrained one-step inversion of spectral CT data. Phys. Med. Biol. **61**(10), 3784 (2016)

14. K. Mechlem, S. Ehn, T. Sellerer, E. Braig, D. Münzel, F. Pfeiffer, P.B. Noël, Joint statistical iterative material image reconstruction for spectral computed tomography using a semi-empirical forward model. IEEE Trans. Med. Imaging **37**(1), 68–80 (2018)
15. W. Zhao, D. Vernekohl, F. Han, B. Han, H. Peng, Y. Yang, L. Xing, J.K. Min, A unified material decomposition framework for quantitative dual- and triple-energy CT imaging. Med. Phys. **45**(7), 2964–2977 (2018)
16. L.-J. Zhang, J. Peng, S.-Y. Wu, Z.J. Wang, X.-S. Wu, C.-S. Zhou, X.-M. Ji, G.-M. Lu, Liver virtual non-enhanced CT with dual-source, dual-energy CT: a preliminary study. Eur. Radiol. **20**, 2257–2264 (2010)
17. L. Yu, J.A. Christner, S. Leng, J. Wang, J.G. Fletcher, C.H. McCollough, Virtual monochromatic imaging in dual-source dual-energy CT: Radiation dose and image quality. Med. Phys. **38**(12), 6371–6379 (2011)
18. L. Yu, S. Leng, C.H. McCollough, Dual-energy CT-based monochromatic imaging. AJR Am. J. Roentgenol. **199**(5 Suppl), 9–15 (2012)
19. A. Macovski, R.E. Alvarez, J.L.-H. Chan, J.P. Stonestrom, L.M. Zatz, Energy dependent reconstruction in x-ray computerized tomography. Comput. Biol. Med. **6**(4), 325–336 (1976)
20. C.H. McCollough, S. Leng, L. Yu, J.G. Fletcher, Dual-and multi-energy CT: principles, technical approaches, and clinical applications. Radiology **276**(3), 637–653 (2015)
21. Y. Zou, M.D. Silver, Analysis of fast kv-switching in dual energy CT using a pre-reconstruction decomposition technique, in *Medical Imaging 2008: Physics of Medical Imaging*, vol. 6913, pp. 392–403 (SPIE, New York, 2008)
22. K. Matsumoto, M. Jinzaki, Y. Tanami, A. Ueno, M. Yamada, S. Kuribayashi, Virtual monochromatic spectral imaging with fast kilovoltage switching: improved image quality as compared with that obtained with conventional 120-kvp CT. Radiology **259**(1), 257–262 (2011)
23. T.R.C. Johnson, B. Krauss, M. Sedlmair, M. Grasruck, H. Bruder, D. Morhard, C. Fink, S. Weckbach, M. Lenhard, B. Schmidt, et al., Material differentiation by dual energy CT: initial experience. Eur. Radio. **17**, 1510–1517 (2007)
24. G.T. Barnes, R.A. Sones, M.M. Tesic, D.R. Morgan, J.N. Sanders, Detector for dual-energy digital radiography. Radiology **156**(2), 537–540 (1985)
25. L. Yu, S. Leng, C.H. McCollough, Dual-energy CT–based monochromatic imaging. Am. J. Roentgenol. **199**(5_supplement), S9–S15 (2012)
26. B. Rutt, A. Fenster, Split-filter computed tomography: a simple technique for dual energy scanning. J. Comput. Assist. Tomogr. **4**(4), 501–509 (1980)
27. A. Euler, A. Parakh, A.L. Falkowski, S. Manneck, D. Dashti, B. Krauss, Z. Szucs-Farkas, S.T. Schindera, Initial results of a single-source dual-energy computed tomography technique using a split-filter: assessment of image quality, radiation dose, and accuracy of dual-energy applications in an in vitro and in vivo study. Invest. Radiol. **51**(8), 491–498 (2016)
28. B. Li, D. Spronk, Y. Luo, C. Puett, C.R. Inscoe, D.A. Tyndall, Y.Z. Lee, J. Lu, O. Zhou, Feasibility of dual-energy cbct by spectral filtration of a dual-focus cnt x-ray source. Plos One **17**(2), e0262713 (2022)
29. H. Chandarana, A.J. Megibow, B.A. Cohen, R. Srinivasan, D. Kim, C. Leidecker, M. Macari, Iodine quantification with dual-energy CT: Phantom study and preliminary experience with renal masses. Am. J. Roentgenol. **196**(6), W693–W700 (2011)
30. C.M. Phan, A.J. Yoo, J.A. Hirsch, R.G. Nogueira, R. Gupta, Differentiation of hemorrhage from iodinated contrast in different intracranial compartments using dual-energy head CT. Am. J. Neuroradiol. **33**(6), 1088–1094 (2012)
31. W.H. Sommer, et al., The value of dual-energy bone removal in maximum intensity projections of lower extremity computed tomography angiography. Invest. Radiol. **44**(5), 285–292 (2009)
32. B. Buerke, G. Wittkamp, H. Seifarth, W. Heindel, S.P. Kloska, Dual-energy cta with bone removal for transcranial arteries: intraindividual comparison with standard CTA without bone removal and TOF-MRA. Acad. Radiol. **16**(11), 1348–1355 (2009)

33. D. Morhard, C. Fink, A. Graser, M.F. Reiser, C. Becker, T.R.C. Johnson, Cervical and cranial computed tomographic angiography with automated bone removal: dual energy computed tomography versus standard computed tomography. Invest. Radiol. **44**(5), 293–297 (2009)
34. B. Schulz, et al., Automatic bone removal technique in whole-body dual-energy CT angiography: performance and image quality. Am. J. Roentgenol. **199**(5), W646–W650 (2012)
35. A.N. Primak et al., Noninvasive differentiation of uric acid versus non–uric acid kidney stones using dual-energy CT. Acad. Radiol. **14**(12), 1441–1447 (2007)
36. G. Ascenti et al., Stone-targeted dual-energy CT: a new diagnostic approach to urinary calculosis. Am. J. Roentgenol. **195**(4), 953–958 (2010)
37. M. Qu et al., Urinary stone differentiation in patients with large body size using dual-energy dual-source computed tomography. Eur. Radio. **23**(5), 1408–1414 (2013)
38. S. Leng et al., Feasibility of discriminating uric acid from non–uric acid renal stones using consecutive spatially registered low-and high-energy scans obtained on a conventional CT scanner. Am. J. Roentgenol. **204**(1), 92–97 (2015)
39. C.-K. Liu, C.-C. Liu, C.-H. Yang, H.-M. Huang, Generation of brain dual-energy CT from single-energy CT using deep learning. J. Digit. Imaging **34**(1), 149–161 (2021)
40. T. Henzler, C. Fink, S.O. Schoenberg, U.J. Schoepf, Dual-energy CT: radiation dose aspects. AJR-Am. J. Roentgenol. **199**(5), S16 (2012)
41. J.R. Grajo, D.V. Sahani, Dual-energy CT of the abdomen and pelvis: radiation dose considerations. J. Am. Coll. Radiol. **15**(8), 1128–1132 (2018)
42. J.R. Wortman, J.Y. Shyu, J. Dileo, J.W. Uyeda, A.D. Sodickson, Dual-energy CT for routine imaging of the abdomen and pelvis: radiation dose and image quality. Emerg. Radiol. **27**, 45–50 (2020)
43. T. Lyu, W. Zhao, Y. Zhu, Z. Wu, Y. Zhang, Y. Chen, L. Luo, S. Li, L. Xing, Estimating dual-energy CT imaging from single-energy CT data with material decomposition convolutional neural network. Med. Image Anal. **70**, 102001 (2021)
44. J. Lyu, Y. Fu, M. Yang, Y. Xiong, Q. Duan, C. Duan, X. Wang, X. Xing, D. Zhang, J. Lin, et al., Generative adversarial network–based noncontrast CT angiography for aorta and carotid arteries. Radiology **309**(2), e230681 (2023)
45. C. Duan, X. Bian, K. Cheng, J. Lyu, Y. Xiong, S. Xiao, X. Wang, Q. Duan, C. Li, J. Huang, et al., Synthesized 7t mprage from 3t mprage using generative adversarial network and validation in clinical brain imaging: A feasibility study. J. Magn. Reson. Imaging **59**(5), 1620–1629 (2023)
46. S. Li, Y. Wang, Y. Liao, J. He, D. Zeng, Z. Bian, J. Ma, Pseudo dual energy CT imaging using deep learning based framework: initial study. arXiv preprint arXiv:1711.07118 (2017)
47. Y. Liao, Y. Wang, S. Li, J. He, D. Zeng, Z. Bian, J. Ma, Pseudo dual energy CT imaging using deep learning-based framework: basic material estimation, in *Medical Imaging 2018: Physics of Medical Imaging*, vol. 10573 (SPIE, New York, 2018), pp. 1190–1194
48. W. Zhao, T. Lv, P. Gao, L. Shen, X. Dai, K. Cheng, M. Jia, Y. Chen, L. Xing, A deep learning approach for dual-energy CT imaging using a single-energy CT data, in *15th International Meeting on Fully Three-Dimensional Image Reconstruction in Radiology and Nuclear Medicine*, vol. 11072 (SPIE, New York, 2019), pp. 355–359
49. W. Zhao, T. Lv, R. Lee, Y. Chen, L. Xing, Obtaining dual-energy computed tomography (CT) information from a single-energy CT image for quantitative imaging analysis of living subjects by using deep learning, in *Pacific Symposium on Biocomputing 2020* (World Scientific, Singapore, 2019), pp. 139–148
50. W. Zhao, T. Lv, P. Gao, L. Shen, X. Dai, K. Cheng, M. Jia, Y. Chen, L. Xing, Dual-energy CT imaging using a single-energy CT data is feasible via deep learning. arXiv preprint arXiv:1906.04874 (2019)
51. I. Goodfellow, J. Pouget-Abadie, M. Mirza, B. Xu, D. Warde-Farley, S. Ozair, A. Courville, Y. Bengio, Generative adversarial nets. Adv. Neural Inf. Proces. Syst. **27** (2014)
52. L. Yao, S. Li, Z. Wei, Y. Deng, M. Zhu, Z. Bian, J. Huang, Q. Lyu, D. Zeng, J. Ma, Direct energy-resolving CT imaging via energy-integrating CT images using a unified generative adversarial network, in *2019 IEEE Nuclear Science Symposium and Medical Imaging Conference (NSS/MIC)* (IEEE, New York, 2019), pp. 1–3

53. L. Yao, S. Li, D. Li, M. Zhu, Q. Gao, S. Zhang, Z. Bian, J. Huang, D. Zeng, J. Ma, Leveraging deep generative model for direct energy-resolving CT imaging via existing energy-integrating CT images, in *Medical Imaging 2020: Physics of Medical Imaging*, vol. 11312, ed. by H. Bosmans, G.-H. Chen (SPIE, New York, 2020), pp. 175
54. P. Isola, J.-Y. Zhu, T. Zhou, A.A. Efros, Image-to-image translation with conditional adversarial networks, in *Proceedings of the IEEE Conference on Computer Vision and Pattern Recognition* (2017), pp. 1125–1134
55. D. Kawahara, S. Ozawa, T. Kimura, Y. Nagata, Image synthesis of monoenergetic CT image in dual-energy CT using kilovoltage CT with deep convolutional generative adversarial networks. J. Appl. Clin. Med. Phys. **22**(4), 184–192 (2021)
56. J. Jeong, A. Wentland, D. Mastrodicasa, G. Fananapazir, A. Wang, I. Banerjee, B.N. Patel, Synthetic dual-energy CT reconstruction from single-energy CT using artificial intelligence. Abdominal Radiology **48**(11), 3537–3549 (2023)
57. T.-C. Wang, M.-Y. Liu, J.-Y. Zhu, A. Tao, J. Kautz, B. Catanzaro, High-resolution image synthesis and semantic manipulation with conditional gans, in *Proceedings of the IEEE Conference on Computer Vision and Pattern Recognition* (2018), pp. 8798–8807
58. J.-Y. Zhu, T. Park, P. Isola, A.A. Efros, Unpaired image-to-image translation using cycle-consistent adversarial networks, in *Proceedings of the IEEE International Conference on Computer Vision* (2017), pp. 2223–2232
59. M.A. Fink, C. Seibold, H.U. Kauczor, R. Stiefelhagen, J. Kleesiek, Jointly optimized deep neural networks to synthesize monoenergetic images from single-energy CT angiography for improving classification of pulmonary embolism. Diagnostics **12**(5), 1224 (2022)
60. C. Jiang, Y. Pan, T. Wang, Q. Chen, J. Yang, Li Ding, J. Liu, Z. Ding, D. Shen, S2DGAN: Generating dual-energy CT from single-energy CT for real-time determination of intracerebral hemorrhage, in *International Conference on Information Processing in Medical Imaging* (Springer, Berlin, 2023), pp. 375–387
61. A. Vaswani, N. Shazeer, N. Parmar, J. Uszkoreit, L. Jones, A.N. Gomez, Ł. Kaiser, I. Polosukhin, Attention is all you need, in *Advances in Neural Information Processing Systems*, vol. 30 (2017)
62. H. Peiris, M. Hayat, Z. Chen, G. Egan, M. Harandi, A robust volumetric transformer for accurate 3d tumor segmentation, in *International Conference on Medical Image Computing and Computer-Assisted Intervention* (Springer, Berlin, 2022), pp. 162–172
63. Y. Zhang, T. Lv, R. Ge, Q. Zhao, D. Hu, L. Zhang, J. Liu, Y. Zhang, Q. Liu, W. Zhao, et al., CD-NET: comprehensive domain network with spectral complementary for dect sparse-view reconstruction. IEEE Trans. Comput. Imaging **7**, 436–447 (2021)
64. X. Yin, J.-L. Coatrieux, Q. Zhao, J. Liu, W. Yang, J. Yang, G. Quan, Y. Chen, H. Shu, L. Luo, Domain progressive 3D residual convolution network to improve low-dose CT imaging. IEEE Trans. Med. Imaging **38**(12), 2903–2913 (2019)
65. W.-A. Lin, H. Liao, C. Peng, X. Sun, J. Zhang, J. Luo, R. Chellappa, S.K. Zhou, Dudonet: Dual domain network for CT metal artifact reduction, in *Proceedings of the IEEE/CVF Conference on Computer Vision and Pattern Recognition* (2019), pp. 10512–10521
66. K. Chen, G. Ji, C. Wang, Z. Peng, X. Ji, H. Tang, C. Yang, Y. Chen, CT-NET: Cascaded t-shape network using spectral redundancy for dual-energy CT limited-angle reconstruction. Biomed. Signal Process. Control **79**, 104072 (2023)
67. Y. Zhang, D. Hu, Z. Yan, Q. Zhao, G. Quan, S. Luo, Y. Zhang, Y. Chen, TIME-Net: Transformer-Integrated Multi-Encoder Network for limited-angle artifact removal in dual-energy CBCT. Med. Image Anal. **83**(August 2022), 102650 (2023)
68. Y. Zhang, D. Hu, T. Lyu, J. Zhu, G. Quan, J. Xiang, G. Coatrieux, S. Luo, Y. Chen, PIE-ARNET: Prior image enhanced artifact removal network for limited-angle DECT. IEEE Trans. Instrum. Meas. **72**, 1–12 (2023)
69. Y. Zhang, D. Hu, T. Lyu, G. Quan, J. Xiang, G. Coatrieux, S. Luo, Y. Chen, SPIE-DIR: Self-prior information enhanced deep iterative reconstruction using two complementary limited-angle scans for DECT. IEEE Trans. Instrum. Meas. **72**, 1–12 (2023)

70. Y. Zhang, H. Chen, W. Xia, Y. Chen, B. Liu, Y. Liu, H. Sun, J. Zhou, Learn++: recurrent dual-domain reconstruction network for compressed sensing CT. IEEE Trans. Radiat. Plasma Med. Sci. **7**(2), 132–142 (2022)
71. W. Wu, D. Hu, W. Cong, H. Shan, S. Wang, C. Niu, P. Yan, H. Yu, V. Vardhanabhuti, G. Wang, Stabilizing deep tomographic reconstruction: Part A. Hybrid framework and experimental results. Patterns (N Y) **3**(5), 100474 (2022)
72. W. Wu, D. Hu, W. Cong, H. Shan, S. Wang, C. Niu, P. Yan, H. Yu, V. Vardhanabhuti, G. Wang, Stabilizing deep tomographic reconstruction: Part B. Convergence analysis and adversarial attacks. Patterns (N Y) **3**(5):100475 (2022). https://doi.org/10.1016/j.patter.2022.100475. PMID: 35607615; PMCID: PMC9122974
73. Z. Huang, X. Liu, R. Wang, Z. Chen, Y. Yang, X. Liu, H. Zheng, D. Liang, Z. Hu, Learning a deep cnn denoising approach using anatomical prior information implemented with attention mechanism for low-dose CT imaging on clinical patient data from multiple anatomical sites. IEEE J. Biomed. Health Inform. **25**(9), 3416–3427 (2021)
74. W. Zhang, H. Zhang, L. Wang, X. Wang, X. Hu, A. Cai, L. Li, T. Niu, B. Yan, Image domain dual material decomposition for dual-energy CT using butterfly network. Med. Phys. **46**(5), 2037–2051 (2019)
75. Y. Xu, B. Yan, J. Chen, L. Zeng, L. Li, Projection decomposition algorithm for dual-energy computed tomography via deep neural network. J. X-Ray Sci. Technol. **26**(3), 361–377 (2018)
76. G. Wang, Z. Liu, Z. Huang, N. Zhang, H. Luo, Improved GAN : using a transformer module generator approach for material decomposition. Comput. Biol. Med. **149**(March), 105952 (2022)
77. P.V. Granton, S.I. Pollmann, N.L. Ford, M. Drangova, D.W. Holdsworth, Implementation of dual- and triple-energy cone-beam micro-CT for postreconstruction material decomposition. Med. Phys. **35**(11), 5030–5042 (2008)
78. T. Su, X. Sun, J. Yang, D. Mi, Y. Zhang, H. Wu, S. Fang, Y. Chen, H. Zheng, D. Liang, et al., DIRECT-Net: a unified mutual-domain material decomposition network for quantitative dual-energy CT imaging. Med. Phys. **49**(2), 917–934 (2022)
79. J. Zhu, T. Su, X. Zhang, J. Yang, D. Mi, Y. Zhang, X. Gao, H. Zheng, D. Liang, Y. Ge, Feasibility study of three-material decomposition in dual-energy cone-beam CT imaging with deep learning. Phys. Med. Biol. **67**(14), 145012 (2022)
80. J. Deng, W. Dong, R. Socher, L.-J. Li, K. Li, L. Fei-Fei, Imagenet: a large-scale hierarchical image database, in *2009 IEEE Conference on Computer Vision and Pattern Recognition* (IEEE, New York, 2009), pp. 248–255
81. H. Gong, S. Tao, K. Rajendran, W. Zhou, C.H. McCollough, S. Leng, Deep-learning-based direct inversion for material decomposition. Med. Phys. **47**(12), 6294–6309 (2020)
82. W. Fang, D. Wu, K. Kim, M.K. Kalra, R. Singh, L. Li, Q. Li, Iterative material decomposition for spectral CT using self-supervised noise2noise prior. Phys. Med. Biol. **66**(15), 155013 (2021)
83. R. Nadkarni, A. Allphin, D.P. Clark, C.T. Badea, Material decomposition from photon-counting CT using a convolutional neural network and energy-integrating CT training labels. Phys. Med. Biol. **67**(15), 155003 (2022)
84. D. Fujiwara, T. Shimomura, W. Zhao, K.W. Li, A. Haga, L.S. Geng, Virtual computed-tomography system for deep-learning-based material decomposition. Phys. Med. Biol. **67**(15), 155008 (2022)
85. J. Lehtinen, J. Munkberg, J. Hasselgren, S. Laine, T. Karras, M. Aittala, T. Aila, Noise2noise: Learning image restoration without clean data. arXiv preprint arXiv:1803.04189 (2018)
86. M. Sheikhi, S. Sina, M. Karimipourfard, Deep-learned generation of renal dual-energy CT from a single-energy scan. Clin. Radiol. **79**(1), e17–e25 (2024)

Performance Evaluation of Implicit Neural Representations in Diagnostic Fan-Beam CT Imaging

Wenhui Qin, Zhentao Liu, Xiaopeng Yu, Mengqing Su, Yang Yang, Yikun Zhang, Yuyao Zhang, Zhiming Cui, Yang Chen, and Xiaochun Lai

1 Introduction

Computed tomography (CT), a prevalent medical imaging technique, noninvasively probes the internal structure of subjects, finding extensive application in clinical diagnosis [26]. A typical scan geometry is fan-beam CT (FBCT) as illustrated in Fig. 1. The CT image reconstruction aims to transform measurement data into detailed images of the scanned object, a process essentially defined as solving an inverse problem [22, 26]. CT reconstruction methodologies traditionally fall into three main categories: analytic, iterative, and deep-learning-based [27]. Among these, filtered back-projection (FBP) is renowned for its analytical application in FBCT [12]. Iterative reconstruction, employing a numerical forward model and a feedback loop to iteratively update image estimate [2, 7], appreciated for its efficacy in reducing inherent measurement noise and managing the sensitivity of mathematical inversion.

The rise of machine learning has propelled deep learning strategies to the forefront of CT reconstruction [14, 27], offering a novel approach that eschews the need for an exact physical imaging model. Lately, supervised deep learning methods have shown considerable potential, although their success significantly depends on the diversity and scale of training data [11], which limits their practical application in clinics due to their incapacity on scarce paired data [3].

W. Qin · Z. Liu · X. Yu · X. Lai (✉)
School of Biomedical Engineering, ShanghaiTech University, Shanghai, China
e-mail: laixch@shanghaitech.edu.cn

M. Su · Y. Yang · Y. Zhang · Z. Cui · Y. Chen
School of Computer Science and Engineering, Southeast University, Nanjing, China

Y. Zhang
School of Information Science and Technology, ShanghaiTech University, Shanghai, China

K. (K.) Iniewski, L. (K.) Cai (eds.), *Deep Learning for Advanced X-ray Detection and Imaging Applications*, https://doi.org/10.1007/978-3-031-75653-5_4

Implicit neural representation (INR) utilizes multilayer perceptrons (MLP) to establish a continuous function to learn the mapping between coordinates and the attributes of signals, proving effective for solving inverse problems in a self-supervised fashion. Inspired by the effort of INR, numerous INR-based neural reconstruction techniques [18, 23, 25, 29, 30, 33, 34] on FBCT have been recently proposed, yielding remarkable results based on conventional image quality metrics in natural images for evaluating reconstruction accuracy, such as peak signal-to-noise ratio (PSNR) [4] and structural similarity index measure (SSIM) [28]. Furthermore, for INR-based techniques to be viable in clinical applications, they must meet the imaging quality standards approved by the American College of Radiology (ACR) [1], including metrics such as CT number accuracy, low-contrast detectability (LCD), and spatial resolution. Additionally, the capacity of INR-based methods to accurately depict human soft tissues requires further investigation to assess their clinical feasibility and potential limitations.

In this work, we conduct comprehensive evaluation studies to assess the imaging performance of typical INR methods using clinical standard protocols and metrics. Our initial step involves simulating the FBCT scanning process adhering to standard clinical protocols, followed by acquiring data from various phantoms, specifically quality assurance (QA) phantoms [8] and XCAT torso phantoms [21]. We then employ four representative INR methods for reconstruction, benchmarking these against the FBP method. The metrics accredited by the ACR [1] in imaging studies—CT number accuracy, image uniformity, low-contrast detectability, and spatial resolution [8, 9]—are pivotal for clinical utility. We employ these metrics to evaluate the imaging performance of INR-based reconstruction techniques. Our qualitative and quantitative analyses provide valuable insights, showing that INR-based methods deliver promising performance while also revealing opportunities for improvement to meet clinical standards.

2 Method

2.1 Evaluation Pipeline

Our evaluation pipeline is illustrated in Fig. 1 and encompasses three critical stages: FBCT processing, INR-based neural reconstruction, and image studies. We initially employ a toolkit for simulating a commercial FBCT processing, aligning clinical CT geometries, and executing image reconstruction via the FBP method. We subsequently adopt several representative INR-based neural reconstruction techniques for FBCT, including classical function-expansion-based INR methods (e.g., the frequency positional encoding (PE+MLP) [15] and SIREN [24]) and state-of-the-art (SOTA) hash-based INR methods (Instant NGP [16] and DINER [31]). Our final phase involves rigorous imaging studies to assess the performance of these INR-based methods within clinical settings. We utilized QA and XCAT torso phantoms, adhering to clinical protocols, to study if these techniques satisfy key clinical requirements.

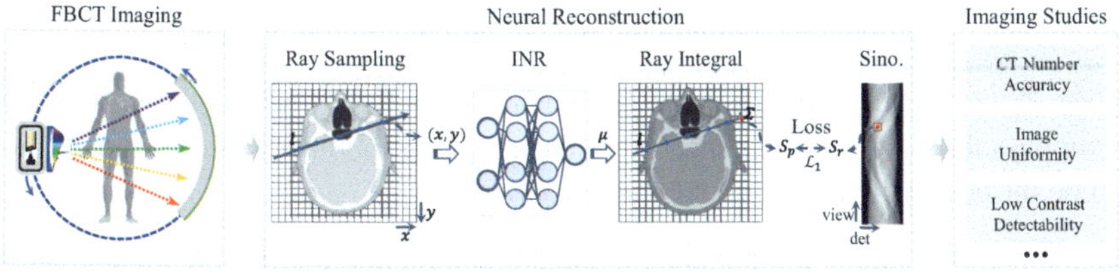

Fig. 1 Imaging evaluation pipeline consists of three components. The sinograms are generated by the FBCT imaging module, and then CT images are reconstructed using INR-based techniques in the neural reconstruction module. Finally, the imaging studies module is designed to evaluate their imaging performance in clinical CT applications

2.2 FBCT Imaging

During CT acquisition, the X-ray photons emitted from the source are sequentially attenuated by a bowtie filter and the scanned object, governed by Lambert-Beer's law [5]. We simulate a commercial clinical diagnostic FBCT process using typical parameters of CT geometry configuration [10] to generate standard datasets for various phantoms. Additionally, since the random number of both detected X-ray photons and optical photons generated by the scintillator follows a Poisson process—where the mean of the detected counts is equal to their variance—we add Poisson noise to each detector cell to replicate the inherent noise in CT measurements accurately. The correction of beam hardening [13] is also employed to process the counts recorded in the detector before image reconstruction. The sinogram S_r for ray l converted from corrected count projection could be formulated as follows:

$$S_r(l) = \int_l \mu(\mathbf{p})d\mathbf{p} = -\ln(N_1/N_0), \tag{1}$$

where $\mathbf{p} = (x, y) \in \mathbb{R}^2$ represents the spatial position along the casting ray, $\mu(\mathbf{p}) \in \mathbb{R}_0^+$ is the corresponding linear attenuation coefficient, and N_0 and N_1 denote initial and attenuated X-ray counts, respectively. These converted sinograms would be used to reconstruct CT images with both the conventional FBP method and various INR-based methods.

2.3 Implicit Neural Representation for CT Reconstruction

INR-based neural reconstruction leverages FBCT geometry to learn attenuation distribution through the comparison of observed and synthesized sinograms. This process optimizes a model that maps spatial positions to linear attenuation coefficients, facilitating CT image reconstruction through voxel querying. Mathematically, we aim to learn a mapping function denoted as follows:

$$\mu(\mathbf{p}) = \mathcal{F}_\Theta(\mathbf{p}) = \mathcal{F}_\Theta(x, y), \tag{2}$$

where $\mathcal{F}_\Theta$ is the learnable implicit neural representation. INR methods typically encompass three key components: ray sampling, INR module, and ray integral. We will be introduce them one by one in the following.

Ray Sampling Each pixel value in a sinogram encapsulates the cumulative attenuation experienced from the X-ray source to the detector across the scanned object. To discretely simulate the attenuation process of X-ray, we uniformly sample a sufficient number of points along each casting ray, ensuring a minimum of one sample per voxel grid in the reconstructed image. These coordinates are subsequently input into the INR module.

INR Module INR utilizes an MLP network to map spatial coordinates to attenuation coefficients in CT images. To address the low-frequency bias of MLP [17], the coordinates would be first mapped into higher-dimensional feature spaces by position encoder before input into the MLP network. Such a mechanism helps the model in learning high-frequency details while mitigating low-frequency overfitting. In this study, we explore several representative position encoding techniques in INR methods, including frequency-based encoding (PE+MLP) [15], SIREN activation [24], multi-resolution hash encoding (Instant NGP) [16], and disorder-invariant hash encoding (DINER) [31]. Then the encoded features would be input into the following MLP network to decode as the attenuation coefficient at a specific position.

Ray Integral The predicted sinogram S_p is numerically synthesized via ray integral along incident ray path l, mathematically expressed as Riemann sum:

$$S_p(l) = \sum_{i=1}^{\mathcal{N}} \mu(\mathbf{p}_i) \cdot \delta, \tag{3}$$

where $\mathbf{p}_i$ is a sample point, $\mathcal{N}$ is the total number of these points, and δ is the fixed distance between any two adjacent points.

Loss Function The INR module is optimized by minimizing L1 discrepancy between the observed sinograms S_r and synthesized ones S_p, expressed as follows:

$$\mathcal{L}(\mathcal{F}_\Theta) = \sum_{l \in \mathbb{L}} \| S_r(l) - S_p(l) \|_1, \tag{4}$$

where $\mathbb{L}$ denotes the set of rays.

CT Reconstruction The final output is a 2D matrix representing a target region of interest (ROI). It is created by querying voxel coordinates of the matrix through the trained INR module. This process estimates the attenuation coefficients of those coordinates.

2.4 Image Studies

Clinical CT applications demand more specialized evaluations of imaging studies when employed on standard phantoms such as QA phantoms and XCAT torso phantoms. The QA phantom includes uniformity phantom, low-contrast phantom, and point phantom in this work.

Human Soft Tissues Imaging Evaluation The performance of INR in clinical CT applications is assessed following standard clinical protocols, with scans covering typical body parts such as the lung, shoulder, teeth, and brain. This evaluation employs simulated data from XCAT torso phantoms, analyzing reconstructed images at specific clinical window levels (WL) and window widths (WW) [8] to scrutinize human soft tissue details.

CT Number Accuracy and Image Uniformity Assessment In adherence to the ACR CT accreditation standards [1], the CT number accuracy and image uniformity are evaluated using a uniformity phantom—a 20 cm cylindrical water phantom. This evaluation involves measuring the mean (CT number accuracy) and standard deviation (image noise) of CT numbers across the center and five peripheral circular ROIs, each with a diameter of 30 mm, in the reconstructed CT image. The image uniformity is measured by the largest difference in mean values among all measured ROIs. The mathematical formula converting the linear attenuation coefficient in each pixel to the CT number (HU) is expressed as:

$$HU = 1000 \cdot \left[\left(\mu - \mu_w \right) / \mu_w \right], \tag{5}$$

where HU, namely Hounsfield unit, denotes the unit of CT number, μ is the linear attenuation at each CT image pixel, and $\mu_w = 0.017$ represents the linear attenuation coefficient of water, specifically calculated in our study.

Low-Contrast Detectability Assessment Low-contrast detectability (LCD), the capability to distinguish materials with closely matched attenuation characteristics [8], is evaluated using a low-contrast phantom. This phantom in our study incorporates four peripheral circular objects, each with a 5 mm radius. These objects exhibit varying contrasts of 1% (with CT number of 10 HU), 3% (30 HU), 5% (50 HU), and 10% (100 HU), thereby facilitating a detailed examination of the reconstruction techniques' sensitivity to subtle differences.

Spatial Resolution Assessment The in-plane spatial resolution is visually demonstrated using a point phantom, characterized by a circular domain with a radius of 0.3 mm. Quantitatively, spatial resolution is further evaluated through the modulation transfer function (MTF) [20]. MTF curves are derived by applying Fourier transformation to the point spread function (PSF) [19], which is obtained from scanning the point phantom.

3 Experiments and Results

3.1 Experimental Settings

FBCT Simulation Details The geometry and parameters of our FBCT simulation closely resemble those commonly used in practice [10]: a source-to-detector distance of approximately 1000 mm, a source-to-isocenter distance of approximately 600 mm, a detector array size of 912, a detector pixel size of 1.0 mm, and a system field of view (FOV) of 50 cm. The FBCT reconstruction configurations are also standard: the FBP method uses a classical ramp filter [6], the reconstructed image matrix is 512 by 512, and the reconstructed image pixel size is 0.5 mm by 0.5 mm, yielding a reconstructed FOV of 25.6 cm, typical for brain imaging.

Datasets Our analysis utilizes QA and XCAT torso phantoms to craft training datasets, employing full-angle sinograms across 360 degrees for CT reconstruction. The uniformity and LCD phantom undergo scanning with clinical protocols of 140 kVp tube voltage, 300 mAs current exposure dose, and body bowtie, excluding the point phantom. The scanning of the point phantom is performed without setting bowtie and Poisson noise addition because point signal is easily annihilated in the count fluctuation. The scans of XCAT phantoms adhere to clinical protocols, ensuring comprehensive evaluation across various clinical scenarios, such as 120 kVp and 300 mAs for tooth and brain phantom, 140 kVp and 180 mAs for shoulder phantom, and 120 kVp and 140 mAs for lung phantom.

Baselines Existing INRs can be broadly categorized into two groups: function-expansion-based INRs and hash-based INRs [35]. For example, PE+MLP [15] utilizes Fourier features as inputs to an MLP to capture high-frequency details in natural signals, while SIREN [24] employs a periodic activation function instead of the conventional ReLU, allowing for finer detail representation in scenes. Both methods treat the INR as a function-expansion problem using different basis functions. Additionally, recent developments have introduced hash-based INRs. These methods replace frequency encoding with learned hash-key indexing, significantly enhancing the models' ability to represent scenes. Specifically, Instant NGP [16] maps the coordinates to compact hash keys using learned hash functions, which are then used to index optimized features for reconstruction. DINER [31] incorporates a hash table into a traditional INR framework, creating a disorder-invariant representation. Our study compares the four most prominent SOTA INR techniques–PE+MLP, SIREN, Instant NGP, and DINER–against the classical FBP baseline, using simulated QA and XCAT phantom datasets.

Metrics Performance metrics, including natural imaging metrics and QA metrics, facilitate both qualitative and quantitative assessments. The natural image metrics include PSNR and SSIM. The QA metrics follow the "Accreditation Phantom Instruction" approved by the ACR [1], which is widely used for clinical diagnostic FBCT acceptance tests and routine maintenance. These metrics include CT number accuracy, image noise, spatial resolution, and LCD.

Implementation Details Four INR models are reproduced and optimized according to the original papers' code. The four INR models employ specific position encoders followed by two hidden-layer MLPs. Each MLP comprises 64 neurons per hidden layer, with ReLU as the activation function for the hidden layers and LeakyReLU for the output layer. Adjustments to the hyperparameters of each model were made to optimize performance. For instance, the PE+MLP model utilizes 12 frequency levels per encoded dimension, while the SIREN model adopts an initial frequency scale parameter of 30 for the sinusoidal activation functions in the hidden layers. Instant NGP integrates 12 hash levels, each with up to 2^{19} entries, where each entry is an eight-dimensional learnable feature vector. It also incorporates a scaling factor of 2 and a base resolution of 2. DINER uses a hash table with dimensions of 1024 by 1024 and 32 feature levels. For more detailed information regarding the position encoders and hyperparameters, please refer to the original papers [15, 16, 24], and [31]. All four INR models were trained using the Adam optimizer with default settings. The learning rate was initially set at 1×10^{-3} and decreased to 5×10^{-4} as the networks converged. The ray batch size was set to $|\mathbb{L}| = 2000$. The training schedule for PE+MLP and SIREN began with a learning rate of 1×10^{-3}, decaying by a factor of 0.5 every 10k iterations, for a total of 20k iterations. Instant NGP followed a similar schedule, starting with a learning rate of 1×10^{-3}, which decayed by a factor of 0.5 every 1k iterations, over a total of 2k iterations. Instant NGP also employed coarse-to-fine progressive training [32], starting with an initial hash level of 4 and activating subsequent levels every 50 iterations. For DINER, the learning rate decayed by a factor of 0.5 every 2k iterations, over a total of 3k iterations. All experiments were conducted on a single RTX 4090 GPU with the memory of 24G.

3.2 Results

3.2.1 Performance on XCAT Phantoms

A qualitative comparison of four INR methods and the FBP method on typical XCAT torso phantoms is presented in Figs. 2 and 3. The results reveal some limitations of INR-based methods in handling human soft tissues, where noticeable distortions or artifacts appear in clinical display settings (e.g., WW/WL = 200/50 HU), while these methods achieve performance comparable to FBP (PSNR and SSIM values) under broader display windows (e.g., WW/WL = 2000/0 HU).

Notably, distinct artifact patterns and amplified noise are observed in clinical display windows, which can be attributed to inherent biases in INR's position encoders. The frequency-based position encoding in the PE+MLP approach introduces grid-like artifacts, while its global representation may lead to a blurrier output with less detail. SIREN's periodic activation function can generate periodic moiré patterns, resulting in a blurred appearance. Instant NGP significantly improves detail performance through its powerful multi-resolution hash encoding, although

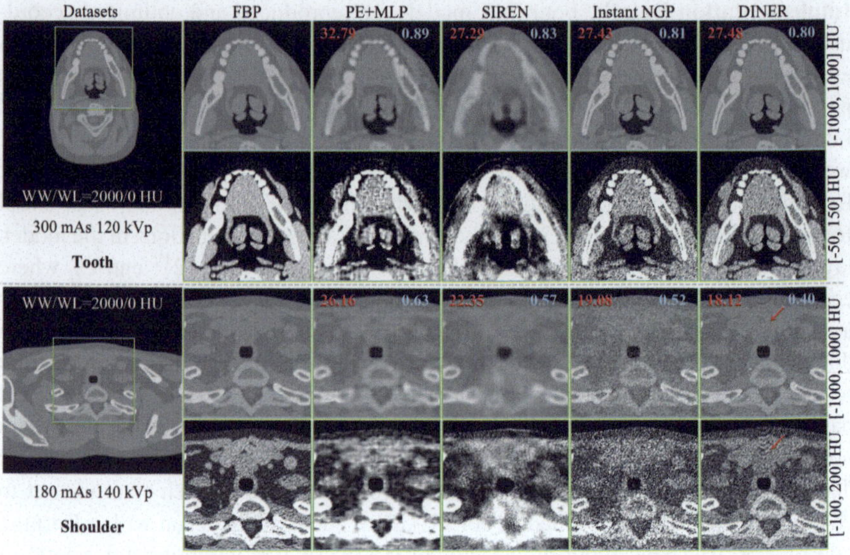

Fig. 2 Qualitative and quantitative results of FBP and four INR methods on tooth and shoulder XCAT phantoms. The red numbers on the left and the blue numbers on the right indicate the PSNR and SSIM values, respectively, calculated between the selected INR-based results and FBP-reconstructed images (used as ground truth)

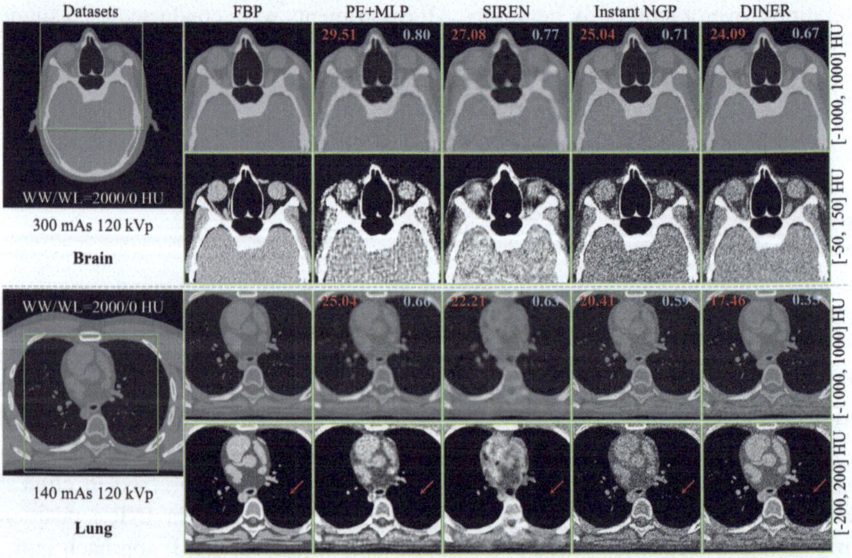

Fig. 3 Qualitative and quantitative results of FBP and four INR methods on brain and lung XCAT phantoms. The red numbers on the left and blue numbers on the right indicate the PSNR and SSIM values, respectively, calculated between the selected INR-based results and FBP-reconstructed images (used as ground truth)

it exhibits noticeable noise due to unstable gradients from hash-index operations. DINER excels with its superior performance, offering excellent detail and minimal noise due to its disorder-invariant feature representation. However, its single-resolution encoding mechanism slightly limits interpolation ability, leading to some noise and artifacts. For example, wavy artifacts can be seen in the upper soft tissue of the shoulder, as indicated by the red arrow.

3.2.2 Performance on QA Phantoms

The qualitative performance of INR methods on QA phantoms is visualized in Fig. 4. These INR-based methods exhibit noise and artifacts similar to those observed in the XCAT phantoms. In the low-contrast phantom test, PE+MLP successfully detects contrasts above 3%, while Instant NGP and DINER perform well at contrasts above 5 %, though they show slightly higher noise levels. SIREN faces challenges in identifying contrast circles due to artifacts. In the point phantom test, INR methods demonstrate difficulties with reconstructing small shapes, leading to grid artifacts in PE+MLP and ring artifacts in DINER. Moreover, the visual quality of SIREN and Instant NGP could benefit from further refinement. MTF curve results are shown in Fig. 5.

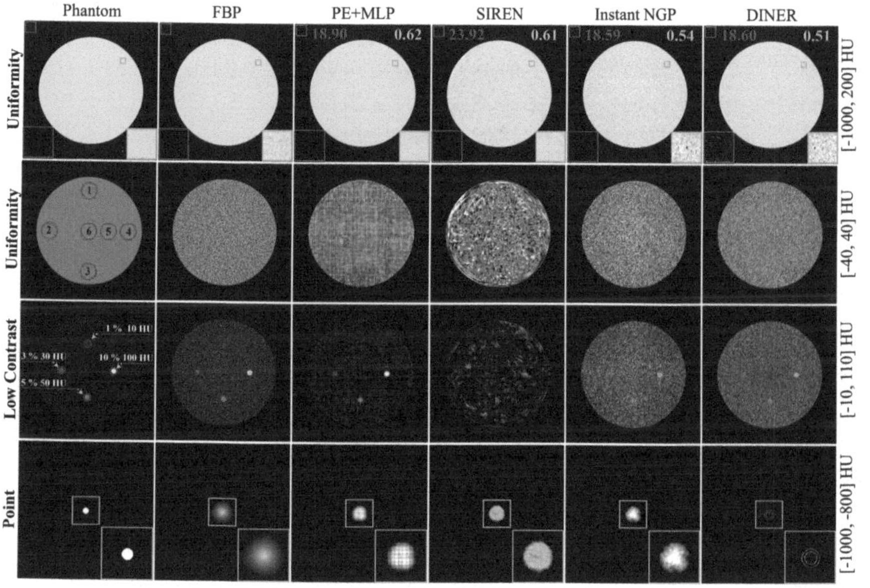

Fig. 4 Qualitative results of FBP and four INR methods on three QA phantoms. The red numbers on the left and the blue numbers on the right represent the calculated PSNR and SSIM values, respectively, between the selected INR-based results and FBP reconstructed images (used as ground truth)

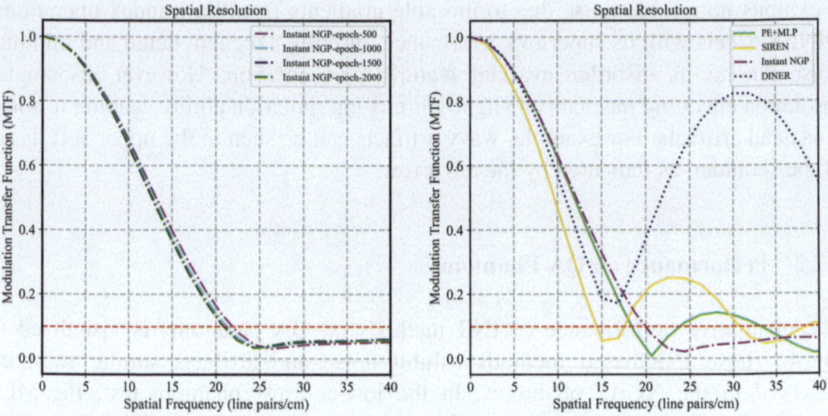

Fig. 5 MTF curves from the point phantom. As shown in the left figure, the spatial resolution will converge within the increase of iteration in Instant NGP. The right figure shows the comparison of MTF curves between INR methods, the jump pattern of DINER is due to the incorrect reconstruction of the point phantom (shown in the Fig. 4)

Table 1 Measured CT number (HU) accuracy and image uniformity of five methods

ROI	FBP	PE+MLP	SIREN	Instant NGP	DINER
ROI-1	-1.80 ± 37.93	0.68 ± 09.85	-1.55 ± 20.02	02.97 ± 114.33	2.98 ± 112.89
ROI-2	-1.48 ± 38.58	-1.67 ± 10.66	2.27 ± 18.72	$\underline{06.09} \pm 113.66$	4.60 ± 114.80
ROI-3	-1.46 ± 37.89	$\underline{-8.28} \pm 10.22$	$\underline{-7.11} \pm 25.56$	$\underline{05.63} \pm 117.07$	4.45 ± 112.53
ROI-4	-1.64 ± 37.83	0.79 ± 09.48	-2.93 ± 19.96	02.23 ± 116.19	3.58 ± 115.60
ROI-5	0.82 ± 41.63	-0.74 ± 10.38	-3.32 ± 20.82	$\underline{14.95} \pm 132.73$	$\underline{6.53} \pm 102.82$
ROI-6	-0.07 ± 43.01	0.70 ± 10.02	-2.34 ± 19.26	$\underline{26.74} \pm 137.58$	$\underline{5.01} \pm 089.69$
Δ(Mean)	**2.62**	9.07	9.38	24.51	3.55

For quantitative analysis, Table 1 provides details on CT number accuracy, image uniformity (Δ), and noise levels. PE+MLP, SIREN, and DINER achieve CT number accuracy that nearly meets the clinical requirement of staying within a ± 5 HU deviation, while Instant NGP does not meet this criterion. Unsatisfactory cases are underlined in the table. Furthermore, the noise levels of PE+MLP and SIREN are comparable to those of FBP, while Instant NGP and DINER display higher noise levels. The image uniformity of INR methods, although improving, remains behind that of FBP.

Through the qualitative and quantitative analysis, the INR-based methods show potential but require further improvements to consistently meet clinical standards in both QA and XCAT torso phantom image studies.

4 Summary and Discussion

In this chapter, we present a comprehensive evaluation of INR imaging performance in clinical CT applications, adhering to the guidelines accredited by the ACR. We simulate the fan-beam clinical CT scanning process and perform standard phantom studies to qualitatively and quantitatively assess the imaging capabilities of representative INR-based methods.

Initially, we simulate a commercial clinical diagnostic FBCT process to generate standard datasets for QA and XCAT phantoms. This simulation involves the physical processes of X-ray photon attenuation and detection, using typical CT geometry and image reconstruction parameters that closely align with standard clinical practices. This well-controlled and ideal simulation provides a reliable framework for fairly comparing the performance potential of INR-based methods.

We then apply four SOTA INR-based techniques (PE+MLP, SIREN, Instant NGP, and DINER) to perform neural reconstruction on various XCAT and QA phantoms. Full-view projections are utilized for image reconstruction, as they avoid artifacts introduced by ill-posed imaging scenarios (e.g., limited-angle or sparse-view) and are more suitable for evaluating the clinical diagnostic potential of INRs according to ACR standards.

Subsequently, we assess the imaging performance using ACR-proposed metrics, including CT number accuracy, LCD, and spatial resolution—metrics widely used in clinical diagnostic FBCT acceptance tests and routine maintenance. While SOTA INR methods show considerable potential for FBCT imaging, further refinement is necessary to fully meet the clinical performance standards set by the ACR.

By analyzing the performance on XCAT phantoms, we qualitatively evaluate the INR's ability to depict human soft tissues. Position encoders, commonly used in typical INRs, address the low-frequency bias in pure MLPs, capturing more details of imaging objects. Although these encoding approaches achieve high PSNR and SSIM values in reconstructed results, amplified noise and induced artifacts remain noticeable in wide display windows (e.g., WW/WL = 2000/0 HU). Moreover, these artifacts (e.g., PE+MLP, SIREN) or noise (e.g., Instant NGP, DINER) become more prominent in narrow clinical display windows (e.g., WW/WL = 80/0 HU, which is a typical brain window used in clinical settings). Similar imaging performances are observed in the reconstruction of QA phantoms, where the results from low-contrast and point phantoms indicate that INR methods need further refinement for imaging human tissues with low contrast and small structures.

We further quantify the clinical diagnostic potential of INR methods using QA test phantoms. CT number accuracy produced by PE+MLP, SIREN, and DINER nearly meets the clinical requirement of being within a ± 5 HU deviation, while Instant NGP struggles to satisfy this standard. The image uniformity of INR methods is generally inferior to FBP. Additionally, the noise levels of PE+MLP and SIREN are comparable to those of FBP, whereas Instant NGP and DINER exhibit significantly higher noise levels. This may be attributed to the hash-index operations employed by Instant NGP and DINER, which create a discontinuous and nonanalytical mapping between coordinates and feature vectors, resulting in broken gradients and decoupled optimization [35]. Introducing continuous, analytical connections between coordinates and feature vectors could potentially improve performance.

However, our study has some limitations. A key limitation is the absence of real-world clinical datasets. Although the FBCT simulations were conducted using standard clinical settings to ensure alignment with real-world practices, it is still crucial to perform imaging studies on experimental datasets. Another limitation is that our study identifies the boundaries of INR techniques without proposing optimization solutions. Despite these limitations, our findings offer valuable insights and underscore the importance of advancing INR techniques to meet the stringent requirements of medical diagnostics. We acknowledge these limitations and propose addressing them in future work. A promising direction for future research is the development of more comprehensive position encoders, which should not only capture subtle, high-frequency details but also effectively mitigate noise. Additionally, developing more precise evaluation metrics is essential for accurately assessing reconstruction quality in clinical applications.

5 Conclusion

In this chapter, we conduct comprehensive studies to evaluate the performance of INR-based CT reconstruction methods in clinical settings, directly utilizing clinical performance metrics accredited by the ACR as a complement to natural imaging metrics such as PSNR and SSIM. We employ XCAT torso and QA phantoms to qualitatively and quantitatively assess clinical diagnostic performance using clinical imaging quality metrics, including CT number accuracy, image uniformity, low-contrast detectability, and spatial resolution. Our findings indicate that while INR-based methods show significant promise for clinical CT reconstruction, further improvements are necessary to address their limitations in medical imaging performance, such as CT number inaccuracies (exceeding ± 5 HU), pronounced image noise, insufficient low-contrast detectability, and induced additional artifacts compared to FBP. These results highlight the need for continued advancements in INR-based techniques to fully realize their potential in clinical settings.

References

1. C. ACR, *Accreditation Phantom Instruction* (American College of Radiology, New York, 2011), pp. 3–10
2. A.H. Andersen, A.C. Kak, Simultaneous algebraic reconstruction technique (SART): a superior implementation of the art algorithm. Ultrason. Imaging **6**(1), 81–94 (1984)
3. V. Antun, F. Renna, C. Poon, B. Adcock, A.C. Hansen, On instabilities of deep learning in image reconstruction and the potential costs of AI. Proc. Natl. Acad. Sci. **117**(48), 30088–30095 (2020)

4. I. Avcibas, B. Sankur, K. Sayood, *Statistical Evaluation of Image Quality Measures* (2001)
5. Beer, Bestimmung der absorption des rothen lichts in farbigen flüssigkeiten. Ann. Phys. **162**(5), 78–88 (1852)
6. A. Brahme, *Comprehensive Biomedical Physics*. Newnes (2014)
7. R. Gordon, R. Bender, G.T. Herman, Algebraic reconstruction techniques (ART) for three-dimensional electron microscopy and x-ray photography. J. Theor. Biol. **29**(3), 471–481 (1970)
8. K. Gulliksrud, C. Stokke, A.C.T. Martinsen, How to measure CT image quality: variations in CT-numbers, uniformity and low contrast resolution for a CT quality assurance phantom. Phys. Med. **30**(4), 521–526 (2014)
9. T. Higaki, Y. Nakamura, J. Zhou, Z. Yu, T. Nemoto, F. Tatsugami, Awai, K., Deep learning reconstruction at CT: phantom study of the image characteristics. Acad. Radiol. **27**(1), 82–87 (2020)
10. J. Hsieh, *Computed Tomography: Principles, Design, Artifacts, and Recent Advances* (2003)
11. K.H. Jin, M.T. McCann, E. Froustey, M. Unser, Deep convolutional neural network for inverse problems in imaging. IEEE Trans. Image Process. **26**(9), 4509–4522 (2017)
12. A.C. Kak, M. Slaney, Principles of computerized tomographic imaging. SIAM (2001)
13. W.A. Kalender, X-ray computed tomography. Phys. Med. Biol. **51**(13), R29 (2006)
14. E. Kang, J. Min, J.C. Ye, A deep convolutional neural network using directional wavelets for low-dose x-ray CT reconstruction. Med. Phys. **44**(10), e360–e375 (2017)
15. B. Mildenhall, P.P. Srinivasan, M. Tancik, J.T. Barron, R. Ramamoorthi, R. Ng, Nerf: representing scenes as neural radiance fields for view synthesis. Commun. ACM **65**(1), 99–106 (2021)
16. T. Müller, A. Evans, C. Schied, A. Keller, Instant neural graphics primitives with a multiresolution hash encoding. ACM Trans. Graph. (ToG) **41**(4), 1–15 (2022)
17. N. Rahaman, A. Baratin, D. Arpit, F. Draxler, M. Lin, F. Hamprecht, Y. Bengio, A. Courville, On the spectral bias of neural networks, in *International Conference on Machine Learning* (PMLR, New York, 2019), pp. 5301–5310
18. A.W. Reed, H. Kim, R. Anirudh, K.A. Mohan, K. Champley, J. Kang, S. Jayasuriya, Dynamic CT reconstruction from limited views with implicit neural representations and parametric motion fields, in *Proceedings of the IEEE/CVF International Conference on Computer Vision* (2021), pp. 2258–2268
19. K. Rossmann, Point spread-function, line spread-function, and modulation transfer function: tools for the study of imaging systems. Radiology **93**(2), 257–272 (1969)
20. M.R. Schroeder, Modulation transfer functions: definition and measurement. Acta Acustica united with Acustica **49**(3), 179–182 (1981)
21. W.P. Segars, G. Sturgeon, S. Mendonca, J. Grimes, B.M. Tsui, 4D XCAT phantom for multimodality imaging research. Med. Phys. **37**(9), 4902–4915 (2010)
22. J.K. Seo, E.J. Woo, *Nonlinear Inverse Problems in Imaging* (Wiley, New York, 2012)
23. L. Shen, J. Pauly, L. Xing, NeRP: implicit neural representation learning with prior embedding for sparsely sampled image reconstruction. IEEE Trans. Neural Networks Learn. Syst. **35**(1), 770–782 (2022)
24. V. Sitzmann, J. Martel, A. Bergman, D. Lindell, G. Wetzstein, Implicit neural representations with periodic activation functions. Adv. Neural Inf. Proces. Syst. **33**, 7462–7473 (2020)
25. Y. Sun, J. Liu, M. Xie, B. Wohlberg, U.S. Kamilov, Coil: Coordinate-based internal learning for tomographic imaging. IEEE Trans. Comput. Imaging **7**, 1400–1412 (2021)
26. G. Wang, H. Yu, B. De Man, An outlook on x-ray CT research and development. Med. Phys. **35**(3), 1051–1064 (2008)
27. S. Wang, G. Cao, Y. Wang, S. Liao, Q. Wang, J. Shi, C. Li, D. Shen, Review and prospect: artificial intelligence in advanced medical imaging. Front. Radiol. **1**, 781868 (2021)
28. Z. Wang, A.C. Bovik, H.R. Sheikh, E.P. Simoncelli, Image quality assessment: from error visibility to structural similarity. IEEE Trans. Image Process. **13**(4), 600–612 (2004)
29. Q. Wu, L. Chen, C. Wang, H. Wei, S.K. Zhou, J. Yu, Y. Zhang, Unsupervised polychromatic neural representation for CT metal artifact reduction. Adv. Neural Inf. Proces. Syst. **36**, 69605–69624 (2024)

30. Q. Wu, R. Feng, H. Wei, J. Yu, Y. Zhang, Self-supervised coordinate projection network for sparse-view computed tomography. IEEE Trans. Comput. Imaging **9**, 517–529 (2023)
31. S. Xie, H. Zhu, Z. Liu, Q. Zhang, Y. Zhou, X. Cao, Z. Ma, Diner: disorder-invariant implicit neural representation, in *Proceedings of the IEEE/CVF Conference on Computer Vision and Pattern Recognition* (2023), pp. 6143–6152
32. J. Yang, M. Pavone, Y. Wang, Freenerf: improving few-shot neural rendering with free frequency regularization, in *Proceedings of the IEEE/CVF Conference on Computer Vision and Pattern Recognition* (2023), pp. 8254–8263
33. R. Zha, Y. Zhang, H. Li, NAF: neural attenuation fields for sparse-view CBCT reconstruction, in *International Conference on Medical Image Computing and Computer-Assisted Intervention* (Springer, Berlin, 2022), pp. 442–452
34. Z. Zhang, S. Yu, W. Qin, X. Liang, Y. Xie, G. Cao, Self-supervised CT super-resolution with hybrid model. Comput. Biol. Med. **138**, 104775 (2021)
35. H. Zhu, F. Liu, Q. Zhang, X. Cao, Z. Ma, Rhino: regularizing the hash-based implicit neural representation. arXiv preprint arXiv:2309.12642 (2023)

Learning-Based Material Decomposition for Spectral X-Ray Imaging

Yanye Lu, Xinliang Zhang, and Jiakui Hu

1 Introduction

"One begets two, Two begets three, Three begets universe." These ancient words, summarizing the fundamental principle that all things in our universe are composed of basic components, were articulated by Laozi, a great philosopher of ancient China (571 BC–500 BC). In the modern world, the exploration of matter's components remains a central pursuit. Various techniques have been developed for this purpose, among which X-ray imaging is notable. X-ray imaging originated from an unexpected discovery. In the late nineteenth century, German physicist Wilhelm Conrad Röntgen was experimenting with cathode ray tubes when he noticed that, under high voltage, the tube emitted a type of radiation that could penetrate certain materials but not others. Initially, Röntgen thought he had discovered a new form of light, but he soon recognized that this radiation was unlike anything previously known. He named it "X-rays" to signify its unknown nature and origin. Röntgen quickly realized the potential of X-rays to transform medicine [1]. By passing X-rays through a patient's body, doctors could visualize internal structures without invasive surgery. Today, traditional computed tomography (CT) imaging, which uses a single energy interval, has largely been replaced by energy spectrum CT. This advanced technology utilizes X-rays across multiple energy intervals. In the context of energy spectrum CT imaging, material decomposition remains a fundamental challenge.

These authors contributed equally to this work.

Y. Lu (✉) · X. Zhang · J. Hu
Institute of Medical Technology, Peking University, Beijing, China
e-mail: yanye.lu@pku.edu.cn

K. (K.) Iniewski, L. (K.) Cai (eds.), *Deep Learning for Advanced X-ray Detection and Imaging Applications*, https://doi.org/10.1007/978-3-031-75653-5_5

2 Spectral X-Ray Material Decomposition

In clinical applications, CT has become the preferred tool for rapidly acquiring high-resolution structural information. However, traditional CT is limited in that it cannot provide molecular or metabolic information about tissues. Different tissues in the human body may have distinct chemical compositions, yet traditional CT often yields similar attenuation coefficients for these tissues, resulting in them being classified within the same Hounsfield unit (HU) value range [2]. This occurs because traditional X-ray CT systems use polychromatic (multi-energy) X-rays [3], producing a continuous energy spectrum. The detectors in these systems convert X-rays from the entire energy spectrum into visible light, performing integral detection that does not differentiate between X-ray photon energies. This approach leads to a loss of valuable X-ray attenuation information, which hampers the accurate assessment of the material's physical properties. In medical applications, this limitation is particularly significant, as it makes it challenging to distinguish between different types of soft tissue based on their contrast in the reconstructed CT images. Consequently, this lack of contrast differentiation can directly impact the interpretation of CT images, reducing diagnostic effectiveness.

Due to the limitations of traditional CT in material identification, researchers have increasingly focused on energy spectrum CT. Unlike traditional CT, where detectors merely measure the overall intensity of X-ray photons, energy spectrum CT employs photon-counting detectors that not only measure photon intensity but also differentiate between photon energies. This additional dimension of information enhances the capabilities of energy spectrum CT, allowing for more precise substance identification and quantitative analysis. In the early development of energy spectrum CT, dual-energy CT (dual-energy computed tomography) was a pioneering technology that utilized X-rays at two distinct energy levels [4–9], as illustrated in Fig. 1. This approach marked an important advancement by providing additional data for improved material characterization.

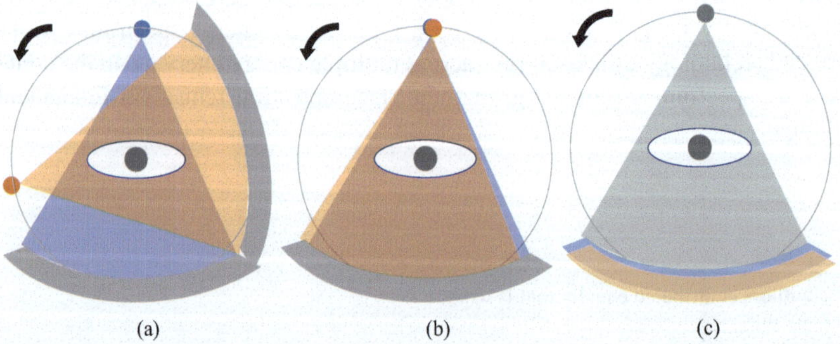

Fig. 1 Illustrations of the various dual-energy CT scanning techniques. (**a**) Dual-source scanning. (**b**) Rapid energy switching. (**c**) Double layer detector

With dual-energy CT, two distinct X-ray attenuation profiles are obtained, providing more detailed information for analyzing the physical properties of materials. This technique has found significant applications in biomedical imaging research [10, 11]. Another cutting-edge approach is the use of photon-counting detectors for X-ray detection [12–14]. Photon counters, typically semiconductor detectors, employ energy gating thresholds to selectively record X-ray photon pulses within different energy ranges. This allows for the differentiation of X-ray photons based on their energies, enabling more accurate material identification and analysis.

The principle of X-ray spectroscopy highlights that different elements exhibit unique X-ray absorption characteristics due to their atomic structures, particularly the K-edge features of materials [15]. The K-edge refers to a distinct jump in the X-ray attenuation coefficient at a specific energy, corresponding to the absorption edge of an element. Unlike traditional X-CT, X-ray spectral CT technology can distinguish the absorption properties of materials at different energy levels. This enables the analysis of K-edge characteristics to identify the composition or condition of materials. Currently, extensive research has been conducted on material identification using X-ray spectral CT, focusing on the K-edge characteristics [16–24]. In this context, spectral CT transcends its role as a mere visualization tool for examining patient anatomy. It evolves into a more powerful tool for material analysis through a process known as material decomposition.

2.1 X-Ray Physics

According to the Beer–Lambert law, if the input X-ray photons are monoenergetic, the intensity of non-attenuated X-ray photons passing through a material can be computed as $I = I^0 \exp^{-\mu(E)l}$, where I^0 represents the intensity of X-ray photons before entering the material, E describes the photon energy, and $\mu(E)$ and l represent the spectral attenuation coefficient of the material at E and the path length of the material, respectively. After applying the minus logarithmic operation, the projection p can be expressed as $p = -ln\frac{I}{I^0} = \mu(E)l$. Normally the scanning object contains various materials $i = 1 \cdots M$, resulting in $p = \sum_{i=1}^{M} p^i$ based on the mixture rule, where the material-specific projection p^i can be expressed as $p = -ln\frac{I_i}{I_i^0} = \mu_i(E)l_i$, in which I_i^0 denotes the material-dependent I^0 of i. The measurements acquired from an X-ray image system yield:

$$q_j = -ln \int [S_j(E)], \tag{1}$$

where $S_j(E)$ represents the distribution function of the X-ray spectrum used in the scan with the index $j = 1 \cdots B$. Note that j counts the independent spectral bins in multi-energy X-ray imaging, and B is the total number of kV settings in multiplekV detection, or the total number of energy bins in energy-resolved spectral detection. $X(E)$ represents the error function of the measurement, which includes several phenomena, such as photon scattering, detector distortion, and signal noise. In practice, $X(E)$ is difficult to model due to nonlinearity. As shown in the functions, we can obtain multi-energy measurements $q_1; \cdots; q_B$ from the same object from a spectral X-ray imaging system.

2.2 Image-Based Material Decomposition

The important matter decomposition idea of CT is that the decay coefficient of a matter can be expressed as a linear combination of the decay coefficients of several other reference substances [4]:

$$\mu(E, \vec{r}) = \sum_{i=1}^{N} f_i(E)c_i(\vec{r}) = c_1(\vec{r})f_1(E) + c_2(\vec{r})f_2(E) + ... + c_N(\vec{r})f_N(E), \tag{2}$$

where $\mu(E, \vec{r})$ represents the attenuation coefficient of the unknown object at position $\vec{r}$) when the ray beam energy is E. $f_1(E), f_2(E), ...f_N(E)$ represents the attenuation coefficient of the known reference substance when the X-ray photon energy is E, $c_1(\vec{r}), c_2(\vec{r}), ..., c_N(\vec{r})$represents the content of various substances when the unknown matter is decomposed into the linear sum of the reference substance, where to acquire its explicit analytical solution is our ultimate goal. According to Eq. 2, the energy and the position of the decay coefficient of matter are successfully separated. In order to acquire its explicit analytical solution, let us define:

$$\overline{\mu_i}(\vec{r}) = \int_{E_i} w(E_i)\mu(E_i, \vec{r})dE_i, \tag{3}$$

where $\overline{\mu_i}(\vec{r})$ represents the equivalent attenuation coefficient of the substance to be tested over the energy interval E_i. Substituting the Eq. 2 into Eq. 3, we get:

$$\overline{\mu_i}(\vec{r}) = \int_{E_i} w(E_i) \sum_{i=1}^{N} f_i(E)c_i(\vec{r})dE_i, \tag{4}$$

where E_i represents the i-th energy interval. For convenient engineering, we have to use the matrix to describe Eq. 4, denoting the left part $\overline{\mu_i}(\vec{r})$ as U:

$$U = \begin{pmatrix} \overline{\mu_1}(\vec{r}) \\ \overline{\mu_2}(\vec{r}) \\ \vdots \\ \overline{\mu_M}(\vec{r}) \end{pmatrix}, \tag{5}$$

and denoting the right part as $Q \times C$:

$$Q = \begin{bmatrix} K_{1,1}, \cdots, K_{1,N} \\ \vdots \cdots \vdots \\ K_{M,1}, \cdots, K_{M,N} \end{bmatrix}, C = \begin{pmatrix} c_1(\vec{r}) \\ c_2(\vec{r}) \\ \vdots \\ c_N(\vec{r}) \end{pmatrix}, \tag{6}$$

where $K_{i,j} = \int_{E_i} w(E_i) \sum_{i=1}^{N} f_i(E) dE_i$. In general, Eq. 4 can be described in format of matrix as:

$$U = Q \times C, \tag{7}$$

where U and C are two tensors which, respectively, represent the reconstructed images and material images

From Eq. 7, we can know that the key to solving C is the matrix Q and the vector U. The vector U is obtained from the measuring experiments, but from the expression definition of Q, the solution is very complicated because it is very difficult to obtain the corresponding spectral information. Even if the spectral information is obtained, the Q calculated by the formula integration will also have a large error. So we use the idea of engineering, and we measure Q directly through energy calibration. It is defined by the formula that each element in Q represents the equivalent decay coefficient of the reference matter i in the energy interval j, which can be easily measured in experiments.

In general, before conducting the material decomposition experiment, we have to first determine the reference material, then obtain the attenuation characteristics of the reference material through the calibration experiment, then scan the substance to be measured under the same experimental conditions as the calibration experiment, and finally use the measured Q and U information to obtain material content C [25–27] through $C = Q^{-1}U$.

2.3 *Projection-Based Material Decomposition*

The matter decomposition algorithm for the spectral CT projection domain originates from the detector's Poisson model [28] of the number of photons detected in a certain energy interval. In the model setting, the number of photons detected by the spectral CT photon counter in an energy interval b satisfies the Poisson distribution:

$$P(m_b|\lambda_b) = \frac{\lambda_b^{m_b}}{m_b!}e^{-\lambda_b}, \tag{8}$$

where m_b represents the photon count detected by the detector at the energy interval b [12, 29], and λ_b represents the expected value of the number of photons that the detector is able to detect at the energy interval b. The parameter λ_b varies with the X-ray scanning conditions. From the Eq. 2, we know that the decay coefficient of an unknown matter can be expressed as a linear summation of the decay coefficients of a series of reference substances. For convenient expression, define:

$$\mathcal{L}(E, A_1, A_2, \cdots, A_N) = \sum_{i=1}^{N} f_i(E)A_i, \tag{9}$$

where $f_i(E)$ represents the decay coefficient of the matter i at the energy E, and A_i represents the line integral length of the reference material i in a certain path of the unknown matter. When a monochromatic light with an initial photon number I_0 and energy E passes through an unknown material, which is composed of $\mathbf{A} = [A_1, A_2, \cdots, A_N]$, the number of the photons I after attenuation can be calculated by:

$$I(E) = I_0(E)e^{-\mathcal{L}(E, A_1, A_2, \cdots, A_N)}. \tag{10}$$

From Eq. 8, it can be seen that, under the energy interval b, the number of X-ray photons after decaying through the unknown matter $\mathbf{A} = [A_1, A_2, \cdots, A_N]$ is the expected value of the number of photons that the detector can detect under the energy interval b:

$$\lambda_b(A_1, \cdots, A_N) = \int_b I_b(E)e^{-\mathcal{L}(E, A_1, A_2, \cdots, A_N)}dE. \tag{11}$$

In Eq. 11, $I_b(E)$ is the number of photons when the energy in the energy interval b is E. After obtaining the photon expectation value of the detector in a single energy interval b, we can model the photon number distribution detected by the energy spectrum CT at all energy channels:

$$P([m_1, \cdots, m_{N_B}]|\lambda_1(A), \lambda_2(A), \cdots, \lambda_{N_B}(A)) = \prod_b^{N_b} \frac{\lambda_b^{m_b}}{m_b!}e^{-\lambda_b}, \tag{12}$$

where $\mathbf{m} = [m_1, \cdots, m_{N_B}]$ represents the number of photons detected by the photon counter at each energy channel. In this manner, the unknown material $\mathbf{A} = [A_1, A_2, \cdots, A_N]$ can be obtained by maximum likelihood method, maximizing the value of Eq. 8. There are many iterative optimization methods [30–32] for the solution of Eq. 12.

3 Advanced Material Decomposition

3.1 Dictional Regularization in Image-Domain Methods

3.1.1 Model-Based Image-Domain Material Decomposition

From the analysis presented in Sect. 2.2, it can be concluded that image-domain material decomposition is essentially an inverse problem. In order to calculate the content C from the measured Q and U information, iterative optimization [33, 34] and least-squares algorithm [35, 36] is required to solve this inverse calculation. To recover material images from Eq. 7, we have the following least square linear program problem:

$$U^* = \text{argmin}\frac{1}{2}||C - Q^{-1} \times U||_F^2, \tag{13}$$

where $||\cdot||_F$ is Frobenius norm.

To improve the decomposed image quality, regularization is a feasible strategy to constrain the solution during iteration process, and Eq. 13 can be reformulated as:

$$U^* = \text{argmin}\left(\frac{1}{2}||C - Q^{-1} \times U||_F^2 + \lambda R(U)\right), \tag{14}$$

where λ is a regularization factor to balance the data fidelity term $\frac{1}{2}||C - Q^{-1} \times U||_F^2$ and regularization term $R(U)$.

Numerous methods have employed regularization techniques [34, 37–39] with the objective of enhancing the signal-to-noise ratio (SNR) of the decomposed results C. They include total variation (TV) minimization [40], spectral prior image constrained compressed sensing (SPICCS) [41, 42], prior rank, intensity and sparsity model (PRISM) [43], tensor PRISM [44], nonlocal low-rank cube-based tensor factorization (NLCTF) [45, 46], spectral-image decomposition with energy-fusion sensing (SIDES) [47], etc. Among these methods, dictionary-based regularization [39, 48–50] methods are widely employed approaches that can further exploit the sparsity of the reconstructed image itself. We will focus on an introduction to this particular method.

3.1.2 Dictionary Learning

Given a set of image patches $x_i \in \mathcal{R}^{s\times s}, i = 1, \cdots, I$, are extracted from the training datasets and unfolded as a vector to train the dictionary $\mathbf{D} \in \mathcal{R}^{S\times T}$, where $S = s \times s$ and T represents the number of atoms. The goal of dictionary learning is to find sparse representation coefficients $\alpha \in \mathcal{R}^{T\times I}$ using the dictionary $\mathbf{D}$. This can be established by solving the following optimization problem [49]:

$$\{\mathbf{D}^*, \alpha^*\} = \operatorname{argmin}\left(\frac{1}{2}\sum_{i=1}^{I}\left(||x_i - \mathbf{D}\alpha_i||_F^2 + v_i||\alpha_i||_0\right)\right), \tag{15}$$

where L_0 is the sparsity level, $||\cdot||_0$ is the quasi-l_0 norm, and v_i is the Lagrange multiplier to control the $||\alpha_i||_0 \le L_0$.

Equation 15 can be solved by utilizing an alternating minimization scheme. The dictionary $\mathbf{D}$ is first fixed and the best coefficient matrix α can be found. As finding the truly optimal α is impossible, we use an approximation pursuit method. Any algorithm such as the orthogonal matching pursuit (OMP) [51] or matching pursuit (MP) [52] algorithm can be used for the calculation of the coefficients. The algorithm will not stop until the coefficient matrix and all the atoms have been updated and the stop criteria are satisfied. In general, the dictionary updating method is based on K-SVD [53] algorithm.

3.1.3 Dictionary Learning in Image-Domain Material Decomposition

Considering the image-domain material decomposition model (Eq. 13) is similar to the dictionary learning for low-dose CT reconstruction [54], the dictionary-learning-based image-domain material decomposition can be constructed as:

$$\{U^*, \{\beta_m^*\}_{m=1}^M\} = \operatorname{argmin}\Bigg(\frac{1}{2}||C - Q^{-1} \times U||_F^2 + \sum_{m=1}^{M}\frac{\lambda_m}{2}\sum_{i=1}^{I}\left(||H_i(U_m) - \hat{\mathbf{D}}\beta_{mi}||_F^2 + v_{mi}||\beta_{mi}||_0\right)\Bigg), \tag{16}$$

where U_m is the mth image U, $\beta_{mi} \in \mathcal{R}^{T\times I}$ are sparse representation coefficients for ith image patch from mth material image and $\beta_m = \{\beta_{mi}\}_{i=1}^{I}$. $H_i(U_m)$ is the ith image patch extraction operator from U_m. $\hat{\mathbf{D}}$ is the trained dictionary. High-quality dictionary is beneficial for sparse representation and accuracy of material decomposition.

To circumvent the challenges associated with the sparse encoding of image features during material decomposition and to enhance the performance of material decomposition, while also optimizing the parallel usage of the dictionaries $\hat{\mathbf{D}}$, [49] proposed that dictionaries be trained for all materials simultaneously. Its experiments also demonstrated that this parallel training method effectively preserves image information and resilience to noise while also reducing the time required for training. Furthermore, this method was furnished with additional physical priors, including the incompatibility of bones and contrast agents. This facilitates more precise material decomposition results.

Nevertheless, due to the low concentration of contrast agents in clinical applications, the iterative process of dictionary learning will result in inconsistent data of the same material under different contrast agents based on dictionary representation. The normalization strategy proposed by Wu et al. [50] was developed as a solution to the data inconsistency issues associated with low-contrast agents. Furthermore, it was suggested that the material image could be expanded along mode-1, with the aim of enhancing the correlation between different materials.

3.2 *Decision Trees in Projection-Domain Methods*

All the abovementioned methods have demonstrated the great performance of material decomposition in image domain. However, there are still some issues worth mentioning. Because performing material decomposition in projection domain is more complicated, most of them have focused on image-based material decomposition, despite that it loses the polychromatic properties of the X-ray spectra, resulting in less accurate decomposition [55]. Meanwhile, the decomposition of materials in the image domain necessitates a significant computational burden. When material images of disparate instruments are encountered, a new coefficient calibration and iterative training regimen must be implemented. Consequently, it is imperative to leverage information in the projection domain and adopt more straightforward and adaptable machine learning techniques to achieve more accurate material decomposition.

3.2.1 Transfer Material Decomposition to Machine Learning Problem

Fortunately, [56] has demonstrated that the material decomposition task can be transferred into a simple machine learning task without the need for calibration of CT imaging parameters and regularization in image domain.

We aim to build material-specific decomposition models to decompose the effective material-specific projection A_i from the multiple energy measurements $\mathbf{A}$ in Sect. 2.3. A target function of the material decomposition model is defined as $\hat{A}_i = D_i(\mathbf{A})$, where $D_i(\mathbf{A})$ is an as-yet-unknown general function estimator of the basis material pi with the multi-energy measurements $\mathbf{A}$. The estimator's goal is to learn a mapping from the measurements to an objective quantity. To learn this mapping, a hypothesis model $H_i(\mathrm{x})$ is defined to approximate $D_i(A_i)$, where x is a feature vector that numerically represents the measurements A_i. Each element in x, a so-called feature, represents a distinct description of A_i. The concept of general function estimation is in accordance with machine learning systems. Thus, a material decomposition task can be transferred to a machine learning problem. After obtaining the features of the projection image, we can train a machine learning model to perform material decomposition. Many reseachers [56–59] chose to use decision tree algorithm to solve this regression task.

3.2.2 Decision Tree

Decision tree is a supervised learning approach that is employed in the fields of machine learning. In this formalism, a classification or regression decision tree is employed as a predictive model to draw inferences about a set of observations. A classification tree is a tree model where the target variable can take a discrete set of values. In these tree structures, the leaves represent class labels, while the branches represent conjunctions of features that lead to those class labels. Decision trees where the target variable can take continuous values (typically real numbers) are referred to as regression trees. In a more general sense, the concept of a regression tree can be extended to any kind of object equipped with pairwise dissimilarities, such as categorical sequences.

3.2.3 Decision Tree in Projection-Domain Material Decomposition

To implement a decision-tree-based learning method for projection-domain material decomposition task, it is first necessary to obtain datasets for training and validation. The data used to generate projection images is derived from the simulation of X-ray images [60]. Motion and non-motion scenes are typically simulated concurrently to ensure greater fidelity of the simulation data [57]. In the motion case, the simulation includes one heart cycle and the lung motion from full exhale to full inhale. A 200-degree vertical rotation around the torso (short scan) is simulated in the non-motion case. For each set of projections, both noisy and clean images were generated. It is also worth noting that some scholars have proposed real-world projection acquisition methods [58]. However, in contrast to simulated data, real-world data exhibits unanticipated motion, e.g., the vibration of the CT Imaging system. Consequently, real-world projection datasets must be rigorously calibrated before utilization.

Once the data has been prepared, we can choose reduced error pruning tree (REPTree) or random tree in boosting [61] and bagging [62] as the basic decision tree, as shown in Fig. 2a and b. It can build feature extractors to extract the spectral information (namely Channel) [63] and polynomial combinations [64] of the spectral information (namely Polynomial). Then, the trainable Weka segmentation (namely WEKA) can be created based on this decision tree to extract structural information from the individual material ED phantom data, from both label projection and training projection.

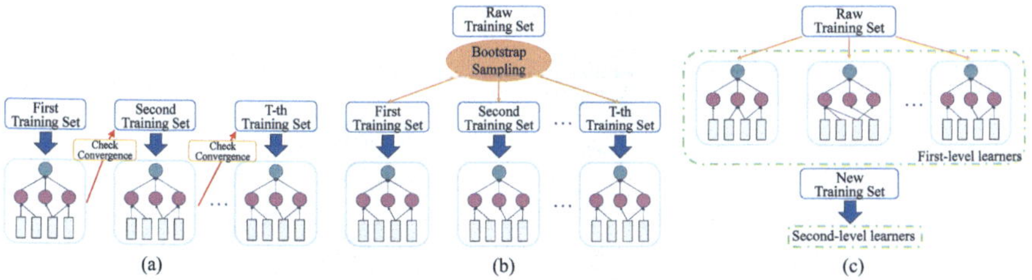

Fig. 2 Schematic description of the three decision tree algorithms. The base learners are either reduced error pruning (REPTree) or random tree in boosting and bagging, while the first-level learners are REPTree and random in stacking. (**a**) boosting, (**b**) bagging, (**c**) stacking

While single decision tree can be a helpful tool for projection-based material decomposition, it still suffers from its overfitting problem, leading to lower robustness in real-world scenes. To alleviate this issue, ensemble decision trees trained by stacking [65] were created, as shown in Fig. 2c. The results in [59] demonstrated that ensemble learning method with appropriate hyperparameters exhibited superior performance compared to two basic decision tree methods. Furthermore, decision-tree based methods also demonstrated that machine learning with gradient descent optimization was more suitable for the material decomposition task.

3.3 *Deep-Learning-Based Methods*

With the successful application of gradient optimization algorithms in the deep learning field, researchers are increasingly inclined to use powerful neural networks [66] for material decomposition in the hope of obtaining material images with less noise and artifacts. Incorporating deep learning with conventional gradient optimization algorithms can bring scalable learnable modeling tools to improve decomposition accuracy.

3.3.1 Direct Decomposition in Projection-Domain

The use of projection-based multi-material decomposition allows for more accurate decomposition, and thus, the incorporation of deep learning into material decomposition tasks represents a significant advancement in this field. As presented in Fig. 3, Geng et al. proposed a novel parallel multi-stream generative adversarial network PMS-GAN [66] that represents a promising approach for multi-material decomposition in the projection domain. The proposed PMS-GAN comprises a parallel multi-stream generator and a discriminator. The discriminator is employed to assist the generator network in learning an implicit loss function [67]. Material decomposition is addressed as a disentanglement problem, with differential maps

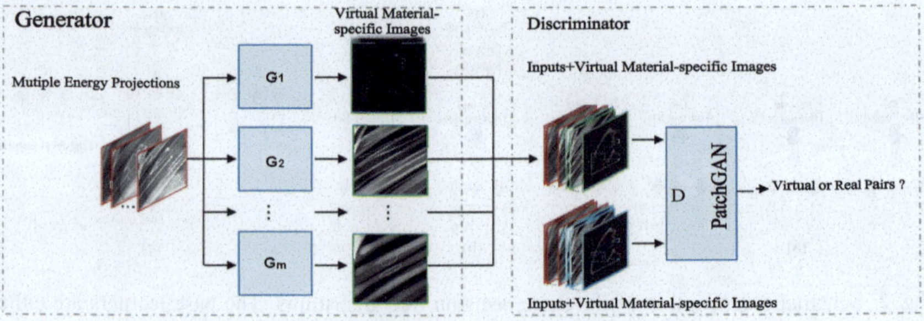

Fig. 3 This figure shows the scheme of projection-domain material decomposition method PMS-GAN [66]. PMS-GAN is composed of two subnetworks: generators $G_1, G_2, \cdots, G_m$ and a discriminator D. First, multi-energy projections $(A_1, A_2, \cdots, A_n)$ are input into the generator. Then, both input and output of the generator are fed into the discriminator which is committed to distinguishing real pairs from the pairs created by the generator

(DMs) devised for the generator to enhance decomposition accuracy. The proposed network addresses material decomposition as an end-to-end task, with the input and output being multi-energy projections and objective material-specific images, respectively. The proposed network was quantitatively evaluated using both simulation and experimental studies, with the results demonstrating that PMS-GAN markedly improves the accuracy of multi-material decomposition.

3.3.2 Extracting Prior Information in Image Domain

However, in projection-domain methods, the raw measured projections are first decomposed to material-specific projection datasets, and then the material-specific images can be obtained independently through conventional reconstruction algorithms. This method also requires a known X-ray emission spectrum. Projection-domain methods depend on accurate spectrum estimation methods ideal results.

In contrast, image-domain methods are capable of utilizing the back projection filtering (BPF) [68] results for material decomposition, a process typically employed in current clinical CTs for the purpose of implementing an analytical and rapid reconstruction algorithm. Moreover, the image domain offers a wealth of prior knowledge that is not available in the projection domain. This allows for the incorporation of a multitude of sophisticated prior regularizations into training a neural network, facilitating noise reduction, image edge protection, and structural recovery, among other applications.

Considering introducing effective prior knowledge to refine the final material-specific images. In the preceding decade, a number of image-domain material decomposition models have been investigated for use in DECT or spectral CT [69–72]. In the context of DECT, Niu et al. [69] developed a penalty least-square estimation, which entailed an investigation of the full variance-covariance properties of the decomposed images. Subsequently, a number of alternative approaches were proposed, including the edge-preserving algorithm with nonlocal mean [70], the similarity-based regularization [73], the entropy minimization [74], the total nuclear norm and L_0 norm [75], the mixed union of learned sparsifying transforms (MULTRA) model [76], and so on. In the context of spectral CT, Tao et al. put forth a novel approach, namely prior image-constrained compressed sensing (PICCS), which employs prior knowledge in an iterative denoising framework (MD-PKAID) [77].

Recently, in light of the redundant similarities between material images, convolutional neural networks (CNNs) have generated considerable interest in the field of computer vision, and a CNN-based approach to image-domain material decomposition has been developed for spectral CT. Concurrently, Chen and Li [78] put forth a robust multi-material decomposition approach for spectral CT utilizing CNNs. Xu et al. [79] used a cascade model and a modified fully convolutional network on material decomposition task. Chen et al. [78] suppressed the influence of ring artifacts on decomposition accuracy based on a visual geometry group (VGG) network and a deep residual network; Feng et al. [80] approached the problem by obtaining virtual monochromatic attenuation maps of scanned objects from polychromatic CT reconstructions; Clark et al. [72] modified U-Net architecture to perform material decomposition; Zhang et al. [81] highlighted the importance of network topological structure and presented a Butterfly-Net, which output tissue and bone images; Wu et al. [82] proposed fully convolutional DenseNets to classify the bone, lung, and soft tissue regions in the image domain. The rapid advancement of neural network architectures has led to the utilization of sophisticated transformers [83] and Graph Neural Networks (GNNs) [84]. These novel approaches aim to harness the formidable long-range information processing capabilities of transformers or GNNs for image-domain material decomposition [85–87].

3.3.3 Transfer Learning: From Simulated Data to Real-World Data

Transfer learning (TL) has emerged as a promising candidate technique to compensate the lack of real-world medical imaging datasets. TL is the ability of a system or model to recognize and apply knowledge and skills learned in previous domains/tasks to novel domains/tasks. TL assumes that the model can generalize to new tasks or new domains by learning features or factors that are common between source and target tasks [88]. Consequently, an increasing number of researchers are utilizing TL technology in the expectation that models trained on simulated data can be applied to real-world data.

Different approaches for TL can be found in [89, 90]; the most common is to pretrain a model on a large-annotated dataset and then to fine-tune the model on a more specific and smaller dataset [91, 92]. Many works proposed related TL approaches based on fine-tuning for material decomposition. Recently, in [93], the authors follow the Sim2Real transfer approach where the knowledge that one learns from synthetic data is transferred to a real-world scenario [94, 95]. The Sim2Real strategy is based on the fact that synthetic data is almost infinitely accessible and diverse, and it becomes pertinent when experimental annotated data is very limited or nonexistent, which is the case of material decomposition. While deep-learning-based approaches perform reasonably well given experimental training data, they may not necessarily work well when trained only on synthetic data. In Sim2Real framework for material decomposition task, numerical phantoms are generated from human CT volumes of the KiTS19 Challenge dataset [96]. Numerical data is then computed by segmenting CT volumes into specific materials (soft tissue and bone), projecting the material volumes and simulating photon counting data, taking into account the energy response of the scanner. Experiments indicated that this meticulously constructed dataset is efficacious in both the projection domain and the image-domain material decomposition and can markedly enhance the performance of the deep learning models in clinical scenes.

4 Summary

In this chapter, we begin by reviewing the fundamental principles of X-ray physics and traditional physics-based methods for material decomposition. Following this, we introduce three main categories of advanced material decomposition methods currently in use: image-domain-based methods, projection-domain-based methods, and deep-learning-based methods. Among these approaches, deep-learning-based methods have demonstrated superior accuracy due to their ability to learn complex patterns and structures directly from data. Moreover, the end-to-end strategy employed by these methods offers ease of use, as it does not rely on predefined mathematical models. Overall, deep-learning-based techniques provide a powerful tool for enhancing the quantitative accuracy of material decomposition.

References

1. A.S. Panchbhai, Wilhelm conrad röntgen and the discovery of x-rays: Revisited after centennial. J. Indian Acad. Oral Med. Radiol. **27**(1), 90–95 (2015)
2. J.J. Schreiber, P.A. Anderson, H.G. Rosas, A.L. Buchholz, A.G. Au, Hounsfield units for assessing bone mineral density and strength: a tool for osteoporosis management. JBJS **93**(11), 1057–1063 (2011)
3. B.J. Heismann, B.T. Schmidt, T.G. Flohr, Spectral computed tomography. (No Title) (2012)
4. R.E. Alvarez, A. Macovski, Energy-selective reconstructions in x-ray computerised tomography. Phys. Med. Biol. **21**(5), 733 (1976)
5. H.K. Genant, D. Boyd, Quantitative bone mineral analysis using dual energy computed tomography. Investig. Radiol. **12**(6), 545–551 (1977)
6. H.K. Choi, A.M. Al-Arfaj, A. Eftekhari, P.L. Munk, K. Shojania, G. Reid, S. Nicolaou, Dual energy computed tomography in tophaceous gout. Ann. Rheum. Dis. **68**(10), 1609–1612 (2009)
7. F. Bamberg, A. Dierks, K. Nikolaou, M.F. Reiser, C.R. Becker, T.R.C. Johnson, Metal artifact reduction by dual energy computed tomography using monoenergetic extrapolation. Eur. Radiol. **21**, 1424–1429 (2011)

8. R. Eliahou, G. Hidas, M. Duvdevani, J. Sosna, Determination of renal stone composition with dual-energy computed tomography: an emerging application, in *Seminars in Ultrasound, CT and MRI*, vol. 31 (Elsevier, Amsterdam, 2010), pp. 315–320
9. S. Dell'Aversana, R. Ascione, M. De Giorgi, D.R. De Lucia, R. Cuocolo, M. Boccalatte, G. Sibilio, G. Napolitano, G. Muscogiuri, S. Sironi, et al., Dual-energy CT of the heart: a review. J. Imag. **8**(9), 236 (2022)
10. J.L. Ducote, Y. Alivov, S. Molloi, Imaging of nanoparticles with dual-energy computed tomography. Phys. Med. Biol. **56**(7), 2031 (2011)
11. A. Graser, T.R.C. Johnson, H. Chandarana, M. Macari, Dual energy CT: preliminary observations and potential clinical applications in the abdomen. Eur. Radiol. **19**, 13–23 (2009)
12. R.M.N. Doesburg, *The Mars Photon Processing Cameras for Spectral CT* (2012)
13. P.M. Shikhaliev, Tilted angle CZT detector for photon counting/energy weighting x-ray and CT imaging. Phys. Med. Biol. **51**(17), 4267 (2006)
14. R. Ballabriga, J. Alozy, G. Blaj, M. Campbell, M. Fiederle, E. Frojdh, E.H.M. Heijne, X. Llopart, M Pichotka, S. Procz, et al., The Medipix3rx: a high resolution, zero dead-time pixel detector readout chip allowing spectroscopic imaging. J. Instrum. **8**(02), C02016 (2013)
15. T.S. Curry, J.E. Dowdey, R.C. Murry, *Christensen's Physics of Diagnostic Radiology*. Lippincott Williams & Wilkins (1990)
16. E.C. Frey, X. Wang, Y. Du, K. Taguchi, J. Xu, B.M.W. Tsui, Investigation of the use of photon counting x-ray detectors with energy discrimination capability for material decomposition in micro-computed tomography, in *Medical Imaging 2007: Physics of Medical Imaging*, vol. 6510 (SPIE, New York, 2007), pp. 119–129
17. S.J. Nik, J. Meyer, R. Watts, Optimal material discrimination using spectral x-ray imaging. Phys. Med. Biol. **56**(18), 5969 (2011)
18. B. Norlin, A. Manuilskiy, H.-E. Nilsson, C. Fröjdh, Material recognition with the medipix photon counting colour x-ray system. Nucl. Instrum. Methods Phys. Res., Sect. A Accelerators, Spectrometers, Detectors and Associated Equipment **531**(1-2), 265–269 (2004)
19. H.Q. Le, S. Molloi, Least squares parameter estimation methods for material decomposition with energy discriminating detectors. Med. Phys. **38**(1), 245–255 (2011)
20. M. Firsching, D. Niederlohner, T. Michel, G. Anton, Quantitative material reconstruction in CT with spectroscopic x-ray pixel detectors–a simulation study, in *2006 IEEE Nuclear Science Symposium Conference Record*, vol. 4 (IEEE, New York, 2006), pp. 2257–2259
21. J.P. Schlomka, E. Roessl, R. Dorscheid, S. Dill, G. Martens, T. Istel, C. Bäumer, C. Herrmann, R. Steadman, G. Zeitler, et al., Experimental feasibility of multi-energy photon-counting k-edge imaging in pre-clinical computed tomography. Phys. Med. Biol. **53**(15), 4031 (2008)
22. E. Roessl, R. Proksa, K-edge imaging in x-ray computed tomography using multi-bin photon counting detectors. Phys. Med. Biol. **52**(15), 4679 (2007)
23. G.A. Waychunas, M.J. Apted, G.E. Brown Jr, X-ray k-edge absorption spectra of fe minerals and model compounds: near-edge structure. Phys. Chem. Miner. **10**(1), 1–9 (1983)
24. T. Yamamoto, Assignment of pre-edge peaks in k-edge x-ray absorption spectra of 3d transition metal compounds: electric dipole or quadrupole? X-Ray Spectrom. Int. J. **37**(6), 572–584 (2008)
25. Z. Li, S. Leng, L. Yu, Z. Yu, C.H. McCollough, Image-based material decomposition with a general volume constraint for photon-counting CT, in *Medical Imaging 2015: Physics of Medical Imaging*, vol. 9412 (SPIE, New York, 2015), pp. 190–197
26. S. Faby, S. Kuchenbecker, D. Simons, H.-P. Schlemmer, M. Lell, M. Kachelrieß, Ct calibration and dose minimization in image-based material decomposition with energy-selective detectors, in *Medical Imaging 2014: Physics of Medical Imaging*, vol. 9033 (SPIE, New York, 2014), pp. 328–339
27. T.G. Schmidt, F. Pektas, Region-of-interest material decomposition from truncated energy-resolved CT. Med. Phys. **38**(10), 5657–5666 (2011)
28. W.T. Eadie, L.D. Kirkpatrick, *Statistical Methods in Experimental Physics* (1973)
29. R. Aamir, A. Chernoglazov, C.J. Bateman, A.P.H. Butler, P.H. Butler, N.G. Anderson, S.T. Bell, R.K. Panta, J.L. Healy, J.L. Mohr, et al., Mars spectral molecular imaging of lamb tissue: data collection and image analysis. J. Instrum. **9**(02), P02005 (2014)

30. C. Bai, G.L. Zeng, G.T. Gullberg, F. DiFilippo, S. Miller, Slab-by-slab blurring model for geometric point response correction and attenuation correction using iterative reconstruction algorithms. IEEE Trans. Nucl. Sci. **45**(4), 2168–2173 (1998)
31. E.J. Candès, J. Romberg, T. Tao, Robust uncertainty principles: Exact signal reconstruction from highly incomplete frequency information. IEEE Trans. Inf. Theory **52**(2), 489–509 (2006)
32. A.P. Dempster, N.M. Laird, D.B. Rubin, Maximum likelihood from incomplete data via the em algorithm. J. R. Stat. Soc. Ser. B Methodol. **39**(1), 1–22 (1977)
33. P.R.S. Mendonça, P. Lamb, D.V. Sahani, A flexible method for multi-material decomposition of dual-energy CT images. IEEE Trans. Med. Imaging **33**(1), 99–116 (2014)
34. Y. Long, J.A. Fessler, Multi-material decomposition using statistical image reconstruction for spectral CT. IEEE Trans. Med. Imaging **33**(8), 1614–1626 (2014)
35. S. Tilley, W. Zbijewski, J.W. Stayman, Model-based material decomposition with a penalized nonlinear least-squares CT reconstruction algorithm. Phys. Med. Biol. **64**(3), 035005 (2019)
36. Z. Li, S. Ravishankar, Y. Long, J.A. Fessler, Image-domain material decomposition using data-driven sparsity models for dual-energy CT, in *2018 IEEE 15th International Symposium on Biomedical Imaging (ISBI 2018)* (2018), pp. 52–56
37. N. Ducros, J.F. Perez-Juste Abascal, B. Sixou, S. Rit, F. Peyrin, Regularization of nonlinear decomposition of spectral x-ray projection images. Med. Phys. **44**(9), e174–e187 (2017)
38. J.F.P.J. Abascal, N. Ducros, F. Peyrin, Nonlinear material decomposition using a regularized iterative scheme based on the bregman distance. I. Problems **34**(12), 124003 (2018)
39. W. Wu, P. Chen, V.V. Vardhanabhuti, W. Wu, H. Yu, Improved material decomposition with a two-step regularization for spectral CT. IEEE Access **7**, 158770–158781 (2019)
40. Q. Xu, H. Yu, J. Bennett, P. He, R. Zainon, R. Doesburg, A. Opie, M. Walsh, H. Shen, A. Butler, et al., Image reconstruction for hybrid true-color micro-CT. IEEE Trans. Biomed. Eng. **59**(6), 1711–1719 (2012)
41. Z. Yu, S. Leng, Z. Li, C.H. McCollough, Spectral prior image constrained compressed sensing (spectral piccs) for photon-counting computed tomography. Phys. Med. Biol. **61**(18), 6707 (2016)
42. S. Wang, W. Wu, J. Feng, F. Liu, H. Yu, Low-dose spectral CT reconstruction based on image-gradient l0-norm and adaptive spectral piccs. Phys. Med. Biol. **65**(24), 245005 (2020)
43. H. Gao, H. Yu, S. Osher, G. Wang, Multi-energy CT based on a prior rank, intensity and sparsity model (PRISM). Inverse Probl. **27**(11), 115012 (2011)
44. L. Li, Z. Chen, G. Wang, J. Chu, H. Gao, A tensor prism algorithm for multi-energy CT reconstruction and comparative studies. J. Xray Sci. Technol. **22**(2), 147–163 (2014)
45. W. Wu, F. Liu, Y. Zhang, Q. Wang, H. Yu, Non-local low-rank cube-based tensor factorization for spectral CT reconstruction. IEEE Trans. Med. Imaging **38**(4), 1079–1093 (2018)
46. W. Wu, D. Hu, K. An, S. Wang, F. Luo, A high-quality photon-counting CT technique based on weight adaptive total-variation and image-spectral tensor factorization for small animals imaging. IEEE Trans. Instrum. Meas. **70**, 1–14 (2020)
47. S. Wang, H. Yu, Y. Xi, C. Gong, W. Wu, F. Liu, Spectral-image decomposition with energy-fusion sensing for spectral CT reconstruction. IEEE Trans. Instrum. Meas. **70**, 1–11 (2021)
48. K. Mechlem, T. Sellerer, S. Ehn, D. Münzel, E. Braig, J. Herzen, P.B. Noël, F. Pfeiffer, Spectral angiography material decomposition using an empirical forward model and a dictionary-based regularization. IEEE Trans. Med. Imaging **37**(10), 2298–2309 (2018)
49. W. Wu, H. Yu, P. Chen, F. Luo, F. Liu, Q. Wang, Y. Zhu, Y. Zhang, J. Feng, H. Yu, Dictionary learning based image-domain material decomposition for spectral CT. Phys. Med. Biol. **65**(24), 245006 (2020)
50. W. Wu, P. Chen, S. Wang, V. Vardhanabhuti, F. Liu, H. Yu, Image-domain material decomposition for spectral CT using a generalized dictionary learning. IEEE Trans. Radiat. Plasma Med. Sci. **5**(4), 537–547 (2020)
51. S. Chen, S.A. Billings, W. Luo, Orthogonal least squares methods and their application to non-linear system identification. Int. J. Control. **50**(5), 1873–1896 (1989)
52. S.G. Mallat, Z. Zhang, Matching pursuits with time-frequency dictionaries. IEEE Trans. Signal Process. **41**(12), 3397–3415 (1993)

53. M. Aharon, M. Elad, A. Bruckstein, K-SVD: An algorithm for designing overcomplete dictionaries for sparse representation. IEEE Trans. Signal Process. **54**(11), 4311–4322 (2006)
54. Q. Xu, H. Yu, X. Mou, L. Zhang, J. Hsieh, G. Wang, Low-dose x-ray CT reconstruction via dictionary learning. IEEE Trans. Med. Imaging **31**(9), 1682–1697 (2012)
55. M. Touch, D.P. Clark, W. Barber, C.T. Badea, A neural network-based method for spectral distortion correction in photon counting x-ray CT. Phys. Med. Biol. **61**(16), 6132 (2016)
56. Y. Lu, M. Kowarschik, X. Huang, Y. Xia, J.-H. Choi, S. Chen, S. Hu, Q. Ren, R. Fahrig, J. Hornegger, et al., A learning-based material decomposition pipeline for multi-energy x-ray imaging. Med. Phys. **46**(2), 689–703 (2019)
57. Y. Lu, J. Geret, M. Unberath, M. Manhart, Q. Ren, R. Fahrig, J. Hornegger, A. Maier, Projection-based material decomposition by machine learning using image-based features for computed tomography, in *The 13th International Meeting on Fully Three-Dimensional Image Reconstruction in Radiology and Nuclear Medicine* (2015), pp. 448–451
58. Y. Lu, M. Berger, M. Manhart, J.-H. Choi, M. Hoheisel, M. Kowarschik, R. Fahrig, Q. Ren, J. Hornegger, A. Maier, Bridge to real data: empirical multiple material calibration for learning-based material decomposition, in *2016 IEEE 13th International Symposium on Biomedical Imaging (ISBI)* (IEEE, New York, 2016), pp. 457–460
59. Y. Lu, M. Kowarschik, X. Huang, S. Chen, Q. Ren, R. Fahrig, J. Hornegger, A. Maier, Material decomposition using ensemble learning for spectral x-ray imaging. IEEE Trans. Radiat. Plasma Med. Sci. **2**(3), 194–204 (2018)
60. A. Maier, H.G. Hofmann, C. Schwemmer, J. Hornegger, A. Keil, R. Fahrig, Fast simulation of x-ray projections of spline-based surfaces using an append buffer. Phys. Med. Biol. **57**(19), 6193 (2012)
61. Y. Freund, R.E. Schapire, A decision-theoretic generalization of on-line learning and an application to boosting. J. Comput. Syst. Sci. **55**(1), 119–139 (1997)
62. L. Breiman, Bagging predictors. Mach. Learn. **24**, 123–140 (1996)
63. M. Firsching, P.T. Talla, T. Michel, G. Anton, Material resolving x-ray imaging using spectrum reconstruction with medipix2. Nucl. Instrum. Methods Phys. Res., Sect. A Accelerators, Spectrometers, Detectors and Associated Equipment **591**(1), 19–23 (2008)
64. N. Maaß, S. Sawall, M. Knaup, M. Kachelrieß, Empirical multiple energy calibration (EMEC) for material-selective CT, in *2011 IEEE Nuclear Science Symposium Conference Record* (IEEE, New York, 2011), pp. 4222–4229
65. D.H. Wolpert, Stacked generalization. Neural Netw. **5**(2), 241–259 (1992)
66. M. Geng, Z. Tian, Z. Jiang, Y. You, X. Feng, Y. Xia, K. Yang, Q. Ren, X. Meng, A. Maier, et al., PMS-GAN: Parallel multi-stream generative adversarial network for multi-material decomposition in spectral computed tomography. IEEE Trans. Med. Imaging **40**(2), 571–584 (2020)
67. P. Isola, J.-Y. Zhu, T. Zhou, A.A. Efros, Image-to-image translation with conditional adversarial networks, in *Proceedings of the IEEE Conference on Computer Vision and Pattern Recognition* (2017), pp. 1125–1134
68. W. Wu, H. Yu, S. Wang, F. Liu, BPF-type region-of-interest reconstruction for parallel translational computed tomography. J. Xray Sci. Technol. **25**(3), 487–504 (2017)
69. T. Niu, X. Dong, M. Petrongolo, L. Zhu, Iterative image-domain decomposition for dual-energy CT. Med. Phys. **41**(4), 041901 (2014)
70. W. Zhao, T. Niu, L. Xing, Y. Xie, G. Xiong, K. Elmore, J. Zhu, L. Wang, J.K. Min, Using edge-preserving algorithm with non-local mean for significantly improved image-domain material decomposition in dual-energy CT. Phys. Med. Biol. **61**(3), 1332 (2016)
71. C. Maaß, M. Baer, M. Kachelrieß, Image-based dual energy CT using optimized precorrection functions: a practical new approach of material decomposition in image domain. Med. Phys. **36**(8), 3818–3829 (2009)
72. D.P. Clark, M. Holbrook, C.T. Badea, Multi-energy CT decomposition using convolutional neural networks, in *Medical Imaging 2018: Physics of Medical Imaging*, vol. 10573 (SPIE, New York, 2018), pp. 415–423

73. J. Harms, T. Wang, M. Petrongolo, T. Niu, L. Zhu, Noise suppression for dual-energy CT via penalized weighted least-square optimization with similarity-based regularization. Med. Phys. **43**(5), 2676–2686 (2016)
74. M. Petrongolo, L. Zhu, Noise suppression for dual-energy CT through entropy minimization. IEEE Trans. Med. Imaging **34**(11), 2286–2297 (2015)
75. Q. Ding, T. Niu, X. Zhang, Y. Long, Image-domain multimaterial decomposition for dual-energy CT based on prior information of material images. Med. Phys. **45**(8), 3614–3626 (2018)
76. Z. Li, S. Ravishankar, Y. Long, J.A. Fessler, Dect-multra: Dual-energy CT image decomposition with learned mixed material models and efficient clustering. IEEE Trans. Med. Imaging **39**(4), 1223–1234 (2019)
77. S. Tao, K. Rajendran, C.H. McCollough, S. Leng, Material decomposition with prior knowledge aware iterative denoising (MD-PKAID). Phys. Med. Biol. **63**(19), 195003 (2018)
78. Z. Chen, L. Li, Robust multimaterial decomposition of spectral CT using convolutional neural networks. Opt. Eng. **58**(1), 013104–013104 (2019)
79. Y. Xu, B. Yan, J. Zhang, J. Chen, L. Zeng, L. Wang, Image decomposition algorithm for dual-energy computed tomography via fully convolutional network. Comput. Math. Methods Med. **2018**(1), 2527516 (2018)
80. C. Feng, K. Kang, Y. Xing, A multi-energy material decomposition method for spectral CT using neural network, in *Medical Imaging 2018: Physics of Medical Imaging*, vol. 10573 (SPIE, New York, 2018), pp. 1160–1167
81. W. Zhang, H. Zhang, L. Wang, X. Wang, X. Hu, A. Cai, L. Li, T. Niu, B. Yan, Image domain dual material decomposition for dual-energy CT using butterfly network. Med. Phys. **46**(5), 2037–2051 (2019)
82. X. Wu, P. He, Z. Long, X. Guo, M. Chen, X. Ren, P. Chen, L. Deng, K. An, P. Li, et al. Multi-material decomposition of spectral CT images via fully convolutional densenets. J. Xray Sci. Technol. **27**(3), 461–471 (2019)
83. A. Vaswani, N. Shazeer, N. Parmar, J. Uszkoreit, L. Jones, A.N. Gomez, Ł. Kaiser, I. Polosukhin, Attention is all you need, in *Advances in Neural Information Processing Systems*, ed. by I. Guyon, U. Von Luxburg, S. Bengio, H. Wallach, R. Fergus, S. Vishwanathan, R. Garnett, vol. 30 (Curran Associates, Inc., New York, 2017)
84. F. Scarselli, M. Gori, A.C. Tsoi, M. Hagenbuchner, G. Monfardini, The graph neural network model. IEEE Trans. Neural Netw. **20**(1), 61–80 (2008)
85. G. Wang, Z. Liu, Z. Huang, N. Zhang, H. Luo, L. Liu, H. Shen, C. Che, T. Niu, D. Liang, et al., Improved gan: using a transformer module generator approach for material decomposition. Comput. Biol. Med. **149**, 105952 (2022)
86. X. Guo, P. He, X. Lv, X. Ren, Y. Li, Y. Liu, X. Lei, P. Feng, H. Shan, Material decomposition of spectral CT images via attention-based global convolutional generative adversarial network. Nucl. Sci. Tech. **34**(3), 45 (2023)
87. Z. Shi, F. Kong, M. Cheng, H. Cao, S. Ouyang, Q. Cao, Multi-energy CT material decomposition using graph model improved CNN. Med. Biol. Eng. Comput. **62**(4), 1213–1228 (2024)
88. Y. Bengio, A. Courville, P. Vincent, Representation learning: A review and new perspectives. IEEE Trans. Pattern Anal. Mach. Intell. **35**(8), 1798–1828 (2013)
89. H.-C. Shin, H.R. Roth, M. Gao, L. Lu, Z. Xu, I. Nogues, J. Yao, D. Mollura, R.M. Summers, Deep convolutional neural networks for computer-aided detection: CNN architectures, dataset characteristics and transfer learning. IEEE Trans. Med. Imaging **35**(5), 1285–1298 (2016)
90. G. Wilson, D.J. Cook, A survey of unsupervised deep domain adaptation. ACM Trans. Intell. Syst. Technol. (TIST) **11**(5), 1–46 (2020)
91. N. Tajbakhsh, J.Y. Shin, S.R. Gurudu, RT. Hurst, C.B. Kendall, M.B. Gotway, J. Liang, Convolutional neural networks for medical image analysis: full training or fine tuning? IEEE Trans. Med. Imaging **35**(5), 1299–1312 (2016)
92. R.K. Samala, H.-P. Chan, L. Hadjiiski, M.A. Helvie, J. Wei, K. Cha, Mass detection in digital breast tomosynthesis: Deep convolutional neural network with transfer learning from mammography. Med. Phys. **43**(12), 6654–6666 (2016)

93. J.F.P.J. Abascal, N. Ducros, V. Pronina, S. Rit, P.-A. Rodesch, T. Broussaud, S. Bussod, P.C. Douek, A. Hauptmann, S. Arridge, et al., Material decomposition in spectral CT using deep learning: a Sim2real transfer approach. IEEE Access **9**, 25632–25647 (2021)
94. C. Doersch, A. Zisserman, Sim2real transfer learning for 3d human pose estimation: motion to the rescue, in *Advances in Neural Information Processing Systems*, vol. 32 (2019)
95. A. Pashevich, R. Strudel, I. Kalevatykh, I. Laptev, C. Schmid, Learning to augment synthetic images for sim2real policy transfer, in *2019 IEEE/RSJ International Conference on Intelligent Robots and Systems (IROS)* (IEEE, New York, 2019), pp. 2651–2657
96. N. Heller, F. Isensee, K.H. Maier-Hein, X. Hou, C. Xie, F. Li, Y. Nan, G. Mu, Z. Lin, M. Han, et al., The state of the art in kidney and kidney tumor segmentation in contrast-enhanced CT imaging: Results of the kits19 challenge, in *Medical Image Analysis* (2020), p. 101821

Scatter Correction Using Deep Learning for Cone-Beam Computed Tomography

Xin Ge, Tianye Niu, and Peng Jin

1 Introduction

Computed tomography (CT) is a type of imaging diagnostic examination. It employs X-ray measurements from multiple angles to reconstruct three-dimensional structures and obtain information about internal spatial structure without causing damage to the objects being scanned [1, 2]. Since 1972, when a biomedical engineer Godfrey Hounsfield invented the computed axial tomography scan and was awarded the Nobel Prize in 1979 for his contribution to the diagnosis of neurological and other disorders [3], all types of CT equipment and related imaging technologies have been developed rapidly, especially in the medical field, where it has become an important diagnostic tool and is still the gold standard for diagnosing various diseases [4–6].

Cone beam CT (CBCT) is an expansion of CT imaging technology that allows for accurate and three-dimensional imaging. Due to the internal structure of the detection object, the system's mechanical design, the reconstruction algorithm, etc., CBCT often has severe artifacts that reduce measurement accuracy [7]. Scattering artifacts and beam-hardening artifacts result from the physical processes involved in CT data acquisition and are constrained by imaging principles. Shadow and stripe artifacts degrade the image quality of CBCT. Siewerdsen and Jaffray [8] found that scattering artifacts account for a significant proportion in cone beam CT images and that scattering reduces image contrast and increases image noise. For large-sized objects such as large phantoms and patient chest and abdomen scans, the cone angle will be large, resulting in a scattering proportion more than 100%. Due to the modulation of the scattering signal, imaging complicated anatomical

X. Ge · T. Niu (✉) · P. Jin
Shenzhen Bay Laboratory, Shenzhen, China
e-mail: niuty@szbl.ac.cn

K. (K.) Iniewski, L. (K.) Cai (eds.), *Deep Learning for Advanced X-ray Detection and Imaging Applications*, https://doi.org/10.1007/978-3-031-75653-5_6

structures with CBCT significantly degrades image quality when these structures are present. These issues restrict the quality and quantity of CBCT imaging in clinical practice. To effectively eliminate the scattering artifacts in CBCT, it is necessary to explore the causes of scattering and the mechanism of interaction between X-ray and matter, so as to provide theoretical guidance for the correction of scattering artifacts. The interaction between X-ray and matter mainly includes photoelectric effect, Compton effect, and Rayleigh scattering. The attenuation of X-ray intensity after passing through the object comes from scattering and absorption. The former means that the X-ray photons collide with matter and change the propagation direction. In the latter case, photoelectric absorption with matter captures X-ray photons. Scattering and absorption will cause the attenuation of X-ray intensity in the original propagation direction [9]. In general, the intensity attenuation resulting from scattering is considerably less than that resulting from absorption. However, the attenuation ratio of scattering is significantly increased for low-energy X-rays, especially the X-rays of medical cone beam CT equipment in the diagnostic energy range, resulting in severe contamination of CT images. In the CBCT imaging system, a primary ray is a ray that passes through an object without changing its propagation direction. A scattered ray is a ray that interacts with the object once or multiple times and arrives at the detector after changing its initial propagation direction. During the interaction between photons and objects, the principles of Compton scattering and Rayleigh scattering dictate that some rays will change their initial propagation direction and propagation energy and enter a pixel of the detector alongside effective rays, thereby increasing the number of photons of that pixel. As shown in Fig. 1, the linear attenuation coefficient of the object will be less than its true value in the reconstructed CT image, resulting in a shadow region. To eliminate shadow artifacts in the CT image, it is necessary to design a scatter correction method that removes scattering signals while retaining only primary ray signals.

2 Scatter Correction Using Deep Learning

Since scattering noise is inherent in the incident X-ray fluence, removing scattering artifacts effectively is crucial for improving the image quality of CBCT. There are two primary approaches to reduce the effect of scattered X-rays: The first method is physical rejection or suppression with the use of additional hardware, such as an air gap or an anti-scatter grid in the ray path [10, 11]. Anti-scatter grids are commonly used in most of X-ray projection imaging systems, but they also reduce the detected signal from the primary photons, resulting in a radiation dose penalty if equivalent image quality is desired [12]. The second method is post-processing. Several algorithms can be categorized as learning-based and non-learning-based methods, depending on whether or not learning techniques are utilized. Measuring-based methods and calculating-based methods are two cate-

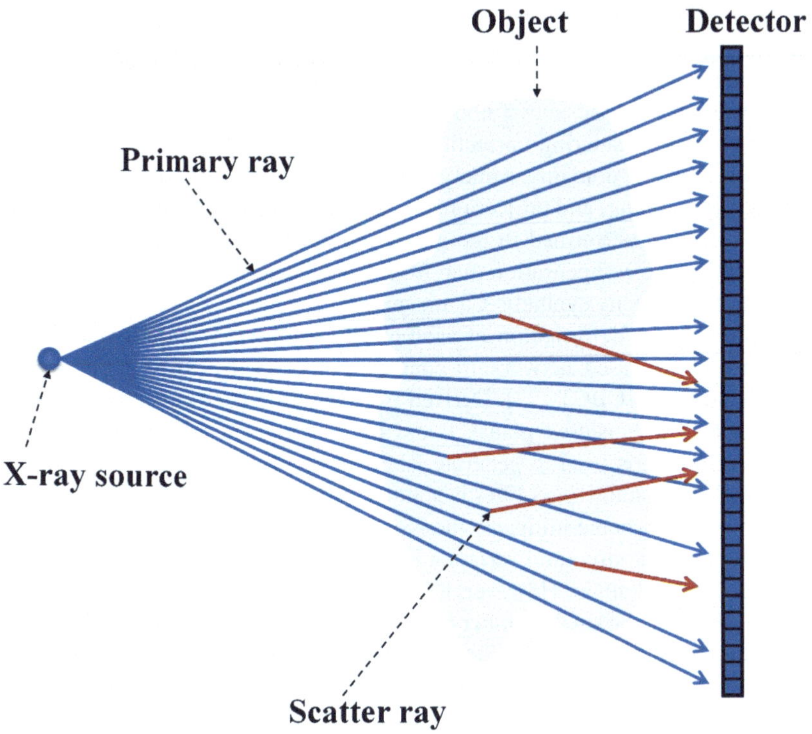

Fig. 1 The process of forming scattering artifacts

gories of non-learning-based approaches. To estimate scatter samples in the blocked projection area, measurement-based methods insert a primary beam-blocker array between the X-ray source and the object [13]. The primary modulation methods are developed by extracting low-frequency scatter components from the Fourier domain [14]. These methods require hardware modification that may not be practical in clinical equipment. The calculation-based methods include the Monte Carlo (MC)-based methods and the analytical scatter kernel-based methods. The MC-based method achieves accurate scatter estimation based on the accurate modeling of the CBCT imaging systems and simulation of the potential physical interaction between photons and matter while at the expense of heavy computational burden [15]. To combat the low computational efficiency, analytical scatter kernel-based methods provide a faster solution by simulating the scattering process between the primary beam and the imaged object through convolutional operations [16, 17]. These methods work well for relatively uniform objects. Due to the nonlinear scattering process, they may not perform effectively for objects with heterogeneous inner structures [18]. Consequently, the corrected CBCT images still have residual artifacts.

Learning-based methods can capture complex scattering distribution through feature extraction and representation learning. The deep convolutional neural network (CNN) can be used to deal with the nonlinear scattering process to overcome the shortcomings of the above non-learning-based methods [19]. The learning-based methods of scattering correction can be implemented in three ways based on different working domains, namely, direct image-to-image mapping, projection-to-projection mapping, and projection to scattering transition. Unlike the scattering correction method performed in the projection domain, image-to-image mapping directly studies the compensation map from CBCT images with scattering-induced artifacts to high-quality synthetic CT images [20, 21].

To minimize the deformation of anatomical structures, David C. Hansen et al. proposed a U-net-based network to convert projection data from uncorrected to scattering-corrected CBCT [22]. Derived from the difference between the forward projection of the registered plan CT and the uncorrected CBCT projection, the scatter distribution is used to generate labels in the training dataset. To eliminate registration errors, a low-pass filter is applied to the difference [23]. The projection domain mapping scheme mitigates the issue of anatomy deformation in the image domain, because the structural variation in the projection domain is less pronounced than in the image domain. However, it is difficult to eliminate anatomical structure deformation due to the small number of patients in the training dataset, which cannot account for the anatomical variations observed in the patient population [22].

To address the problem of inconsistent anatomy, the projection-to-scatter transition has been proposed to estimate the scatter signal as opposed to the projection data. The estimated scatter signal is subtracted from the raw projection to correct for scatter contamination without affecting the anatomical structure of the projection. Yusuke Nomura et al., for instance, utilized the U-net to obtain the mapping from the measured CBCT projection to the corresponding scatter distribution generated by MC simulation in non-anthropomorphic phantoms [19]. Since scatter signal is predominately low frequency and has minimal impact on relatively high-frequency anatomical structures, this scheme avoids the inconsistent anatomy issue of indirect mapping methods.

2.1 Network Architecture

The following is a brief summary of two of our recent works, DRCNN and PixPix GAN. A depth residual convolutional neural network (DRCNN) image domain method employs U-net depth convolution neural network and shadow compensation. In order to train the mapping function from uncorrected CBCT to high-quality corrected CBCT, the depth residual convolution neural network employs the MC simulation method. Due to the supervised learning property of the deep residual convolution neural network, the CBCT images in the training and test data sets must have the same physiological structure. Due to nonideal physics, shading artifacts in the uncorrected CBCT image significantly degrade image quality. As the difference between the corrected and uncorrected images, the shading compensation map effectively compensates for low-frequency shading artifacts. In the image domain, the corrected image can be generated by directly adding the shading compensation map to the uncorrected image. In scatter correction, the proposed DRCNN is used to learn the shading compensation map. The DRCNN mapping function is expressed as follows:

$$f^{corr} = M\left(f^{un}\right) + f^{un}, \tag{1}$$

where f^{corr} is the corrected image, f^{un} is the uncorrected image, and M is an end-to-end shading compensation mapping function.

The proposed DRCNN trains the mapping function using supervised learning with the output size matching the input size. We utilize a deep CNN as the primary component due to its superior learning ability. The deep CNN collaborates with the residual learning framework (RLF) to reduce the difficulty of training, as the deep CNN typically experiences the gradient vanishing issue. Two residual network modules (RNMs) are designed based on the RLF to mitigate the gradient vanishing problem. The residual block is the primary component of the residual contracting path (the left side) and the residual expansive path (the right side). The second residual network module is applied to learn the shading compensation map by connecting the network's input to its output via a global shortcut. Since the number of collected CBCT images is limited and the distributions of shading artifacts vary, the DRCNN runs the risk of overfitting. Dropout operations are utilized in the DRCNN to avoid the overfitting issue [24]. In each network training iteration, the dropout operation randomly eliminates the network layer's feature channels and corresponding parameters, generating new networks with simpler network structures. In contrast to the U-net, batch normalization (BN) operations are used to enhance computational efficiency by resolving the internal covariate shift [25]. To normalize the variables of the activation function, BN operations are placed between the convolution and activation operations. These diverse networks are efficiently combined to increase the DRCNN's robustness. The architecture and kernel parameters of the proposed DRCNN are depicted in Fig. 2.

To overcome the constraint of consistent anatomy requirement in DRCNN, the cycle generative adversarial network (GAN) is combined with residual block to investigate the mapping function from CBCT to its counterpart, such as the paired planning CT [21]. GAN can generate high-quality samples, because the updated parameters in the generator are driven by the back propagation of the discriminator and not by the data samples themselves [26, 27]. Cyclic GAN enforces inverse transformation and consistency to facilitate greater precision in learning [21]. Inconsistent anatomical training in the cyclic GAN may alter the anatomical structure of the corrected CBCT image, resulting in suspicious results [13].

Estimate accurate scatter distributions from CBCT projections acquired at different imaging volume sizes and X-ray energy spectra to achieve the generalized image quality improvement while preserving anatomy. We employ the pixel-to-pixel (Pix2pix) GAN to achieve the generalized scatter estimation, because the Pix2pix GAN enhances the original GAN's ability to achieve pixel-wise image-to-image transition by the correspondence construction between the sample and label. The

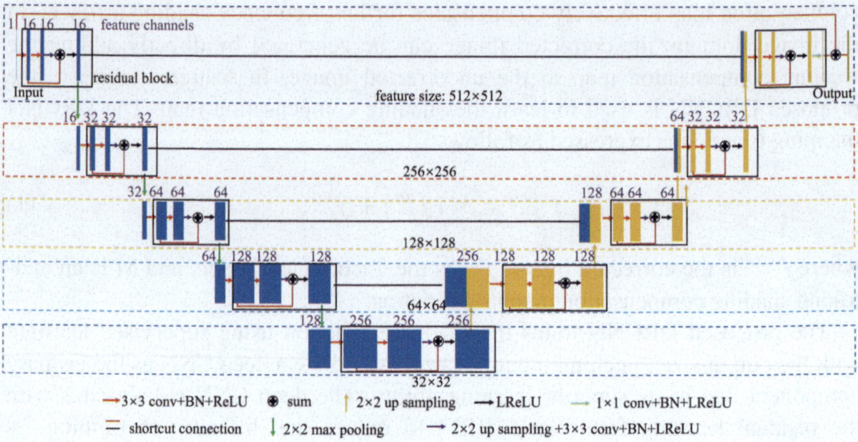

Fig. 2 The proposed DRCNN architecture. The network layers are represented by blue or yellow boxes. Using the purple font, the number of features of each layer is displayed on top of the box. The feature dimensions extracted by the network layers in the dashed rectangle of the same color are identical and are denoted by black fonts. The residual blocks are represented by the black solid rectangles. This figure is adopted from reference [20]

network architecture and loss function of Pix2pix GAN are optimized for the best performance during the projection-to-scatter transition. CT data from the abdomen and head of patients are incorporated into the training set. The trained projection to scatter mapping model is applied to CBCT projection data of various anatomy and X-ray energy spectra in order to estimate scatter distributions. In the testing procedure, both head phantom and abdominal patient CBCT projections acquired at various energy spectra are utilized.

The training dataset can be generated using any scatter estimation method that does not rely on a learning scheme, such as the MC-based or analytical scatter kernel-based methods. Since scatter-induced artifacts are uncommon in multidetector CT (MDCT) images due to well-designed scatter suppression schemes over the past decades [23, 28], the MDCT data are used as one example of implementing the framework in this work.

The workflow of the proposed scatter correction method is depicted in Fig. 3. The area enclosed by the dashed line represents the training dataset generation and model training process. On the multiple spectral MDCT images, the forward projection algorithm and MC simulation are performed to generate scatter-free projection data and scatter signal, respectively. The scatter signal is added to the scatter-free projection in order to simulate the scatter-contaminated projection using the same scatter-to-primary ratio (SPR) as the measured CBCT projection. For the training of the scatter estimation model, the Pix2pix GAN employs the mimicked scatter contaminated projection and scatter signal as the sample and label, respectively. To achieve CBCT scatter correction, the trained model is applied to the CBCT projection data to predict the scatter distribution.

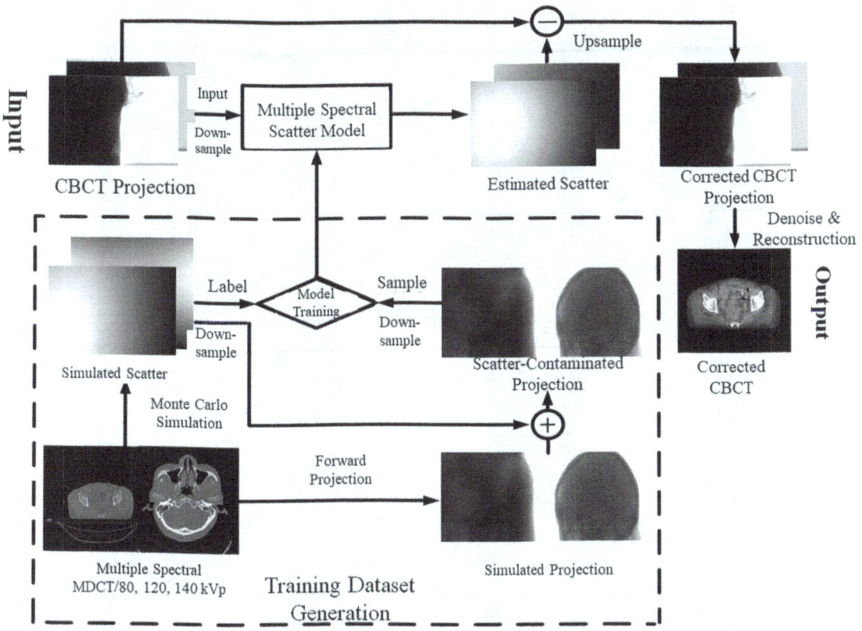

Fig. 3 The workflow of the scatter estimation using PixPix GAN. This figure is adopted from reference [29]

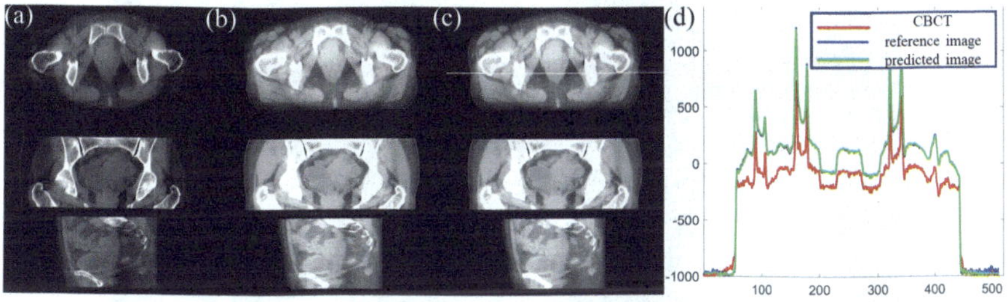

Fig. 4 CBCT images reconstructed of the testing dataset using (**a**) FDK without scatter correction; (**b**) the trained model; and (**c**) the MC simulation as the reference. (**d**) 1D horizontal profiles along the solid line drawn in (**c**). The coronal and sagittal views of images (**a**)–(**c**) are shown in the second and third rows. Displace window: [−250,300] HU. This figure is adopted from reference [20]

2.2 *Network Performance*

To evaluate the performance of the DRCNN network, we use the testing datasets that are not included in the training or validation datasets. Figure 4 depicts the 3D views of the CBCT images of the testing dataset. Figures 4a–c depict, respectively, the uncorrected image, the corrected image, and the reference image. Figure 4a demonstrates that the trained network successfully removes shading artifacts from

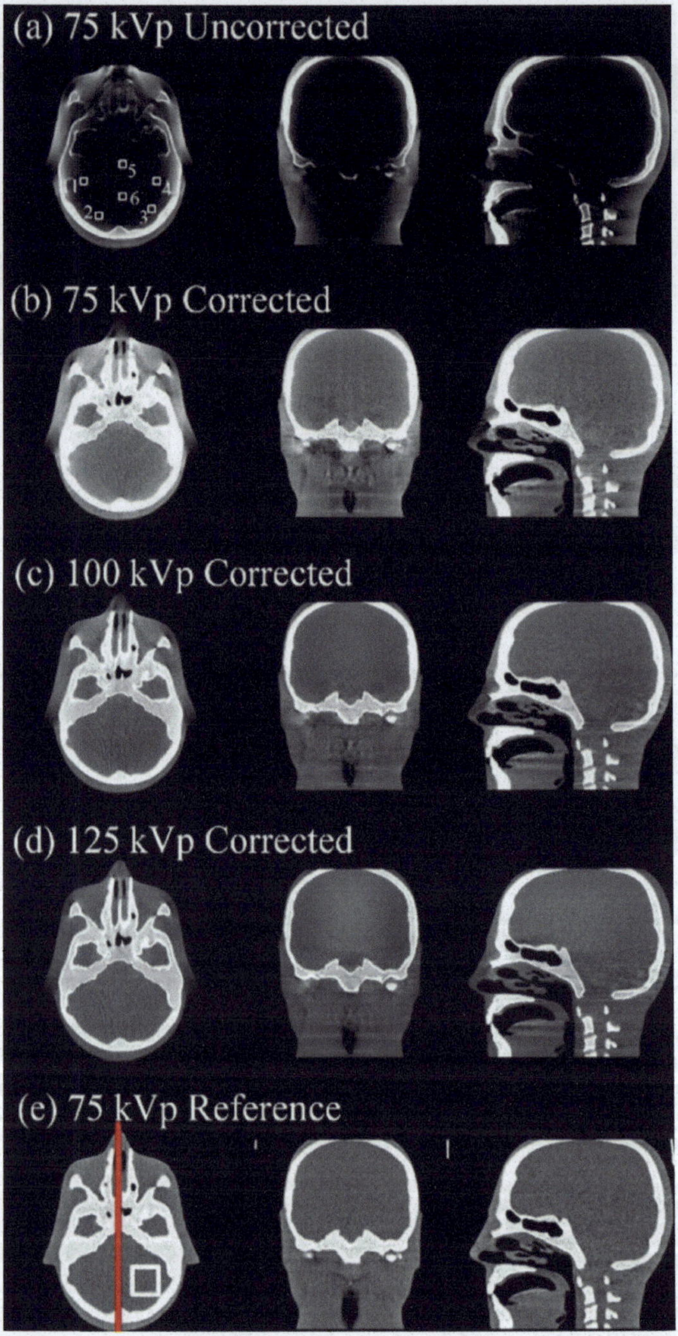

Fig. 5 The axial, coronal, and sagittal views of the head phantom CBCT data: (**a**) uncorrected image at 75 kVp, (**b**)–(**d**) corrected image at 75 kVp, 100 kVp, 125 kVp, and (e) the reference MDCT image at 75 kVp. Displace window is [−150 400] HU for all. This figure is adopted from reference [29]

the uncorrected CBCT images. A quantitative evaluation is performed with respect to the reference image (Fig. 4c).

Figure 5 depicts the axial, coronal, and sagittal views of the uncorrected, corrected, and reference CBCT images of a head phantom scanned at 75, 100, and 125 kVp. At 75 kVp, we only display the uncorrected and reference images. In the head phantom experiment, we use the registered MDCT as the reference image for comparison. Results from correcting the above three sets of scanning spectra suggest that the proposed method can remove most correction artifacts and bring CBCT image quality closer to that of high-quality MDCT.

In summary, we mainly established two methods to reduce the impacts of scatter artifacts in CT imaging technology. One is the feature-based method in DRCNN, and the other one is the usage of multiple spectral CT labels in Pix2pix GAN. Both of the methods are realized through deep learning and can improve CT image quality and more suitable for clinical usage.

References

1. G. Hounsfield, A method of and apparatus for examination of a body by radiation such as X or gamma radiation. The Patent Office, London, Patent Specificiation, vol. 1283915
2. J. Hsieh, *Computed Tomography: Principles, Design, Artifacts, and Recent Advances* (SPIE Press, 2003)
3. R. Paxton, J. Ambrose, The EMI scanner. A brief review of the first 650 patients. Br. J. Radiol. **47**(561), 530–565 (1974)
4. W. Roberts, J. Bax, L. Davies, Cardiac CT and CT coronary angiography: Technology and application. Heart **94**(6), 781–792 (2008)
5. S.J. Schambach, S. Bag, L. Schilling, C. Groden, M.A. Brockmann, Application of micro-CT in small animal imaging. Methods **50**(1), 2–13 (2010)
6. S. Tashman, W. Anderst, In-vivo measurement of dynamic joint motion using high speed biplane radiography and CT: Application to canine ACL deficiency. J. Biomech. Eng. **125**(2), 238–245 (2003)
7. R. Schulze, U. Heil, D. Groβ, D.D. Bruellmann, E. Dranischnikow, U. Schwanecke, E. Schoemer, Artefacts in CBCT: A review. Dentomaxillofac. Radiol. **40**(5), 265–273 (2011)
8. J.H. Siewerdsen, D.A. Jaffray, Cone-beam computed tomography with a flat-panel imager: Magnitude and effects of x-ray scatter. Med. Phys. **28**(2), 220–231 (2001)
9. T. Niu, L. Zhu, Overview of x-ray scatter in cone-beam computed tomography and its correction methods. Curr. Med. Imaging **6**(2), 82–89 (2010)
10. Y. Kyriakou, W. Kalender, Efficiency of antiscatter grids for flat-detector CT. Phys. Med. Biol. **52**(20), 6275 (2007)
11. J.A. Sorenson, J. Floch, Scatter rejection by air gaps: An empirical model. Med. Phys. **12**(3), 308–316 (1985)
12. C. Yang, P. Wu, S. Gong, J. Wang, Q. Lyu, X. Tang, T. Niu, Shading correction assisted iterative cone-beam CT reconstruction. Phys. Med. Biol. **62**(22), 8495 (2017)
13. T. Niu, L. Zhu, Scatter correction for full-fan volumetric CT using a stationary beam blocker in a single full scan. Med. Phys. **38**(11), 6027–6038 (2011)
14. L. Zhu, N.R. Bennett, R. Fahrig, Scatter correction method for X-ray CT using primary modulation: Theory and preliminary results. IEEE Trans. Med. Imaging **25**(12), 1573–1587 (2006)
15. K. He, X. Zhang, S. Ren, J. Sun, Deep residual learning for image recognition, pp. 770–778

16. M. Baer, M. Kachelrieß, Hybrid scatter correction for CT imaging. Phys. Med. Biol. **57**(21), 6849 (2012)
17. W. Zhao, D. Vernekohl, J. Zhu, L. Wang, L. Xing, A model-based scatter artifacts correction for cone beam CT. Med. Phys. **43**(4), 1736–1753 (2016)
18. Y. Nomura, Q. Xu, H. Peng, S. Takao, S. Shimizu, L. Xing, H. Shirato, Modified fast adaptive scatter kernel superposition (mfASKS) correction and its dosimetric impact on CBCT-based proton therapy dose calculation. Med. Phys. **47**(1), 190–200 (2020)
19. Y. Nomura, Q. Xu, H. Shirato, S. Shimizu, L. Xing, Projection-domain scatter correction for cone beam computed tomography using a residual convolutional neural network. Med. Phys. **46**(7), 3142–3155 (2019)
20. Y. Jiang, C. Yang, P. Yang, X. Hu, C. Luo, Y. Xue, L. Xu, X. Hu, L. Zhang, J. Wang, Scatter correction of cone-beam CT using a deep residual convolution neural network (DRCNN). Phys. Med. Biol. **64**(14), 145003 (2019)
21. J. Harms, Y. Lei, T. Wang, R. Zhang, J. Zhou, X. Tang, W.J. Curran, T. Liu, X. Yang, Paired cycle-GAN-based image correction for quantitative cone-beam computed tomography. Med. Phys. **46**(9), 3998–4009 (2019)
22. D.C. Hansen, G. Landry, F. Kamp, M. Li, C. Belka, K. Parodi, C. Kurz, ScatterNet: A convolutional neural network for cone-beam CT intensity correction. Med. Phys. **45**(11), 4916–4926 (2018)
23. T. Niu, M. Sun, J. Star-Lack, H. Gao, Q. Fan, L. Zhu, Shading correction for on-board cone-beam CT in radiation therapy using planning MDCT images. Med. Phys. **37**(10), 5395–5406 (2010)
24. N. Srivastava, G. Hinton, A. Krizhevsky, I. Sutskever, R. Salakhutdinov, Dropout: A simple way to prevent neural networks from overfitting. J. Mach. Learn. Res. **15**(1), 1929–1958 (2014)
25. S. Ioffe, C. Szegedy, Batch normalization: Accelerating deep network training by reducing internal covariate shift. pp. 448–456
26. I. Goodfellow, J. Pouget-Abadie, M. Mirza, B. Xu, D. Warde-Farley, S. Ozair, A. Courville, Y. Bengio, Generative adversarial networks. Commun. ACM **63**(11), 139–144 (2020)
27. Y. Zhang, N. Yue, M.Y. Su, B. Liu, Y. Ding, Y. Zhou, H. Wang, Y. Kuang, K. Nie, Improving CBCT quality to CT level using deep learning with generative adversarial network. Med. Phys. **48**(6), 2816–2826 (2021)
28. Z.C. Li, G. Zhai, J. Zhang, Z. Wang, G. Liu, G.Y. Wu, D. Liang, H. Zheng, Differentiation of clear cell and non-clear cell renal cell carcinomas by all-relevant radiomics features from multiphase CT: A VHL mutation perspective. Eur. Radiol. **29**(8), 3996–4007 (2019)
29. Y. Jiang, Y. Zhang, C. Luo, P. Yang, J. Wang, X. Liang, W. Zhao, R. Li, T. Niu, A generalized image quality improvement strategy of cone-beam CT using multiple spectral CT labels in Pix2pix GAN. Phys. Med. Biol. **67**(11), 115003 (2022)

Correcting Charge Sharing Distortions in Photon Counting Detectors Utilizing a Spatial-Temporal CNN

Aaron Smith, James Atlas, Ali Atharifard, Corin Simcock, Sikiru A. Adebileje, Steven D. Alexander, Maya R. Amma, Marzieh Anjomrouz, Fatemeh Asghariomabad, Anthony P. H. Butler, Philip H. Butler, Pierre Carbonez, Claire Chambers, Krishna M. Chapagain, Alexander I. Chernoglazov, Jennifer A. Clark, Frances Colgan, Jonathan S. Crighton, Jérôme Damet, Theodorus Dapamede, Niels J. A. de Ruiter, Devyani Dixit, Robert M. N. Doesburg, Karen Dombroski, Neryda Duncan, Steven P. Gieseg, Joseph L. Healy, Luke Holmes, Kevin Jonker, Tracy Kirkbride, V. B. H. Mandalika, Aysouda Matanaghi, Mahdieh Moghiseh, Manoj Paladugu, David Palmer, Raj K. Panta, Peter Renaud, Yann Sayous, Nanette Schleich, Jereena S. Sheeja, Lieza Vanden Broeke, V. S. Vivek, E. Peter Walker, and Manoj Wijesooriya

1 Introduction

Photon counting computerised tomography (CT) scanners are becoming an increasingly useful area of medical imaging [6]. The ability of the photon-counting CT scanner to generate medical scans with high spatial resolutions and with lower

A. Smith (✉) · J. Atlas · D. Dixit
MARS Bioimaging Limited, Christchurch, New Zealand

University of Canterbury, Christchurch, New Zealand
e-mail: aaron.smith@pg.canterbury.ac.nz

A. Atharifard · S. D. Alexander · M. R. Amma · M. Anjomrouz · F. Asghariomabad · C. Chambers · R. M. N. Doesburg · K. Dombroski · N. Duncan · J. L. Healy · A. Matanaghi · J. S. Sheeja · L. V. Broeke · V. S. Vivek · M. Wijesooriya
MARS Bioimaging Limited, Christchurch, New Zealand

C. Simcock · K. M. Chapagain · K. Jonker · M. Moghiseh
MARS Bioimaging Limited, Christchurch, New Zealand

University of Otago Christchurch, Christchurch, New Zealand

S. A. Adebileje · A. I. Chernoglazov · V. B. H. Mandalika
MARS Bioimaging Limited, Christchurch, New Zealand

Human Interface Technology Laboratory New Zealand, University of Canterbury, Christchurch, New Zealand

K. (K.) Iniewski, L. (K.) Cai (eds.), *Deep Learning for Advanced X-ray Detection and Imaging Applications*, https://doi.org/10.1007/978-3-031-75653-5_7

doses when compared to traditional CT scanners [10] enables patient care benefits for medical imaging. Photon-counting detectors' (PCD) discretised energy values enable enhanced material decomposition [14]. However, the decreased pixel size used to detect individual photons at increased resolutions introduces distortion issues such as charge sharing. Charge sharing introduces spectrum distortion via a single incident photon being recorded as two or more lower-energy photons. This also results in an increase in noise present in the scan [13].

A. P. H. Butler · P. H. Butler
MARS Bioimaging Limited, Christchurch, New Zealand

University of Canterbury, Christchurch, New Zealand

University of Otago Christchurch, Christchurch, New Zealand

European Organisation for Nuclear Research (CERN), Geneva, Switzerland

Human Interface Technology Laboratory New Zealand, University of Canterbury, Christchurch, New Zealand

P. Carbonez
University of Otago Christchurch, Christchurch, New Zealand

European Organisation for Nuclear Research (CERN), Geneva, Switzerland

J. A. Clark
MARS Bioimaging Limited, Christchurch, New Zealand

University of Otago Christchurch, Christchurch, New Zealand

Ara Institute of Canterbury, Christchurch, New Zealand

F. Colgan · J. S. Crighton · T. Dapamede · E. P. Walker
University of Otago Christchurch, Christchurch, New Zealand

J. Damet
University of Otago Christchurch, Christchurch, New Zealand

European Organisation for Nuclear Research (CERN), Geneva, Switzerland

Institute of Radiation Physics, Lausanne University Hospital, Lausanne, Switzerland

N. J. A. de Ruiter
MARS Bioimaging Limited, Christchurch, New Zealand

University of Canterbury, Christchurch, New Zealand

University of Otago Christchurch, Christchurch, New Zealand

Human Interface Technology Laboratory New Zealand, University of Canterbury, Christchurch, New Zealand

S. P. Gieseg
University of Canterbury, Christchurch, New Zealand

University of Otago Christchurch, Christchurch, New Zealand

European Organisation for Nuclear Research (CERN), Geneva, Switzerland

L. Holmes
Canterbury District Health Board, Christchurch, New Zealand

T. Kirkbride
Ara Institute of Canterbury, Christchurch, New Zealand

Some photon-counting chips, such as the Medipix3RX, have extra built-in circuitry to combat the effects of charge sharing by enabling charge summing mode (CSM) [3]. This charge summing circuitry consumes valuable space on the board and as such is only implemented in four out of the seven regular energy channels. The other three channels are currently unusable in reconstruction due to the artifacts they introduce. Finding an accurate correction method for these three channels would increase spectral resolution and the ability to differentiate between materials in material decomposition. Other approaches to correcting the charge sharing problem have also been suggested at the hardware level, such as the implementation of multi-energy inter-pixel coincidence counters [15], which are used to count the charge sharing occurrences and subsequently correct the scan afterwards. Some other work has developed analytical methods in correcting spectrum distortions in photon-counting detectors [5].

Machine learning is increasingly becoming an area of interest for tomographic image reconstruction in the medical imaging domain [17]. Some machine learning applications to challenging medical imaging problems include reducing motion artefacts in 4D cone-beam CT [19], increasing small feature preservation on CT reconstruction [16] and enabling operators to manage the trade-offs of denoising CT images with user driven denoisers [2].

Previous methods to correct charge sharing effects include neural networks trained on simulated data [11] and the use of U-Nets to correct spectral distortion issues encountered in single-pixel mode (SPM)when using energy integrated images as targets [9]. This chapter proposes what we believe to be a novel solution to the charge sharing problem which is based on the unique design of the Medipix3RX. Our solution is to use a U-Net based on the CycN-Net design [19], which is fed spatial-temporal correlated data. The data contains projection images with the three single-pixel mode (SPM) channels, the arbitration counter as a prior, and an N time series window to generate charge sharing corrected SPM channels. The corrected SPM channels are compared to target charge summing mode channels set to the same energy threshold. The aim is to correct the charge sharing issues while still maintaining the high level of spatial resolution made possible with PCD,

M. Paladugu · Y. Sayous
University of Canterbury, Christchurch, New Zealand

D. Palmer
Lincoln University, Lincoln, New Zealand

R. K. Panta
MARS Bioimaging Limited, Christchurch, New Zealand

European Organisation for Nuclear Research (CERN), Geneva, Switzerland

P. Renaud
University of Canterbury, Christchurch, New Zealand

University of Otago Christchurch, Christchurch, New Zealand

N. Schleich
University of Otago Wellington, Wellington, New Zealand

since image features are present in the time adjacent projections. This would allow the CSM channels to be set to different thresholds on a future scan with the net result being four CSM channels, three SPM corrected channels, and one arbitration counter channel. The rest of this chapter is organised as follows: In Sect. 2, we describe the charge sharing problem and the corrections performed at the hardware level by the Medipix3RX chip. Section 3 provides an outline of the proposed framework and implementation. Section 4 provides an outline of the experimental set-up. Section 5 presents results obtained and comparisons to current state-of-the-art methods when performed on a variety of datasets. Section 6 analyses current issues with our implementation and is followed by the conclusion of the chapter in Sect. 7.

2 Background

2.1 Charge Sharing

Charge sharing occurs in photon-counting detectors due to a combination of factors. First, the pixel size provides high spatial resolution and is small when compared to integrated energy CT detectors. Second, photon-counting detectors measure photons through an electron cloud produced when the photon strikes the surface of the detector. The electron cloud has a high probability of overlapping significant areas of multiple pixels. This may lead to multiple pixels registering a lower-energy photon from a single higher-energy incident photon. A counter will not register the photon event when the charge shared between pixels is less than the lowest energy threshold of the counter. The pixel thus misses a count at the correct counter and may have erroneous counts in lower threshold counters. When image reconstruction algorithms are performed later in the pipeline, multiple photon ray paths are computed that never existed, leading to increased noise and, possibly, an incorrect material decomposition. Equation 1 [5] describes the expected width at the anode σ_t of a charge cloud with a Gaussian charge density distribution from an initial electron cloud width σ_i (taken to be 5μm) [8]:

$$\sigma_t = \sqrt{2\frac{k_B T z d}{qU} + (\frac{z d N q}{10\pi\epsilon U})\frac{1}{\sqrt{5\sigma_i}} + \sigma_i^2}, \tag{1}$$

where d is the depth of the pixel's crystal layer, U is the bias voltage, and z is the electron cloud's height above the anode. The number of charges per photon is given as $N = E_e/\Delta_e$, where Δ_e is the energy per electron hole pair, which for a CdTe detector is 4.43 eV/ehp [12]. The dominant factor in variation of the charge cloud is dependent on the original photon collision point because the variables defined here should not change greatly between projections.

Charge sharing is not a uniform occurrence across all sensors. The charge clouds produced can vary due to manufacturing variation that affects the pixel electric fields and how the charge is dissipated for the electron cloud.

2.2 *Medipix3RX*

The Medipix3RX is a photon-counting detector which contains a 256 x 256 matrix of 55 μm x 55 μm pixels [3]. When spectroscopic mode is enabled, every group of 2 x 2 pixels forms a single larger pixel which reduces the matrix to 128 x 128 pixels with an area of 110 μm x 110 μm. The Medipix3RX detector has seven normal energy channels plus an arbitration counter. Four of these channels are capable of mitigating the charge sharing issue when set to charge summing mode, and the arbitration counter can facilitate this mode. This mode utilises summing circuits located in the corner boundaries between pixels in a 2 x 2 cluster to detect charge sharing events. A pixel records a photon count when it has the largest charge deposition with respect to its neighbours in the grid and the summing circuit recorded value exceeds a given threshold. Figure 1 provides a diagram of this charge summing circuit.

Charge summing allows the Medipix3RX chip to correct the spectrum distortion by collecting all the charge for a single photon. Due to hardware space limitations, only four out of seven channels have this feature. The remaining three channels are limited to single-pixel mode and cannot be directly used in reconstruction without inducing errors. This decreases the ability to distinguish between materials because the remaining channels must cover larger energy ranges.

Reference [7] provides a detailed analysis of the Medipix3's charge summing correction circuits for the later Medipix3RX version and shows that the algorithm is effective in suppressing charge sharing events.

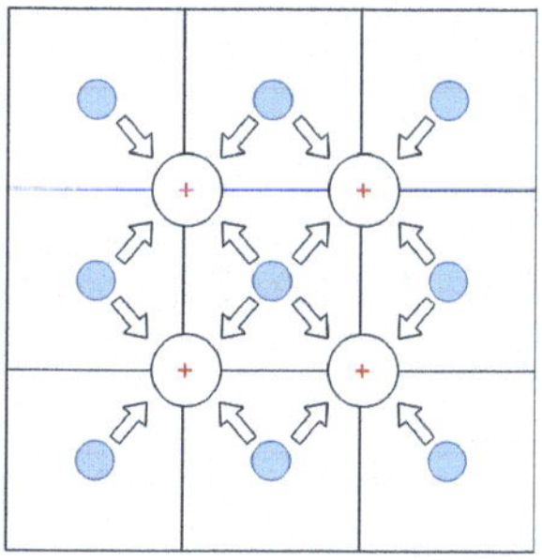

Fig. 1 Charge summing circuit: charge is deposited into the material which is connected to a comparator. This is used to determine the largest charge in each 2 x 2 grid so that the sum of charges can be assigned to a single pixel

3 Proposed Framework

3.1 Network

We propose a machine learning model to correct acquired single-pixel mode PCD data to have non-distorted, ideal, spectral, and spatial data. In Sect. 2.2, we determined that the Medipix3RX hardware charge summing algorithm is an approximate correction of the charge sharing distortions to the spectrum and thus is an approximate inverse of equation one: it assigns all energy of the electron cloud to the pixel which had the majority of the initial electron cloud σ_i in its region. As a result, we can use this data as the target for training the model. Our model should, therefore, be a function that maps our distorted input x to the target charge sharing corrected data y. To do this, the following minimisation loss function is applied to a neural network:

$$\min_{f} mean \frac{||f(x) - y||}{y + \epsilon} * 100, \tag{2}$$

where f is the network forward pass and ϵ is a small value to prevent zero division. This loss function defines a mean absolute percentage error (MAPE) and is used instead of mean squared error because higher-energy threshold channels will count the same or fewer photons than the lower-energy counterpart for the same exposure.

3.2 U-Net

A typical neural network approach to this type of correction problem is to implement a U-Net. However, a U-Net fails to fully utilise all the features in the dataset. Our arbitration counter data contains a total count of all photons recorded above the lowest energy threshold. This data is not an ideal input for a U-Net since we are not trying to correct it. It also follows its own distributions different from the SPM data. However, for comparison to the later network, we implement a standard U-Net outlined in Fig. 2.

3.3 Modified CycN-Net

Because of the issues outlined with the utilisation of the arbitration counter, we modify the standard U-Net. We add another compression pathway for the arbitration counter data and connect it to the up-sampling pathway. The arbitration counter in this way is treated as a prior. If we consider a stack of projections from frame

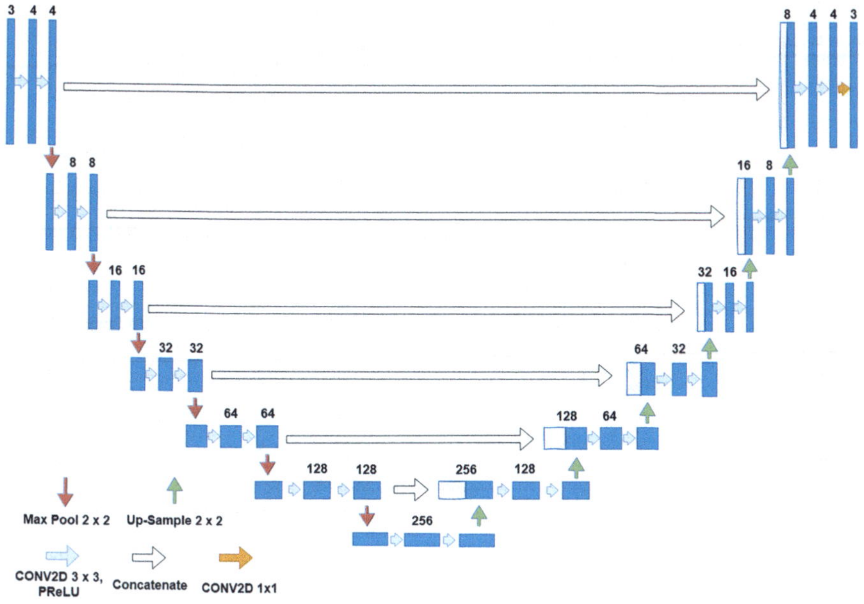

Fig. 2 Comparison U-Net architecture: input is the three single-pixel mode images and the arbitration counter. The network contains blocks of Conv2D operations with a 3x3 window per level followed by a 2x2 max pool for the compression process. Upsampling pathways utilise Conv2D with 3x3 windows. All Conv2D operations with 3x3 windows are followed by PReLu activation functions. Batch normalisation is used before each Conv2D operation. The beginning 2D convolution block for each upsampling layer is provided with the previous upsampled result concatenated with the adjacent down-sampling result. The final upsampling block contains an extra Conv2D operation with a 1x1 window

$n - i$ to $n + i$ distorted by a charge sharing distribution, we have a case where objects observed by the scanner appear shifted based on a function of n. However, since charge sharing distribution has no time-dependent variables between multiple frames, we can utilise a stack of these images to increase the total number of counts in order to measure from the distribution and stabilise the corrected projection. This is done by allowing the network to account for object movement and reduce the variance in the charge sharing distribution.

From this, we derive a network structure that can process the time series data based on CycN-Net [19] which was used for stabilising images distorted by respiratory cycles in 4D cone beam CT scans. We do make a key change to the authors' design in the up-sampling pathway. The CycN-Net design summed the current level of concatenated blocks with the past TConv output before the current TConv block. We removed this step and added an additional concatenation step to keep all the filters and found this increased the performance of the network on the datasets studied in this chapter. The network design is described in Fig. 3.

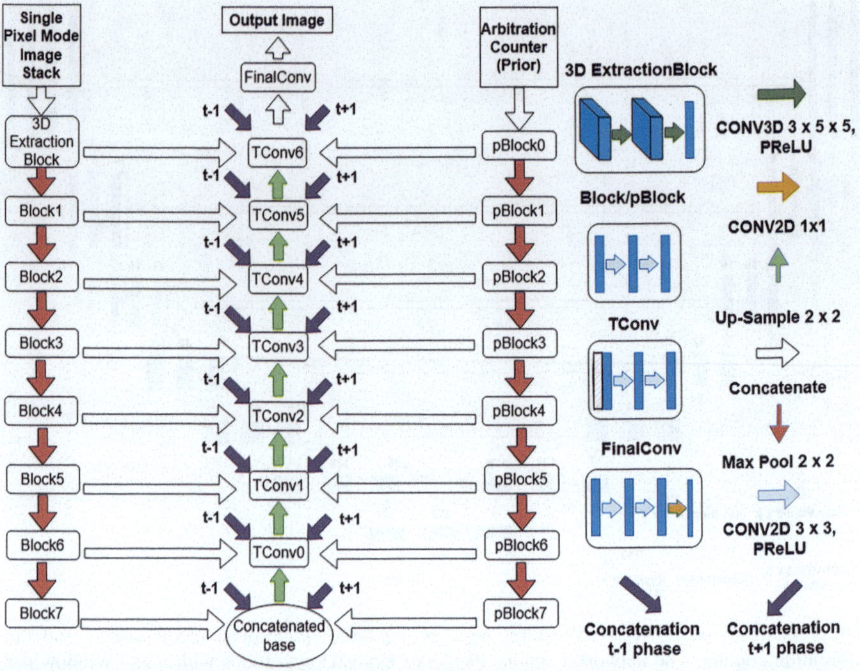

Fig. 3 Modified CycN-Net architecture: the network consists of three copies stacked together. Each copy receives a block of five projections, with the middle projection being the projection to be corrected. The extra copies of the network are used to provide additional time stabilisation for the target frame. The 3D extraction block utilises two Conv3D blocks each with a 3x5x5 window. Blocks and pBlocks contain the same operations, including batch normalisation and PReLu activation functions, described in the U-Net. The concatenated base consists of the concatenated results of Block7, pBlock7, and the results from the $t - 1$ and $t + 1$ versions of these blocks. TConv blocks are the same as the U-Net's, but with additional filters concatenated at the start of each block. These concatenated filters are the results from the previous TConv block joined with the result of the corresponding p/block and the $t - 1$ and $t + 1$ p/blocks. The FinalConv block is the same as the described U-Net

4 Experiment One Set-up

4.1 *Training and Testing Data*

The purpose of the following experiments is to evaluate the ability of the modified CycN-Net model to correct charge sharing distortions at set energy thresholds for the Medipix3RX with different crystal sensor layers. We compare the corrections to the original single-pixel mode projections, projections corrected through an analytical correction approach we define as scaled single-pixel mode (SSPM), and the U-Net we defined previously. For training and evaluation of the modified CycN-Net model and the comparison U-Net, we conducted two scans of different whole lamb legs with two different MARS V6 wrist scanners. Both scanners contained 12

Medipix3RX PCDs arranged in parallel, with spectroscopic mode enabled for a total detector size of 128 x 1536. The first scanner utilised Medipix3RX detectors with a 2mm thick CdZnTe crystal sensor layer. The second scanner utilised Medipix3RX detectors with a 0.75mm thick CdTe sensor layer. These sensor layers have different detection properties, including their absorption efficiency, with 0.75mm CdTe being transparent to some higher-energy photons which the 2mm CdZnTe absorbs. For both scanners, the energy bin thresholds were set to 40, 50, 60, and 70kev for the charge-summing-mode-enabled channels. The single-pixel mode channels were paired with the 40, 50, and 60 kev channels. The arbitration counter was set to a threshold of 7kev. The tube current was set to 30 μA, the tube voltage was set to 120kVp, and the exposure time was set to 160ms.

The lamb leg was used as a training phantom for a few key reasons. First, it comprises biological materials and contains a variety of materials typically found in human wrist CT scans, such as muscle, fat, and bone, in human-similar locations. Second, it is large and has a varied, asymmetrical shape. This means different materials will appear in multiple regions of the scan and should reduce over-fitting which was observed on more homogeneous samples. Image morphology methods fail on this dataset due to variances in charge sharing behaviour in different pixels.

The first scan produced a total of 16,832 projection images. The second scan produced 13,795 projection images. Both scans had groups of 100 removed at varying time intervals of the scan at approximately every 1000 images. This was to ensure the test dataset was an accurate subset of the scan and did not contain a small subset of repeating scan angles. Dataset one had 1692 test images and dataset two had 1410. In total, 14,918 images were used for training the network for dataset one, and dataset two had 12,199 images for training. The remaining projections were lost due to overlap with the testing and training set.

For both scans, an air scan was conducted after the scanning of the lamb legs. These air scans contain only air, the 1.6mm thick aluminium guard tube, which the beam passes through twice, and the 0.125mm brass filter contained in the scanner. From these air scans, a mask file was generated by masking any pixel that recorded photon counts that were more than two standard deviations from the mean count recorded on the air scan for the CdZnTe detectors and any more than three standard deviations for the CdTe detectors. This was done to remove inconsistent pixels from the training pool. The differences in the masking method were due to the apparent increased inconsistent response of the CdTe detector, with two standard deviation mask removing too great of a percentage of pixels for training. Finally, an extra mask was generated per projection to remove any pixels that failed to register a single count for each energy channel.

4.2 Network Training and Testing

For training both the U-Net and the modified CycN-Net, an RTX 3080 10Gb was used with TensorFlow version 2.7.0. The loss function used was the previously described MAPE and was minimised using the ADAM optimiser. The starting learning rate was set to 10^{-3} and decreased by half after four epochs and then decreased by half again after another four epochs. Both the U-Net and the modified CycN-Net were trained for 15 epochs on dataset one and 20 epochs on dataset two, with the epoch with the best score on the metrics being used for the results. The batch size for the U-Net training was set to 16. However, due to increased network and input data size, the batch size for CycN-Net was limited to only four.

5 Experiment Two Set-up

The purpose of this section is to explore the effectiveness of the outlined correction methods in a practical setting using a phantom with known material concentrations. We also analyse the impact of the design decisions made for the network correction approaches. The features tested include the use of scaled single-pixel mode (SSPM) images and the arbitration counter as priors to the network, temporal stabilisation, and the use of masking. Each of these aspects was theorised to bring a different value to the network.

5.1 Training and Testing Data

Each study was trained on the same dataset comprised of eight separate scans. Each scan contained a 40mm section of a material phantom filled with 13 material vials: a vial of water and four different materials where each material has three different concentrations. This resulted in approximately 2250 images per scan and around 18,000 images total. This was split into 900 subsections of 20 images; from each group, two images were randomly selected to test the model throughout the entire scan. However, this reduces the total number of projections trained on to approximately 11,500 due to the requirement that any projection within three time steps of a test is removed from training.

The scans were taken using keV energy thresholds of 65, 75, and 85 with the SPM paired to a CSM threshold, plus a base threshold of 7 keV for the arbitration counter. All scans used the same detector layout outlined in the previous study and CdZnTe as the detector crystal material with a thickness of 2mm. The same computer system and settings as outlined above was used to train the models.

5.2 Follow-up Study: Material Decomposition

To measure the effectiveness of the correction methods for material analysis, we used the trained CycN-Net and U-Net and the calibrated SSPM approach to correct a scan with the materials in Table 1.

Table 1 Materials used for comparison of correction methods

Material	Conc. (mg/ml) 1	Conc. 2 (mg/ml)	Conc. 3 (mg/ml)	K-edge (keV)
Au	2	4	8	80.7
I	4	8	16	33.2
Gd	2	4	8	50.2
Ca	35	70	140	4.03

The scans were taken using keV energy thresholds of 25, 35, 45, and 55 for the CSM counters and 65, 75, and 85 for the SPM counters plus a base threshold of 7 keV for the arbitration counter. After reconstruction, we calibrated a set of material basis vectors and covariance matrices from ten slices. We used a logistic regression classifier and an image domain material decomposition algorithm [4] to examine separability of material classes. Each material class included all three vial concentrations for that material. The Logistic regression classifier is useless in practice because it does not generalise to unknown material quantities as it tightly fits the training data, but it provides a reasonable assessment of separability information. We trained each classifier on ten slices, applied them to another ten slices. We performed statistical analysis of the calibration and comparison of the outputs with known ground truth regions of interest in each material vial. For comparison, we also include outputs from a base model, which uses the same dataset as SPM without the last three channels (it includes only CSM-derived channels).

5.3 *Ablation Study: Effect of Masking*

With any detector system, there exists pixels which act significantly different from the average pixel. We hypothesise that these pixels will negatively impact the training process of both neural network correction approaches. To test the impact of these pixels, we created a mask to remove these pixels from training. The mask creation uses an open beam scan to identify pixels with a large inter-channel variance, non-concordant with the average pixel variance. Pixels further than 2.5 standard deviations away from the mean were masked out. This mask includes pixels that are on the edge of the detectors, pixels that were obscured by the guard tube, and also faulty pixels (which were about 3%). This mask is then used for training both neural networks and withheld during training for both neural networks as a comparison. For analysis on the test set, the mask is used.

5.4 *Ablation study: Effect of Arbitration Counter as Prior*

The arbitration counter is set to a low energy threshold of 7keV. Because of this, a large amount of counts are recorded in this energy channel. This channel is distinct from both the CSM channels and the SPM channels in its distribution. As such, it may contain extra information that better allows the models to correct charge sharing errors. For this experiment, we removed the arbitration counter in both the U-Net and the CycN-Net by supplying an array of all zeroes instead of the arbitration counter. We compare the correction results on the test projections to the standard U-net and the CycN-Net.

5.5 Ablation Study: Effect of SSPM as Prior

The purpose of this ablation study was to analyse the impact of including analytic correction methods with the neural networks. Previous testing had shown that an analytic approach could outperform the neural networks at lower-energy thresholds. As such, the inclusion of the SSPM images should allow the model to be at least equivalent to the SSPM approach at the low energy thresholds. To investigate this, the model was re-trained with arrays initialised with zeroes in the place of the corrected images. Due to the poor performance of the SSPM correction approach in the material dataset caused by the use of high energy thresholds, the lamb leg dataset with the CdZnTe detectors is revisited.

5.6 Ablation Study: Effect of Time Offset Projections

To analyse the importance of time-offset projections, the CycN-net was retrained with the same projection stacked seven times where there would otherwise be time-offset projections. We hypothesise that the inclusion of the time offset projections is important, particularly in areas with lower counts which include high material density and high energy thresholds. However, due to small differences observed between the U-net and CycN-net in performance for correcting the material dataset, this study was also conducted on the lamb leg dataset with the CdZnTe detectors. For this study, we kept the same CycN-net structure as the no time series information network.

6 Results

6.1 Metrics and Evaluation

To evaluate our approach to correcting charge sharing distortions, we measure the distortion induced in the single-pixel mode channels by calculating mean absolute percentage error (MAPE) against the corresponding charge summed channels. We also utilise the peak signal-to-noise ratio (PSNR) as a metric for performance as defined in Eq. 3:

$$PSNR = 10\log_{10}(\frac{Maxv^2}{MSE}), \tag{3}$$

where $Maxv$ is the maximum pixel value for the projection in the charge summed image and MSE is the mean squared error. Mean squared error is defined in Eq. 4:

$$MSE = \frac{1}{N}\sum_{i=1}^{N}(x_i - y_i)^2, \tag{4}$$

where N is the number of pixels in our projection image, x_i is the corrected photon count for pixel index i, and y is the original photon counts for pixel index i.

We also chose to use structural similarity [18] (SSIM) as an additional metric for analysis as defined in Eq. 5:

$$SSIM(x, y) = \frac{(2\mu_x\mu_y + C_1)(2\sigma_{xy} + C_2)}{(\mu_x^2 + \mu_y^2 + C_1)(\sigma_x^2 + \sigma_y^2 + C_2)}, \tag{5}$$

where x is defined as our corrected projection; y is our ground truth charge summed mode channel; μ_x and μ_y are the window means; σ_x and σ_y are the variances of the x and y sample, and σ_{xy} is their covariance; and C_1 and C_2 are small values to stabilise division. We calculate the SSIM in a two-dimensional window and calculate the SSIM for each energy channel independently.

We generate comparison results for three approaches to correct charge sharing distortions. To provide a baseline analytical approach for scaled single-pixel mode (SSPM), we calculate a per pixel mean scaling transformation matrix for the original single-pixel mode channels to match the target mean value. This is described in Eq. 6:

$$z_{jc} = \frac{1}{M}\sum_{i=1}^{M}\frac{(x_{jc} + \epsilon)}{(y_{jc} + \epsilon)}, \tag{6}$$

where z is the scalar matrix, M is the size of air scan dataset, x is the original single-pixel mode projection, and y is the target charge summed mode projection. j is the pixel index, c is the energy channel, and ϵ is a small value to prevent zero division.

Following this, the matrix is then used to scale the single-pixel mode counts to be the approximate correct value in the following equation:

$$sspm_{jc} = \frac{x_{jc}}{z_{jc}}, \tag{7}$$

where $sspm$ is our scaled single-pixel mode projection. This is to account for the consistent reduced counts detected by the single-pixel mode channels by artificially increasing the counts to be closer to the corresponding charge summed channels for each projection. This takes into account different per pixel charge sharing distributions and allows a direct comparison between methods. The SSPM method will still contain errors due to charge sharing distortion. We calculate metrics for projections corrected by the SSPM method, the U-Net described in Sect. 3.2, and the modified CycN-Net described in Sect. 3.3.

In addition, for material analysis, we adopt standard classification and regression metrics for CT imaging: root mean squared error (RMSE), RMSE as a percentage of the pseudo Houndsfield unit (RMSE%), sensitivity, specificity, and area under the receiver operating curve (AUROC). We also report the mean cosine angular distance in degrees (μ_θ) from the calibrated basis vector to each voxel in the calibration material to measure spectral variance.

6.2 Results for Experiment One

Figure 4 displays our results for experiment one. We display a heat map of the MAPE for the 60 keV energy channel thresholded to 20% to better show the differences between the output and desired channel.

We can see that both the SSPM and U-Net have high errors within the denser materials around the bone and tissue. The U-Net also has high error at material edges. However, the CycN-Net has consistently lower error in this region.

Table 2 outlines our results for the MAPE for experiment one. We see that the U-Net offers some improvement over the SSPM approach, but only in the highest energy channel. CycN-Net results in some improvement at the 50keV threshold, and it significantly reduces MAPE in the highest energy channels. However, the SSPM approach is more consistent in the lowest energy channel.

Table 3 displays our results for the computed PSNR on the test dataset. Again, we see the same trend: the scaled approach works best in the lowest energy bins, and the CycN-Net performs best in the remaining two higher-energy bins and comparable performance at 50keV. The U-Net still performs better than the SSPM approach at 60kev.

Table 4 shows our computed SSIM results. Here, the CycN-Net performs equivalently to the SSPM approach in the lowest energy channel and significantly better in the remaining channels. However, the U-Net is worse than the SSPM approach in all channels.

6.3 Results for Experiment Two

Here, the results are presented for experiment two, with the MAPE in Table 5.

Similar to experiment one, we again see the SSPM approach performing best in the lower-energy channels and the U-Net and CycN-Net better in the higher-energy

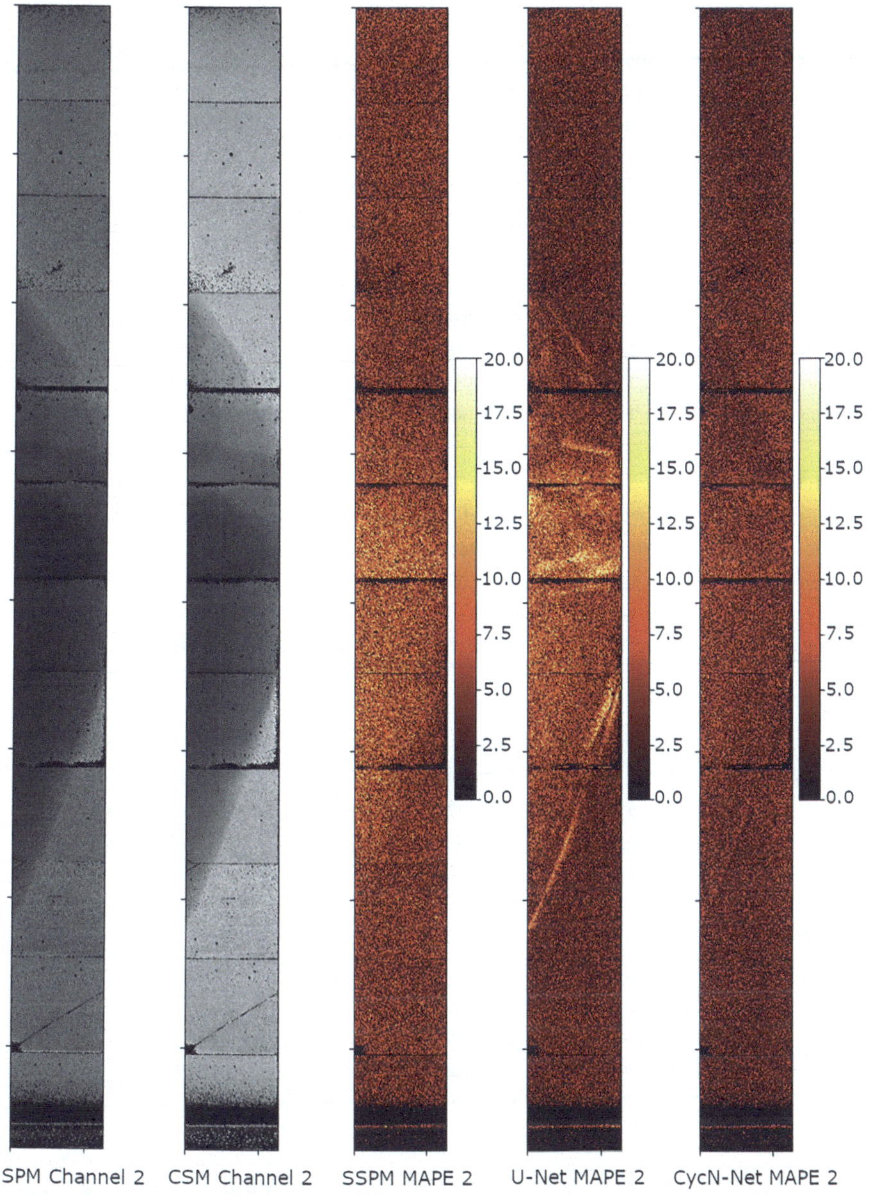

Fig. 4 MAPE comparison of the three correction methods at 60keV threshold. From left: original SPM channel, corresponding ground truth CSM channel, MAPE heat map for the SSPM approach, U-Net MAPE heat map, and CycN-Net MAPE heat map. Here we see that all methods do a reasonable job of predicting the air, but both the SSPM and the U-Net struggle with the darker regions

Table 2 Mean absolute percentage error

Energy channel	SPM	SSPM	U-Net	Modified CycN-Net
40kev	26.78272	**3.34848**	4.99544	3.61354
50kev	41.49372	4.83571	5.38180	**3.95082**
60kev	50.80966	7.45931	6.47575	**5.13708**
Average	39.69537	5.21450	5.06868	**4.23381**

Table 3 PSNR

Energy channel	SPM	SSPM	U-Net	Modified CycN-Net
40kev	22.6004	**41.38833**	36.86978	38.73384
50kev	17.76955	37.07403	35.84707	**37.64204**
60kev	15.77009	33.19969	34.43373	**35.95446**
Average	18.71335	37.22068	35.71686	**37.44345**

Table 4 SSIM

Energy channel	SPM	SSPM	U-Net	Modified CycN-Net
40kev	0.70738	**0.93941**	0.81533	0.93825
50kev	0.5144	0.88105	0.79107	**0.92092**
60kev	0.38553	0.79035	0.75456	**0.86632**
Average	0.53577	0.87027	0.78699	**0.90850**

Table 5 MAPE

Energy channel	SPM	SSPM	U-Net	Modified CycN-Net
40keV	31.21718	**3.70097**	4.95682	5.02096
50keV	33.47278	**5.4297**	5.83362	5.803
60keV	21.05811	10.01085	8.6161	**8.43344**
Average	28.58269	**6.38050**	6.46885	6.41913

Table 6 PSNR

Energy channel	SPM	SSPM	U-Net	Modified CycN-Net
40kev	18.18138	**37.85414**	34.52062	34.57402
50kev	17.30738	**34.08984**	33.38491	33.33244
60kev	20.49841	29.34949	30.55597	**30.61967**
Average	18.66239	**33.76449**	32.82050	32.84204

channels. However, we do not see the MAPE overtake the SSPM until 60keV for the networks. We also see an increase in MAPE for all channels for the SSPM approach when compared to experiment one with the largest increase in the higher-energy channels.

Table 6 shows the computed PSNR values. The best PSNR values are seen in the SSPM channel at 40keV and 50keV with the CycN-Net performing better at 60keV. The overall computed PSNR values are lower than the results obtained from experiment one.

Table 7 SSIM

Energy channel	SPM	SSPM	U-Net	Modified CycN-Net
40keV	0.49543	**0.87897**	0.6717	0.66845
50keV	0.44756	**0.81241**	0.70577	0.70642
60keV	0.37888	0.70297	0.71571	**0.71833**
Average	0.44062	**0.79812**	0.69773	0.69773

Table 8 Material calibration

Metric	Material	Base	SPM	SSPM	U-Net	Modified CycN-Net
RMSE%	Water	11.858	12.475	**11.426**	11.911	11.947
	Au 8	10.467	10.524	10.154	**10.129**	10.145
	Ca 140	9.815	10.369	**9.726**	10.012	10.001
	Gd 8	11.149	10.239	10.723	**9.803**	9.838
	I 16	9.883	9.394	9.781	9.102	**9.027**
μ_θ	Water	**1.071**	1.754	1.615	1.494	1.528
	Au 8	**0.958**	1.469	1.459	1.233	1.253
	Ca 140	**0.912**	1.446	1.375	1.167	1.338
	Gd 8	**1.012**	1.467	1.554	1.231	1.480
	I 16	**0.901**	1.232	1.294	1.028	1.124
Accuracy	All	0.837	0.788	0.817	0.812	**0.842**
AUROC	All	**0.971**	0.956	0.964	0.961	0.969

Table 7 displays our calculated SSIM. For the SSIM values, the performance trend repeats. However, there is a notable exception: for all past metrics, the CycN Net has performed better or on par with the U-Net, but for the 40keV energy threshold, in this case, the U-Net performs better.

6.4 *Results for Material Analysis*

Table 8 shows analysis of the calibration. This gives a good indication of how well the material voxel values form separable clusters in high-dimensional spectral space.

For the calibration, we can clearly see that using any correction method with the additional 3 SPM channels lowers the error associated with overall magnitude of each voxel (RMSE%) compared to the baseline. However, due to going from five to eight dimensions, the angular variance increases. The angular variance is also clearly improved with the corrections, and we see U-Net and CycN-Net outperforming the SSPM approach. AUROC is similar between all approaches with the uncorrected SPM performing worse because of the extra noise.

In decomposition analysis (Table 9), there are a few differences of note. First, AUROC is the preferred metric because it is not misled by small changes to basis vectors which cause swings in the other classification metrics (sensitivity and

Table 9 Material decomposition

Metric	Material	Base	SPM	SSPM	U-Net	Modified CycN-Net
RMSE	Au	2.696	2.922	3.186	2.667	**2.597**
	Ca	**20.698**	28.632	29.599	28.582	28.364
	Gd	1.975	1.899	1.942	**1.714**	1.975
	I	**1.541**	2.181	2.341	2.290	2.361
Sensitivity	Au 2	0.346	0.339	0.299	**0.377**	0.299
	Au 8	0.603	0.740	0.777	**0.832**	0.807
	Ca 45	0.394	0.481	0.484	0.498	**0.669**
	Ca 140	0.691	0.916	**0.961**	0.933	0.545
	Gd 2	0.709	0.617	0.687	0.845	**0.988**
	Gd 8	0.997	0.986	0.980	0.999	**1.000**
	I 4	**0.894**	0.844	0.860	0.826	0.808
	I 16	**1.000**	**1.000**	0.999	**1.000**	**1.000**
Specificity	Au	0.840	0.934	**0.935**	0.929	0.922
	Ca	**0.918**	0.884	0.879	0.900	0.893
	Gd	0.961	**0.981**	0.977	0.973	0.954
	I	0.925	0.951	**0.952**	0.943	0.946
AUROC	Au 2	0.578	0.631	0.613	**0.647**	0.601
	Au 8	0.731	0.844	0.866	**0.893**	0.852
	Ca 45	0.653	0.680	0.682	0.696	**0.781**
	Ca 140	0.832	0.951	**0.975**	0.961	0.747
	Gd 2	0.843	0.802	0.836	0.915	**0.991**
	Gd 8	0.998	0.993	0.990	0.999	**1.000**
	I 4	**0.942**	0.914	0.922	0.904	0.893
	I 16	**1.000**	**1.000**	**1.000**	**1.000**	**1.000**

specificity). We see that overall the corrected approaches achieve higher AUROC. They perform significantly better for the two materials with high k-edges (Au and Gd) with the U-Net being the best with Au and CycN-Net the best with Gd. This makes a lot of sense given that the differentiation between Au and Gd would be difficult for the base five-channel model as both k-edges fall within the same channel. Using additional corrected energy information seems to be a promising approach to differentiate high k-edge materials.

6.5 Results from Ablation Studies

Results from ablation studies can be seen in the appendix. We see that all correction approaches significantly improve the results of metrics on the material phantom dataset. We also observe what we expected with the SSPM method significantly degrading at higher-energy levels due to the sparsity of photon counts at higher-energy thresholds. However, the U-Net and the CycN-Net have equivalent performance, and the U-Net appears to converge faster than the CycN-Net to its optimal performance. For this reason, the CycN-Net was trained for an additional eight epochs. This closely matched performance is of note as it suggests that time series information is not beneficial in the prediction of charge sharing distortions for the material phantom dataset.

For the mask ablation study, we observed what was expected where training with no mask decreased the performance of both neural networks. This effect is more clearly seen in the CycN-Net. This confirms our hypothesis that the pixels that significantly fall outside the typical response of the detector negatively impact the training process of the neural networks and should be masked during training.

Our hypothesis that the arbitration counter would provide benefit to the correction process appears to be inconsistent on the material phantom dataset. It appears to provide value to the U-net with all metrics observing slight improvements with the arbitration counter included in the training data. However, the CycN-Net instead was detrimentally affected by the use of the arbitration counter with all metrics significantly decreased.

With our time series ablation study, we can clearly see the benefit the inclusion of time adjacent projections provides with significant improvements in both the MAPE and the PSNR across all energy channels. It is interesting to see that the SSIM does not greatly vary between the two correction methods.

The inclusion of the SSPM as prior information is detrimental to all of the used metrics, across all energy bins. This indicates that the method used to include the SSPM energy bins may not be the best method available.

7 Discussion

7.1 Experiments One and Two

As shown by our results, the implementation of a U-Net reduces the impact of charge sharing in single-pixel mode only energy channels for energy bins at the 60keV threshold. We can further improve the results by reducing the charge sharing distortions through the use of the CycN-Net. This allows the CycN-Net to outperform the SSPM approach consistently at the 50kev and above energy range on the 2mm CdZnTe detector dataset. These performance gains are made possible through a network structure allowing for the utilisation of spatial-temporal features present in projection images, which stabilises the charge sharing corrections. This approach tends to provide the most improvement over the single-pixel mode images at the higher-energy ranges and in regions with reduced photon counts. There is increased variance in charge sharing distribution per projection, so a single scalar matrix is only able to reduce charge sharing distortions in a limited capacity. Figure 4 supports this hypothesis, showing that most of the MAPE is concentrated in

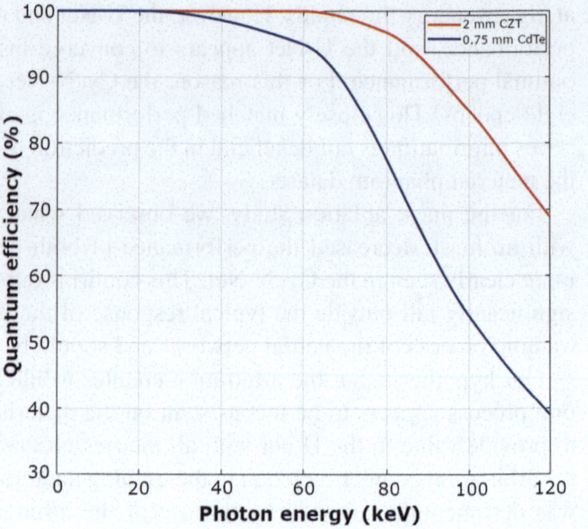

Fig. 5 Quantum efficiency for sensor layers over increasing photon energy: red shows 2mm CdZnTe and blue shows 0.75mm CdTe

the darker regions of the projection. It also shows that CycN-Net is able to produce a result with reduced error in this region.

An area of interest is the performance differences between the two different datasets seen by all explored correction methods. Although the CycN-Net improves the image quality by all used metrics when compared to the SSPM images at the 60kev range, there is a large difference in the average metrics for all correction methods. Material phantom performance generally is improved in both U-Net and CycN-Net approaches for some materials with some better in the baseline five-channel model. Further testing will need to be done to determine the cause of these performance differences. One such potential cause of this is the difference in detector response curve observed between the 2mm CdZnTe and the 0.75mm CdTe sensor displayed in Fig. 5 [1].

We see that the CdTe sensor has a response curve that begins to decrease in the 30 keV range, compared to the 2mm CdZnTe which has 100 percent quantum efficiency until a photon energy of approximately 65keV. This reduced counting of higher-energy photons could be distorting the charge sharing distribution, making it more difficult to predict.

7.2 Ablation Studies

Our ablation studies support the hypothesis that neural network correction approaches can decrease the error induced by uncorrected charge sharing in single-pixel mode detectors. We demonstrate through a material analysis study the potential application of using these corrected energy channels to accurately

Fig. 6 MAPE comparison of the three correction methods at 85kev threshold. From left: original SPM channel, ground truth CSM channel, MAPE heat map for the SSPM approach, U-Net MAPE heat map, CycN-Net MAPE heat map. The U-Net and CycN-Net have similar heat maps indicating they learned similar features of the dataset. The SSPM method performs poorly at this high keV range

identify and quantify materials present. We demonstrate the importance of a mask to reduce the impact of pixels that have a significantly different response function for photon-counting during the training of the neural network correction methods. We also show some of the value that the arbitration counter can provide, but suggest how it decreases the rate of convergence for the neural networks. We also show how the inclusion of the SSPM corrected energy bins as a prior to the CycN-Net is detrimental to performance and requires further study into determining an effective method of utilising this information. It is of note that we observed very little difference in performance between the U-Net and the CycN-Net in the correction of the material phantom dataset. We suspect that this is due to the relative simplicity of the dataset compared to the lamb leg as the time-series information may provide less value. This is supported by Fig. 6 that shows both the U-Net and the CycN-Net appear to have learned similar features of the dataset. We also see why the SSPM performs so poorly on this dataset: the SPM projection contains a large variation in counts compared to the CSM which produces a noisy image.

Because of this, we returned to the lamb dataset for the time series ablation study and SSPM as an added prior. The time series study saw an improvement of MAPE and PSNR in all energy channels when using the time series projections. This shows that the value comes from the inclusion of time series data, not the additional parameters, because we kept all weights active in the non-time series test. However, there was a decrease in all metrics for the CycN-Net compared to the first experiment. We believe this is due to the need to reduce training data from approximately 15,000 to 11,500 because of the more frequent test projection sampling outlined in Sect. 5.1.

8 Conclusion

In conclusion, we introduce a novel approach to correct charge sharing distortions by customising a CycN-Net architecture and training it to correct charge sharing distortions on the Medipix3RX photon-counting detector single-pixel mode channels in post processing. Our study suggests the viability of correcting charge sharing distortions in this manner. Our network design reduces the mean absolute percentage error, increases the peak signal-to-noise ratio, and increases the structural similarity index measurement on the 2mm CdZnTe dataset at the 50 and 60keV thresholds. The 0.75mm CdTe dataset showed the network was capable of improving these metrics at the 60keV range. We also show through multiple ablation studies the value and risks of including time series data, arbitration counter data, and an effective pixel mask. Our approach allows the single-pixel mode energy channels to be utilised in tomographic reconstruction and the material decomposition processes. This is expected to increase the accuracy of material identification and quantification for larger groups of materials which opens up the possibility of new applications in medical imaging.

Acknowledgments This project was funded by the Ministry of Business, Innovation and Employment (MBIE), New Zealand, under contract number UOCX1404, by MARS Bioimaging Ltd and the Ministry of Education through the MedTech CoRE. The authors would like to acknowledge the Medipix2, Medipix3, and Medipix4 collaborations. Also, we would like to take this opportunity to acknowledge the generous support of the MARS Collaboration.

Appendix

Ablation Studies

See Tables 10, 11, 12, 13, 14, 15, 16, 17, 18, 19, 20, 21, 22, 23, 24.

Table 10 Adjustment methods MAPE

Energy channel	SPM	SSPM	U-Net (13 epochs)	CycN-Net (13 epochs)	CycN-Net (21 epochs)
65kev	50.36847	10.37076	**4.13246**	5.06688	4.67292
75kev	54.28265	19.12228	**5.45602**	6.34633	6.00479
85kev	62.43968	37.19895	**7.84918**	8.67104	8.38968
Average	55.69693	22.23066	**5.81255**	6.69475	6.3558

Table 11 Adjustment methods PSNR

Energy channel	SPM	SSPM	U-Net (13 epochs)	CycN-Net (13 epochs)	CycN-Net (21 epochs)
65kev	15.32299	34.14054	**35.84165**	33.9876	34.62289
75kev	14.68201	27.53387	**33.20046**	31.94956	32.30521
85kev	13.85647	21.05622	**29.69878**	28.92591	29.05058
Average	14.62049	27.57688	**32.91363**	31.62102	31.99289

Table 12 Adjustment methods SSIM

Energy channel	SPM	SSPM	U-Net (13 epochs)	CycN-Net (13 epochs)	CycN-Net (21 epochs)
65kev	0.38915	0.85584	**0.88543**	0.87605	0.8789
75kev	0.29329	0.6494	**0.82359**	0.81602	0.81786
85kev	0.23816	0.47167	**0.7831**	0.7769	0.77767
Average	0.30687	0.65897	**0.83071**	0.82299	0.82481

Table 13 MAPE with Mask Removed

Energy channel	U-Net	CycN-Net
65kev	5.22506	10.94909
75kev	6.53711	11.82251
85kev	8.94361	13.5757
Average	6.90193	12.11577

Table 14 PSNR with Mask Removed

Energy channel	U-Net	CycN-Net
65kev	30.81998	25.80608
75kev	29.74253	25.48542
85kev	27.47223	24.48854
Average	29.34491	25.26001

Table 15 SSIM with Mask Removed

Energy channel	U-Net	CycN-Net
65kev	0.8687	0.81482
75kev	0.80448	0.76793
85kev	0.76667	0.73767
Average	0.81328	0.77347

Table 16 MAPE without arbitration counter

Energy channel	U-Net	CycN-Net
65kev	4.63069	4.36298
75kev	5.86815	5.7021
85kev	8.0926	8.15077
Average	6.19715	6.07195

Table 17 PSNR without arbitration counter

Energy channel	U-Net	CycN-Net
65kev	35.46065	35.00454
75kev	32.84665	32.51249
85kev	29.37287	29.09916
Average	32.56006	32.2054

Table 18 SSIM without arbitration counter

Energy channel	U-Net	CycN-Net
65kev	0.88283	0.88338
75kev	0.82051	0.82074
85kev	0.77948	0.77942
Average	0.82761	0.82785

Table 19 Temporal feature inclusion (MAPE)

Energy channel	CycN-Net with	CycN-Net without
65kev	5.35668	7.91813
75kev	5.80704	8.06101
85kev	6.90012	8.58811
Average	6.02128	8.18908

Table 20 Temporal feature inclusion (PSNR)

Energy channel	CycN-Net with	CycN-Net without
65kev	40.78072	37.79852
75kev	40.09325	37.22485
85kev	38.35594	36.15272
Average	39.7433	37.0587

Table 21 Temporal feature inclusion (SSIM)

Energy channel	CycN-Net with	CycN-Net without
65kev	0.77654	0.7817
75kev	0.74625	0.7551
85kev	0.70754	0.71454
Average	0.74344	0.75045

Table 22 SSPM as prior (MAPE)

Energy channel	CycN-Net with	CycN-Net without
65kev	7.5306	5.35668
75kev	7.40866	5.80704
85kev	8.13617	6.90012
Average	7.69181	6.02128

Table 23 SSPM as prior (PSNR)

Energy channel	CycN-Net with	CycN-Net without
65kev	37.25876	40.78072
75kev	37.10649	40.09325
85kev	36.15838	38.35594
Average	36.84121	39.7433

Table 24 SSPM as prior (SSIM)

Energy channel	CycN-Net with	CycN-Net without
65kev	0.7610	0.77654
75kev	0.73394	0.74625
85kev	0.69316	0.70754
Average	0.72937	0.74344

References

1. A. Atharifard, J. Healy, B. Goulter, M. Ramyar, L.V. Broeke, M. Walsh, C. Onyema, R. Panta, R. Aamir, D. Smithies, et al., Per-pixel energy calibration of photon counting detectors. J. Instrument. **12**(03), C03085 (2017)
2. T. Bai, B. Wang, D. Nguyen, B. Wang, B. Dong, W. Cong, M.K. Kalra, S. Jiang, Deep interactive denoiser (did) for x-ray computed tomography. IEEE Trans. Med. Imag. **40**(11), 2965–2975 (2021)
3. R. Ballabriga, J. Alozy, G. Blaj, M. Campbell, M. Fiederle, E. Frojdh, E. Heijne, X. Llopart, M. Pichotka, S. Procz, et al., The medipix3rx: a high resolution, zero dead-time pixel detector readout chip allowing spectroscopic imaging. J. Instrument. **8**(02), C02016 (2013)
4. C. Bateman, D. Knight, B. Brandwacht, J.M. Mahon, J. Healy, R. Panta, R. Aamir, K. Rajendran, M. Moghiseh, M. Ramyar, D. Rundle, J. Bennett, N. de Ruiter, D. Smithies, S. Bell, R. Doesburg, A. Chernoglazov, V. Mandalika, M. Walsh, M. Shamshad, M. Anjomrouz, A. Atharifard, L.V. Broeke, S. Bheesette, T. Kirkbride, N. Anderson, S. Gieseg, T. Woodfield, P. Renaud, A. Butler, P. Butler, Mars-md: rejection based image domain material decomposition. J. Instrument. **13**(05), P05020 (may 2018). https://doi.org/10.1088/1748-0221/13/05/P05020
5. E.S. Dreier, J. Kehres, M. Khalil, M. Busi, Y. Gu, R. Feidenhans, U.L. Olsen, Spectral correction algorithm for multispectral cdte x-ray detectors. Opt. Eng. **57**(5), 054117 (2018)
6. T. Flohr, M. Petersilka, A. Henning, S. Ulzheimer, J. Ferda, B. Schmidt, Photon-counting CT review. Physica Medica **79**, 126–136 (2020)
7. E. Gimenez, R. Ballabriga, M. Campbell, I. Horswell, X. Llopart, J. Marchal, K. Sawhney, N. Tartoni, D. Turecek, Study of charge-sharing in medipix3 using a micro-focused synchrotron beam. J. Instrument. **6**(01), C01031 (2011)
8. L. Hamel, M. Benoit, B. Donmez, J.R. Macri, M.L. McConnell, T. Narita, J.M. Ryan, Optimization of single-sided charge-sharing strip detectors, in *2006 IEEE Nuclear Science Symposium Conference Record*, vol. 6, pp. 3759–3761 (2006). https://doi.org/10.1109/NSSMIC.2006.353811

9. M. Holbrook, D. Clark, C. Badea, Deep learning based spectral distortion correction and decomposition for photon counting CT using calibration provided by an energy integrated detector, in *Medical Imaging 2021: Physics of Medical Imaging*, vol. 11595, p. 1159520 (International Society for Optics and Photonics, 2021)
10. S. Leng, Z. Yu, A. Halaweish, S. Kappler, K. Hahn, A. Henning, Z. Li, J. Lane, D.L. Levin, S. Jorgensen, et al., Dose-efficient ultrahigh-resolution scan mode using a photon counting detector computed tomography system. J. Med. Imag. **3**(4), 043504 (2016)
11. M. Li, D.S. Rundle, G. Wang, X-ray photon-counting data correction through deep learning. arXiv preprint arXiv:2007.03119 (2020)
12. D. McGregor, H. Hermon, Room-temperature compound semiconductor radiation detectors. Nucl. Instrum. Methods Phys. Res. Sect. A Accelerators Spectrometers Detectors Associated Equip. **395**(1), 101–124 (1997). https://doi.org/10.1016/S0168-9002(97)00620-7. https://www.sciencedirect.com/science/article/pii/S0168900297006207, proceedings of the Fourth International Workshop on GaAs Detectors and Related Compounds
13. G. Pellegrini, M. Maiorino, G. Blanchot, M. Chmeissani, J. Garcia, M. Lozano, R. Martinez, C. Puigdengoles, M. Ullan, Direct charge sharing observation in single-photon counting pixel detector. Nucl. Instrum. Methods Phys. Res. Sect. A Accelerators Spectrometers Detectors Associated Equip. **573**(1-2), 137–140 (2007)
14. R. Symons, B. Krauss, P. Sahbaee, T.E. Cork, M.N. Lakshmanan, D.A. Bluemke, A. Pourmorteza, Photon-counting CT for simultaneous imaging of multiple contrast agents in the abdomen: an in vivo study. Med. Phys. **44**(10), 5120–5127 (2017)
15. K. Taguchi, Multi-energy inter-pixel coincidence counters for charge sharing correction and compensation in photon counting detectors. Med. Phys. **47**(5), 2085–2098 (2020)
16. X. Tao, Y. Wang, L. Lin, Z. Hong, J. Ma, Learning to reconstruct CT images from the vvbp-tensor. IEEE Trans. Med. Imag. **40**(1), 3030–3041 (2021)
17. G. Wang, M. Jacob, X. Mou, Y. Shi, Y.C. Eldar, Deep tomographic image reconstruction: Yesterday, today, and tomorrow–editorial for the 2nd special issue "machine learning for image reconstruction". IEEE Trans. Med. Imag. **40**(11), 2956–2964 (2021)
18. Z. Wang, A. Bovik, H. Sheikh, E. Simoncelli, Image quality assessment: from error visibility to structural similarity. IEEE Trans. Image Process. **13**(4), 600–612 (2004). https://doi.org/10.1109/TIP.2003.819861
19. S. Zhi, M. Kachelrieß, F. Pan, X. Mou, Cycn-net: A convolutional neural network specialized for 4d cbct images refinement. IEEE Trans. Med. Imag. **40**(11), 3054–3064 (2021)

Machine Learning Approaches for CdZnTe/CdTe Radiation Detectors

Srutarshi Banerjee, Miesher Rodrigues, Manuel Ballester, Alexander Hans Vija, and Aggelos K. Katsaggelos

1 Introduction

RTSDs such as CdZnTe/CdTe are widely used for homeland security, medical imaging, astronomy, high-energy physics, and other applications [1–5], which requires high-quality crystals at reasonable cost, with uniform and optimized charge transport properties. They are often used as compact radiation detection units with highly segmented anode and cathode patterns.

These detectors have high-energy resolution spectral responses without cooling requirements. However, the performance and yield of such high-quality detector-grade materials for real-world applications are limited by the presence of high concentrations of defects which are randomly distributed. The manufacturing process of these crystals is often nonuniform, and this leads to random distribution of defects in the material. These RTSDs suffer from major detrimental defects [6] such as compositional inhomogeneity due to non-unity segregation coefficient of

S. Banerjee (✉)
Electrical and Computer Engineering, Northwestern University, Evanston, IL, USA

Data Science and Learning Division, Argonne National Laboratory, Lemont, IL, USA
e-mail: srutarshibanerjee2022@u.northwestern.edu

M. Rodrigues
Vital Materials, Buffalo Grove, IL, USA
e-mail: miesher@umich.edu

M. Ballester · A. K. Katsaggelos
Electrical and Computer Engineering, Northwestern University, Evanston, IL, USA
e-mail: manuelballestermatito2021@u.northwestern.edu; a-katsaggelos@northwestern.edu

A. H. Vija
Molecular Imaging, Hoffman Estates, IL, USA
e-mail: hans.vija@siemens-healthineers.com

K. (K.) Iniewski, L. (K.) Cai (eds.), *Deep Learning for Advanced X-ray Detection and Imaging Applications*, https://doi.org/10.1007/978-3-031-75653-5_8

Zn [7], the presence of high concentration of secondary phases, Te inclusions, and sub-grain boundaries/dislocation walls in high concentration in CZT array. These defects act as trapping centers, hindering localized charge transport and impose spatial nonuniformity in charge transport properties, thereby adversely affecting the detector performance [8–13]. These trapping centers can be nonuniform over the volume of the material and typically differs for electrons and holes.

The efforts to characterize these detectors have been done over the last several years. For instance, using thermoelectric emission spectroscopy (TES) and thermally stimulated conductivity (TSC) measurements, the thermal ionization energies of the electron and hole traps were measured [14]. Determination of the trap lifetime was done using a microwave cavity perturbation method in CdZnTe and HgI_2 [15]. In [16], the electron and hole traps of CdZnTe samples were radially irradiated with five MeV-focused proton beams to generate electron-hole pairs and fill traps, which were released later by thermal reemission. Electron and hole traps were distinguished by excitation near the vicinity of the appropriate electrodes. Analyzing simultaneous multiple peaks on TSC measurements allowed characterization of deep trap levels in CdZnTe [17]. In [18], the average hole trapping time τ_h was derived using statistical model of charge collection efficiency based on known electron average trapping time. In [19–21], average hole detrapping time τ_{Dh} was extracted by comparing the measured and simulated signals for holes as measured in the cathode. Imperfections due to mechanical damage or adsorbed chemical species trap charges or increase leakage current. Using pulsed laser microwave cavity perturbation method selectively at the surface and in the bulk region of CZT, these defects have been characterized [22]. The influence of type of metal contacts and deposition techniques on the recombination and trapping defects at the metal-semiconductor interface has also been studied [23]. The uniformity of high flux CdZnTe has also been characterized [13]. In literature, the defects and charge transport properties of electrons and holes are measured using classical approaches considering homogeneous behavior over the detector. This requires cumbersome multiple experiments and technical know-how. However, the homogeneity and repeatability of the defects and charge transport properties within a detector and across multiple detectors spatially and temporally are unknown. Moreover, achieving high-energy resolution below 1.0% at 662 keV and submillimeter position detection accuracy depends on in-depth characterization of the RTSDs. Detailed electrical and material characterization of RTSD will also aid in developing improved reconstruction algorithms. However, this approach traditionally is hugely time-consuming and requires numerous sophisticated experiments and skilled manpower. On the other hand, deployment of such detector arrays requires precise characterization of the individual detectors and knowledge of the defects within the sensors spatially and temporally. Sensors with defects are typically discarded, which significantly reduce manufacturing yield, increase production cost, and result in financial loss.

To solve this important problem, we utilize a machine learning approach. Machine learning and deep learning-based models have not only been popular in the last few years but also have created a paradigm shift in different fields such as material science, drug discovery, and others. Integrating physics-based modeling and machine learning is becoming more popular over the years [24, 25]. The overall objectives of such approaches are to develop inverse models and improve predictions beyond state-of-art physical models, model parameterization, partial differential equations (PDEs) solutions, discover symbolic governing equations, and others. Solving problems in physics governed by PDEs using neural networks has been done [26, 27]. For example, the two-dimensional wave equation is modeled as a recurrent neural network [28]. DeepONets [29] have been demonstrated as a powerful tool to learn nonlinear operators in a supervised data-driven manner. The 2D Poisson equation has been solved with a physics-informed neural networks [30]. In most of these physics-based machine learning approaches, relatively simpler PDEs have been solved. However, the charge transport in a RTSD has multitude of coupled PDEs [31] involving charge drift, trapping, detrapping, and recombination. To the best of our knowledge, *our work is the first novel approach to characterize the radiation detectors using machine learning*.

In our machine learning-based physical models [32–41], we solve the following problems currently plaguing the characterization of radiation detectors with a reasonable detection area for its wide-scale implementation in medical and security applications: (a) micron-level characterization of detector material properties with spatial and temporal uniformity in a fast and efficient way, (b) determination of detector material properties as per industry standards, (c) determination of detector material properties based on reduced amount of data, and (d) micron-level defect identification and characterization.

2 Classical Approach to Detector Modeling

In solid-state detectors, depending on the end application, electron and hole transport properties play a significant role in selecting detectors. Compensation techniques used to increase the resistivity of RTSDs with mostly shallow defect levels introduce deep defect levels in them [42]. The deeper defects degrade the performance of the RTSDs owing to longer trapping centers for charges. The trapping, detrapping, and recombination are governed by Shockley-Read-Hall theory [43, 44]. Rodrigues et al. [45] worked on techniques to measure more detailed properties of these materials using the charge transport equations driven by the charge continuity equations with multiple electron and hole defect levels—trapping centers—coupled with Poisson's equation [46–49]. The macroscopic equations in [31] describe the various phenomena occurring in the material when photons, X-rays, or energetic gamma rays interact with the material usually by photoelectric, Compton, or pair-production type of interaction, creating electron-hole pairs. Once the electron-hole pairs are created, the following phenomena occur: (1) drift of charges (electrons and holes) due to electric field between the electrodes, (2) free charges getting trapped and detrapped in defect levels within the material, and

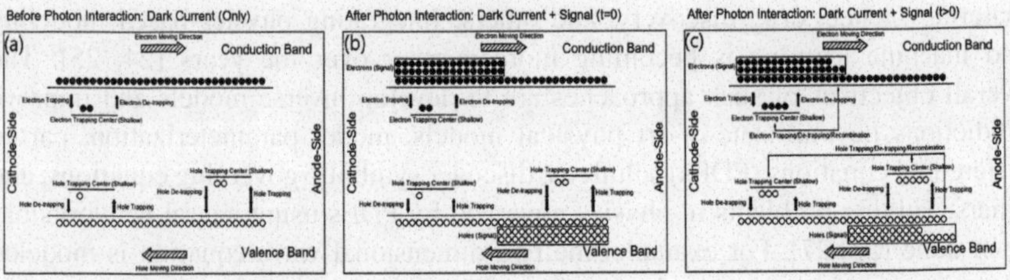

Fig. 1 (**a**) Energy level diagram of a RTSD showing shallow electron trapping centers along with shallow and deep hole trapping centers, before photon interaction; (**b**) electrons-holes in excess of equilibrium generated at $t = 0$; (**c**) transport of electrons and holes by operating electric field along with trapping, detrapping, and recombination for $t > 0$. Figure adapted with permission [32]

(3) recombination of free charges, which are modeled as capture of free electrons followed by capture of free holes in the material [50] (Fig. 1).

Equation (1) describes the electron movement in the RTSD. The dynamic concentration of free electrons is n_e more than equilibrium. There is an increase in electron concentration at a spatial region due to charge creation in the bulk, drift of charges to that region, and detrapping from trapped levels. On the other hand, there is reduction of charges in the same region due to diffusion, trapping, and recombination of charges. The model considers that the trapped charges never saturate the trapping centers, since excess carrier concentrations are small compared to the number of available trapping and recombination centers, and the simplified equations can be used [46]. In Eq. (1), only one trapping level for electrons has been shown, while in principle, there can be several trapping levels,

$$\frac{\partial n_e}{\partial t} + \nabla \bullet (n_e \mu_e \nabla \phi) = -\frac{n_e}{\tau_e} - \frac{n_e}{\tau_{eT1}} + \frac{\tilde{n}_{e1}}{\tau_{eD1}} + \cdots + \delta_e, \tag{1}$$

where μ_e is the mobility of the electrons, ϕ is the voltage, τ_e is the electron lifetime, and $\tilde{n}_{e1}$, τ_{eT1}, and τ_{eD1} are, respectively, the concentration, trapping, and detrapping lifetime of electrons in trapped level 1, and δ_e is the source term.

The increase in concentration of electrons in trapped level 1 is dependent on its own concentration $\tilde{n}_{e1}$, concentration of free electrons n_e, and trapping and detrapping lifetimes τ_{eT1} and τ_{eD1}, respectively, as shown in Eq. (2):

$$\frac{\partial \tilde{n}_{e1}}{\partial t} = \frac{n_e}{\tau_{eT1}} - \frac{\tilde{n}_{e1}}{\tau_{eD1}}. \tag{2}$$

A similar equation applies to holes, as shown in Eqs. (3) and (4) considering two trapping centers for holes and recombination of electrons and holes. Equation (4) shows the increase in holes trapped in trapping centers 1 and 2:

$$\frac{\partial n_h}{\partial t} + \nabla \bullet (n_h \mu_h \nabla \phi) = -\frac{n_h}{\tau_h} - \frac{n_h}{\tau_{hT1}} + \frac{\tilde{n}_{h1}}{\tau_{hD1}} - \frac{n_h}{\tau_{hT2}} + \frac{\tilde{n}_{h2}}{\tau_{hD2}} + \cdots + \delta_h, \tag{3}$$

$$\frac{\partial \tilde{n}_{h1}}{\partial t} = \frac{n_h}{\tau_{hT1}} - \frac{\tilde{n}_{h1}}{\tau_{hD1}}, \frac{\partial \tilde{n}_{h2}}{\partial t} = \frac{n_h}{\tau_{hT2}} - \frac{\tilde{n}_{h2}}{\tau_{hD2}}, \tag{4}$$

where n_h is the dynamic concentration of free holes in excess of equilibrium, μ_h is the mobility of the holes, and τ_h is the hole lifetime. $\tilde{n}_{h1}$, τ_{hT1}, and τ_{eD1} are the concentration, trapping, and detrapping lifetime of holes in trapped level 1, respectively. Similarly, $\tilde{n}_{h2}$, τ_{hT2}, and τ_{hD2} are the concentration, trapping, and detrapping lifetime of holes in trapped level 2, respectively. δ_h is the source term.

The drift of charges is dependent on the electric field E, as shown in Eq. (5), with the voltage satisfying the Poisson equation Eq. (6):

$$E = -\nabla \phi, \tag{5}$$

$$\nabla^2 \phi = -k\frac{q}{\epsilon}\left(n_e + \tilde{n}_{e1} + n_h + \tilde{n}_{h1} + \tilde{n}_{h2} + \cdots\right). \tag{6}$$

where ϵ represents the permittivity of the RTSD material in Eq. (6).

The temporal dynamics of free electrons and holes following the SRH model is affected by the trapping energy states in the bandgap [46]. For free hole concentration p, with trapping and detrapping lifetimes τ_{Ti} and τ_{Di}, respectively for i^{th} trap, and p_{ti} as the concentration of holes in the ith trap, the following equation can be written [51]:

$$\frac{dp}{dt} = \sum\nolimits_i \left(-\frac{p}{\tau_{T1}} + \frac{p_{ti}}{\tau_{Di}}\right). \tag{7}$$

The trapping and detrapping lifetimes dictate whether the defects induce short-term or long-term trapping of charges in the detector. Considering the low probability of transition of charges between the trapping centers, the occurrence of such process [52] is neglected in Eq. (7).

Signals collected at the electrodes arise due to the movement of charges [53–56]. High-energy photons (γ-rays or X-rays) are incident on the detector and create electron-hole pairs. The hole drifts toward the cathode and electrons toward the pixelated anodes as shown in Fig. 2a. The detector setup is shown in Fig. 2b(i, ii) with the nine grid electrodes on the anode side (NW, N, NE, W, C, E, SW, S, SE) and one single large cathode electrode (CAT). Since the mobility of electrons is higher than holes, only single cathode is used, and grid anodes are used. In case hole mobility is also significant, grid cathode can also be used. Figure 3(i) shows two cases of the electric field—uniform and piecewise linear which are plausible inside the RTSD, while Fig. 3(ii) shows the signals observed in the cathode electrode and

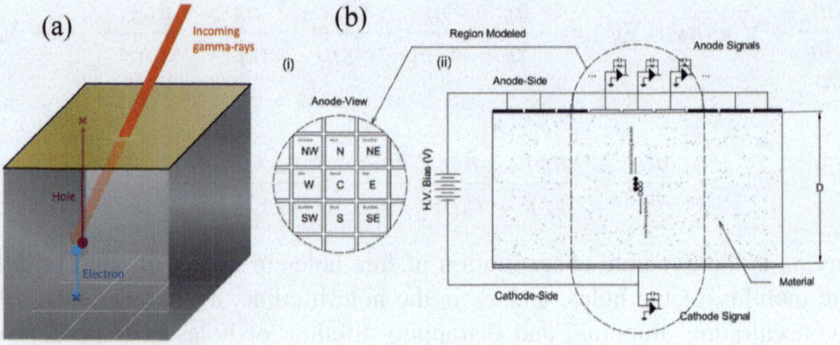

Fig. 2 (**a**) Incoming high-energy photons (γ-rays or X-rays) interacting with the detector having pixelated anode and a single cathode; (**b**) electron-hole motion in the detector, (i) sensor-pixelated anode pattern, (ii) central region model illustrating pixelated anode design with uniform material and electrons-holes drift in bulk. Figure adapted with permission [32]

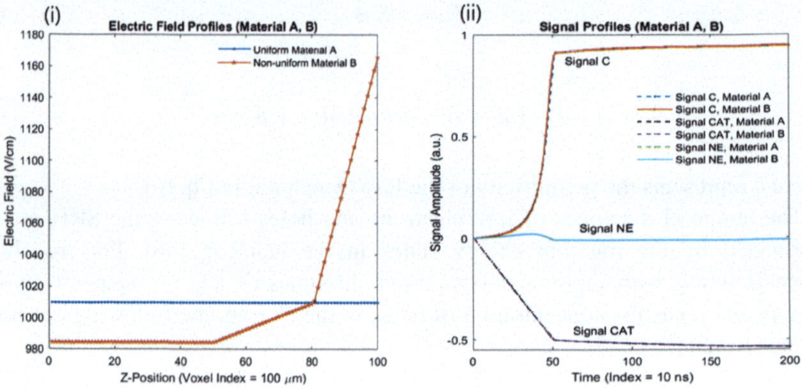

Fig. 3 (i) Electric field profiles for simulated materials A and B, (ii) anode signals C, CAT, and NE simulated for materials A and B (for electron-hole charge injection at Voxel 50). Figure adapted with permission [32]

three adjacent anodes (center or collecting anode, north and north-east or neighbors) for electron-hole pair injection at voxel position 50, where in this case the cathode was at position 0 and anode at position 100. At a quick glance, there is almost no difference in the signals generated at the electrodes due to the two electric fields as shown in Fig. 3(ii). However, with a closer look at the electric field profiles in Fig. 3(i), we can see distinct differences between them. Our learning-based approach can clearly distinguish between these two electric field profiles as shown in Fig. 3(i) from the signals shown in Fig. 3(ii), along with the charge transport in the volume of the RTSD.

The data for training the proposed learning model has been generated using the classical equations, Eqs. (1, 2, 3, 4, 5, 6, and 7). A code based on these equations has been developed to define the charge transport equations in the detector, by defining the trapping, detrapping, and recombination lifetimes of electrons and holes as constant predefined parameters, with varying the electric field along the material. This classical model has been created by spatially discretizing the detector. For a charge input at a voxel position, the signals are generated in the cathode and pixelated anode. Additionally, the free and trapped charges in different spatial pixel locations are computed. These signals and free and trapped charges are computed for each time step, with the total number of time steps defined a priori. The input-output data for training the learning-based model consists of the input electron-hole pair injected at a known voxel position and the output as the signals, free and trapped charges in different spatial voxels of the classical model over different time steps.

3 Machine Learning-Based Full Model of Detector in 1D

We introduce a novel method to determine the material properties such as $\mu_{e,h}$, τ_{eT1}, τ_{eD1}, τ_{hT1}, τ_{hD1}, and recombination coefficients of free and trapped electrons and holes at spatially discretized locations in the RTSD. The detector is subdivided into N voxels, where the division can be in a 1D, 2D, or 3D manner. In each of the discretized volume (or voxel) i, the RTSD properties such as $\mu_{e,h,i}$, $\tau_{eT1,i}$, $\tau_{eD1,i}$, $\tau_{hT1,i}$, and $\tau_{hD1,i}$ are defined. These weights are thus spatially discrete from each other. For instance, the mobility μ_e is converted into N discrete values of $w_{Trpt,e}$. Now, the relation between μ_e and μ_h can be separately modeled as unknowns or be related to each other. This physics-based learning model therefore allows the determination of each of these unknowns in the RTSD. For each of these coefficients (referred here as τ in general), we compute the number of charges particles (electrons or holes) remaining in that particular level as $N_{left} = N_0 e^{-t/\tau}$, where N_0 and N_{left} are the number of charged particles at a particular level time at $t = 0$ and at time t, respectively. Thus, given the time $t = t_1$, (e.g., $t_1 = 10$ ns for the 100-voxel model) and given τ, we can compute the fraction of charges transitioning from an energy level to another and those remaining in the same energy level (Fig. 4).

The trainable model weights are initialized to a value. Based on the electron-hole pair input to this model, the output (signals, free, and trapped charges) is computed over time. The model consists of training and testing phases. In the training phase, the loss is computed using Eq. (15) for every electron-hole input pair based on the outputs from this model and the ground truth output data over time. Since our model is a recurrent network structure over time, backpropagation through time (BPTT) [57, 58] is used to compute the gradients of the loss with respect to the trainable (or tunable) weights of the model. BPTT unfolds the learnable model in time by creating several copies of the model which can be treated as a feed-forward deep network with tied weights. The update of the trainable weights is based on a stochastic gradient descent method—ADAM optimization [59], which is based on adaptive estimation of first- and second-order moments. ADAM optimizer is used with a

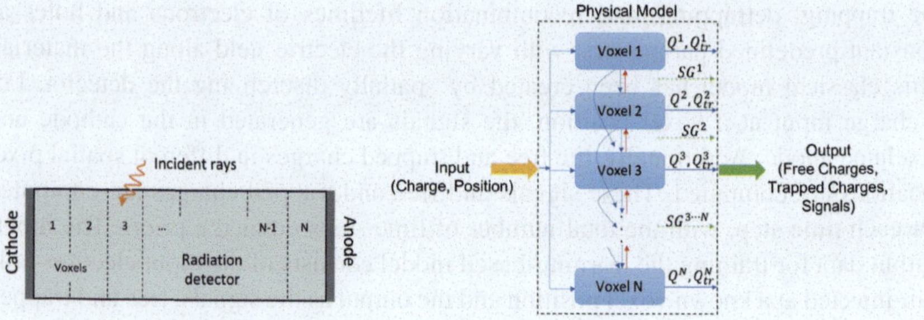

(i) 1D Voxelated Model of RTSD (ii) Overview of learning-based physical model

Fig. 4 RTSD model with (i) 1D voxels and electrodes, and (ii) overview of the learning-based physical model of RTSD. The model consists of trainable weights in each voxel. The charges move dynamically from one voxel to another based on the electric field and polarity (blue for electrons and red for holes). The free and trapped charges in each voxel with the signals generated during movement of charges are used as output of the model. Cathode and anode are on outer ends of voxel 1 and N, respectively. Figure adapted with permission [33]

learning rate of 5×10^{-4} with two momentum terms set as $\beta_1 = 0.9$ and $\beta_2 = 0.999$. Learning rate higher than 5×10^{-4} causes oscillations in the loss function, while a lower learning rate slows down convergence. In each epoch (iteration), the ADAM optimization updates the trainable weights of the model based on the gradients, with the goal to minimize the loss function over the epochs. This learning-based physical model is developed using Tensorflow library [60] in Python. Over the epochs, the weights of the model such as electron transport weights $w_{Trpt,\,e,\,i}$, electron trapping weights $w_{eT,\,i}$, electron detrapping weights $w_{eD,\,i}$, and other weights in voxel i as shown in Fig. 5(i) in the learning-model are trained. Once the model is trained, the model can be tested based on the electron-hole pair input at voxel position. The injection position is different from the ones used for training the model. In the testing phase, there is no computation of loss. Only the signals and charges are obtained in each time step as output from the model.

3.1 Machine Learning-Based Physical Model

A voxelated representation of the detector in one dimension is shown in Fig. 4(i). Linearly, the material is divided into N voxel segments with the electrodes at either end, anode on the right and cathode on the left. We consider a grid anode with M electrodes ($M = 9$ for illustration) and a single cathode. The high-energy photons can interact in any position in the detector. The resultant physical model in turn is a discretized version of the physical phenomena occurring in the RTSD over time.

The model is trained with input-output data pairs. The input data is the position of injected electron-hole pair along with the fraction of charge generated in that

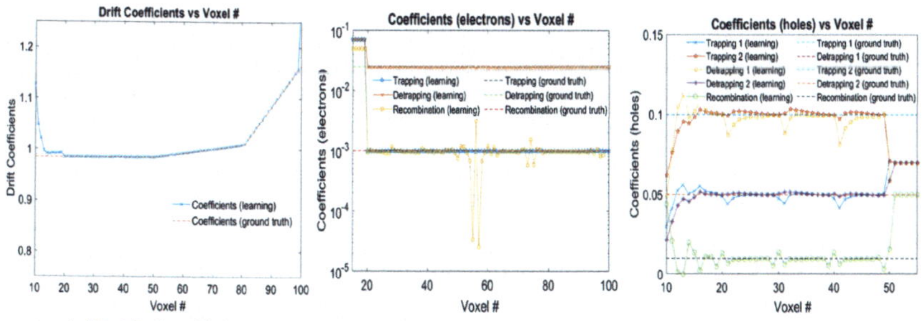

(a)Drift Coefficients (b) Electron Coefficients (c) Hole Coefficients

Fig. 5 (**a**) Drift coefficients (μ_e), (**b**) electron coefficients ($w_{eT,1}, w_{eD,1}, w_{eRec}$), and (**c**) hole coefficients ($w_{hT,1}, w_{hD,1}, w_{hT,2}, w_{hD,2}, w_{hRec}$) for e-h injection at voxels 20,30,40, and 50 with $k = 0.1, l = 100, n = 100$ in Eq. (15). Figure adapted with permission [32]

voxel. The magnitude of the injected charges is normalized to 1. The output data are the signals obtained at the electrodes along with the electrons and holes (free and trapped) in each of the voxels over time. During training of the model, the loss function is computed as the sum of the squared errors between the signals at the electrodes and charges in the voxels compared to the ground truth signals along with the $error^2_{voltage}$. The overall loss function (*LF*) for this model is shown in Eq. (8). The loss function in Eq. (8) is shown for simulation experiments considering CZT detector with two trapping centers for holes and one trapping center for electrons [14, 31]. However, in general, there can be several trapping centers for both electrons and holes, and thus several such terms, whose number is predefined, and the learning-based model needs to be designed accordingly. In the loss function, the errors due to the signals and voltage are grouped together, free and trapped electron charges are grouped together, and free and trapped charges due to holes are grouped together with weighting terms *k*, *l*, and *n*, respectively. The higher the value of the parameters *k*, *l* and/or *n*, the lower the errors associated with those terms. In these error terms, the subscript *gt* for a particular parameter *X* (for instance, *X* is *signal* or q_e) refers to the ground truth data for that parameter *X* generated using the classical model as described and the subscript *L* for the same parameter *X* refers to the data generated by the learning-based model.

$$
\begin{aligned}
LF = & k\left[\left(signal_{gt} - signal_L\right)^2 + error^2_{voltage}\right] \\
& + l\left[\left(q_{e,gt} - q_{e,L}\right)^2 + \left(q_{et,gt} - q_{et,L}\right)^2\right] \\
& + n\left[\left(q_{h,gt} - q_{h,L}\right)^2 + \left(q_{hT_1,gt} - q_{hT_1,L}\right)^2 + \left(q_{hT_2,gt} - q_{hT_2,L}\right)^2\right]
\end{aligned}
\tag{8}
$$

Numerical Experiments with Weighted Loss Function

We performed experimental studies with *k*, *l*, *n* ∈ {0.1, 1.0, 10.0, 100.0} in Eq. (15) using the 100 voxel size model. Now *k*, *l*, *n* can take any finite values. However, in the experimental studies, k, l, n are chosen to be linear in logarithmic scale from 0.1 to 100, which spans a wide range. The experimental data and the model size used for learning the model consist of transport, trapping, and detrapping centers for electrons and holes, along with recombination of charges. We find that for $k = 0.1$, $l = 100$, and $n = 100$, the normalized mean square error (NMSE) [32] has a minimum of 0.1761. This combination of k, l, n values in the learning model is used for injecting charges at multiple voxel positions of 20, 30, 40, 50. Figure 5a–c show the drift, electron, and hole coefficients, respectively, for material B (material B has nonuniform electric field profile as shown in Fig. 3(i)). The weighted model prioritizes signals at the electrodes, charges (free and trapped) due to electrons, and holes differently than the unweighted model. Now, since these hole and electron charges are more fundamental and the signals are generated only due to charge motion, the higher weightage of charges (free and trapped) in the loss function compared to the signals drives the model to its optimized material parameters better than the unweighted model. In both cases, e-h injection is at the same voxel locations of 20, 30, 40, and 50. Clearly, choosing appropriately the regularization parameters in the loss function improved the accuracy in the learned parameters.

The physical model can be designed with voxels other than 100 voxels as well. For a material with the same dimension as before, increasing the number of voxels in the model will reduce the size of each voxel. The decrease in voxel size in the model will improve the response characteristics of the output signals.

3.2 *Machine Learning-Based Physical Model with Reduced Data*

The learning-based full model of the detector uses a loss function taking into consideration the complete data as required by the classical physical equations—signals, voltage distribution in the material, free, and trapped charges. In the real world, each of these data must be obtained from experimental hardware setups with the detector, which not only requires costly equipment but also skilled manpower and time. To address this issue, in this section, we propose learning-based models to learn from fewer data than in the full model (which is dictated by the classical physical model). We explore the models by training with fewer data than what is dictated by the classical physical equations, stepwise removing small fraction of relevant data from the full learning-model (such as charge trapped and trapping centers) and evaluate the performance of the new physics based learning model.

Physical Models with Reduced Data

We develop four physical models based on progressively training with fewer data while developing these models. The Physical Model-1 has been developed using signals at the electrodes, voltage distribution along the detector, free electron charges, and free and trapped hole charges in one trap center as shown in Eq. (9).

We use hole charges corresponding to only one trapping center and no electron charges for its trapping center for CdZnTe detector with two trapping centers for holes and one trapping center for electrons. Our model as shown in Figs. 4 and 5 can be trained using charges corresponding to any one of the hole trapping centers, shallow or deep. However, for illustration purposes, we have used data from shallow hole trapping center. For any other material with NT_h trapping centers for holes and NT_e trapping centers for electrons, we would use the data corresponding to $NT_e - 1$ trapped charges for electrons and $NT_h - 1$ trapped charges for holes.

$$\begin{aligned} LF_1 = k\left[\left(signal_{gt} - signal_L\right)^2 + error_{voltage}^2\right] + l\left[\left(q_{e,gt} - q_{e,L}\right)^2\right] \\ + n\left[\left(q_{h,gt} - q_{h,L}\right)^2 + \left(q_{hT_1,gt} - q_{hT_1,L}\right)^2\right] \end{aligned} \quad (9)$$

In Physical Model-1, we observed from our simulation experiments that the hyperparameters *k*, *l*, *n* in the loss function are biased heavily toward the hyperparameters *l*, *n* are much higher than *k*. Thus, the Physical Model-2 has been developed considering only the free electron charges and free and trapped hole charges—in this case using hole charges in hole trapping center—1 (for illustration purposes). The loss function LF_2 is then defined in Eq. (10):

$$LF_2 = l\left[\left(q_{e,gt} - q_{e,L}\right)^2\right] + n\left[\left(q_{h,gt} - q_{h,L}\right)^2 + \left(q_{hT_1,gt} - q_{hT_1,L}\right)^2\right] \quad (10)$$

In general, for NT_h trapping centers for holes and NT_e trapping centers for electrons, we can use the free hole and electron charges, as well as electron and hole trapped charges for $NT_e - 1$ and $NT_h - 1$ trapping centers, respectively, in training the model.

In the Physical Model-3, we further reduce the dependency on any of the trapped hole charges which were used in the Physical Model-2, and this results in a model which characterizes the trapping centers in an equivalent manner. The equivalent trapping and detrapping lifetimes are the contribution of several trapping and detrapping lifetimes in the detector which contributes to the dynamics of charge motion in the detector as shown in Eq. (10). The properties of the physical detector can be attributed as defect-free properties in addition to equivalent defects in the material. The presence of multiple trapping and detrapping levels can be converted to equivalent trapping and detrapping levels. In such a scenario, for two hole trapping centers of CZT with trapping lifetimes τ_1 and τ_2, the equivalent trapping lifetime is given in Eq. (11):

$$\frac{1}{\tau_{eq}} = \frac{1}{\tau_1} + \frac{1}{\tau_2} \quad (11)$$

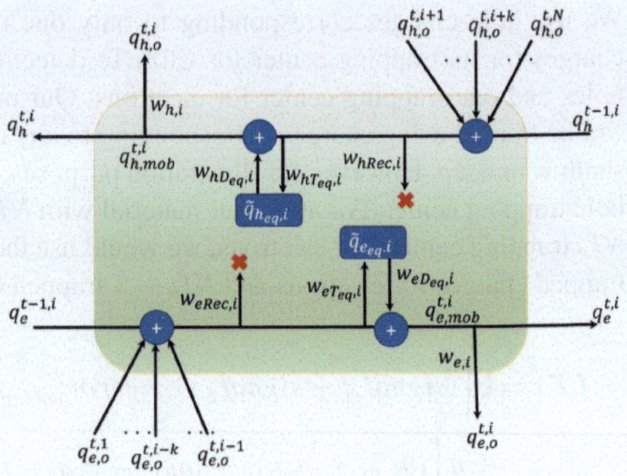

Fig. 6 Equivalent operations in voxel i at time t. Figure adapted with permission [33]

Considering the probability of trapping 1 level as p_{τ_1} and detrapping 1 lifetime as τ_{dt_1}, along with probability of trapping 2 level as $p_{\tau_2}$and detrapping 2 lifetime as τ_{dt_2}, we can calculate the equivalent detrapping lifetime as in Eq. (12):

$$\frac{p_{\tau_{eq}}}{\tau_{eq}} = \frac{p_{\tau_1}}{\tau_{dt_1}} + \frac{p_{\tau_2}}{\tau_{dt_2}}. \tag{12}$$

The Physical Model-3 is designed as a combination of defect-free model and model with equivalent defects. The equivalent computations in a voxel i are shown in Fig. 6. The equivalent trapping and detrapping weights are $w_{hT_{eq},i}$ and $w_{hD_{eq},i}$ for holes and similarly for electrons, and the corresponding trapping and detrapping weights are $w_{eT_{eq},i}$ and $w_{eD_{eq},i}$. The charges in equivalent trap center are $\tilde{q}_{h_{eq},i}$ and $\tilde{q}_{e_{eq},i}$ for holes and electrons, respectively. The loss function consists of only the free electron and free hole charges to train the model, as shown in Eq. (13),

$$LF_3 = l\left[\left(q_{e,gt} - q_{e,L}\right)^2\right] + n\left[\left(q_{h,gt} - q_{h,L}\right)^2\right]. \tag{13}$$

In Physical Model-4, we use only the signals generated at the cathode and anodes due to motion of the charged particles to train. Typically, the signals are generated at the electrodes by the superposition of signals generated individually due to transport of electrons and transport of holes. However, in Physical Model-4, we separate out the signals generated due to the drift of electrons from the signals generated due to the drift of holes and then train the model. The loss function for training the Physical Model-4 is shown in Eq. (14):

$$LF_4 = \left[\left(sg_{e,gt} - sg_{e,L}\right)^2\right] + \left[\left(sg_{h,gt} - sg_{h,L}\right)^2\right] \tag{14}$$

It is also observed from our simulation experiments that by using the loss function in Eq. (14) leads to the trained model weights which fail to converge to the ground truth solution. The solution converges to a local minimum which is different from global minimum, and hence the trained weights differ from the ground truth weights. We use a total variation (T.V.) regularization on the different weights of the model to converge the learned solution to the global minimum solution as observed in our simulation experiments. The loss function gets modified to $LF_{4,\,m}$ as shown in Eq. (15):

$$\begin{aligned} LF_{4,m} = & \left[\left(sg_{e,gt} - sg_{e,L}\right)^2\right] + \lambda_1 \left\| \nabla w_{eT_{eq}} + \nabla w_{eD_{eq}} + \nabla w_{e,Rec} \right\|_2 \\ & + \left[\left(sg_{h,gt} - sg_{h,L}\right)^2\right] + \lambda_2 \left\| \nabla w_{hT_{eq}} + \nabla w_{hD_{eq}} + \nabla w_{h,Rec} \right\|_2 \end{aligned} \tag{15}$$

The T.V. regularization ensures smoothness in the trained weights of the model. The optimal values of λ_1 and λ_2 are determined through simulation experiments by finding the minimum error between the ground truth weights and the trained weights.

The trained weights for the different physical models have been evaluated by computing an error metric for each of the weights. For example, for electron trapping weight (w_{eT}) with $w_{eT,\,gt}$ and $w_{eT,\,lr}$ as the ground truth and trained weights, respectively, the error is expressed as in Eq. (16):

$$Err\left(w_{eT}\right) = \sqrt{\frac{1}{N_{fin} - N_{inj} + 1} \sum\nolimits_{i=N_{inj}}^{N_{fin}} \left\{ \frac{w_{eT,lr,i} - w_{eT,gt,i}}{w_{eT,gt,i}} \right\}^2} \tag{16}$$

The error is computed over the injection positions of the electrons/holes and the number of voxels over which the weight gets trained over the epochs. The difference between the trained weights and ground truth weights for the trained region is normalized by the ground truth weights, to take into account the different ranges of weights in the model and put equal emphasis on the different model weights.

Numerical Experiment with Physical Model-3

The electron-hole charge pairs are injected at voxel positions 9 with stride of 5 voxels until voxel 59 during training the Physical Model-3 as well. The learned physical properties of the model—drift coefficients, electron coefficients, and hole coefficients are shown in Fig. 7a–c, respectively. Equation (13) is used as the loss function with weights $l = 10$ and $n = 1$. Clearly, the electron drift, trapping, detrapping, and recombination coefficients follow the ground truth values. The learned recombination coefficients for the holes follow the ground truth values as well. For multiple trapping centers for holes (two in this case), the learning-based model finds the equivalent trapping center, with equivalent trapping and detrapping lifetimes following Eqs. (11) and (12), respectively.

The ground truth values of trapping 1 lifetime are 0.195 μs and trapping 2 lifetime is 0.094 μs corresponding to probability of trapping holes in trap center 1 and 2 to be 0.05 and 0.10, respectively. The ground truth equivalent trapping lifetime (τ_{eq}) is calculated using Eq. (12) to be 0.0634 μs. The fraction of holes getting trapped in the equivalent trapping center is $1 - 0.8541 = 0.1459$.

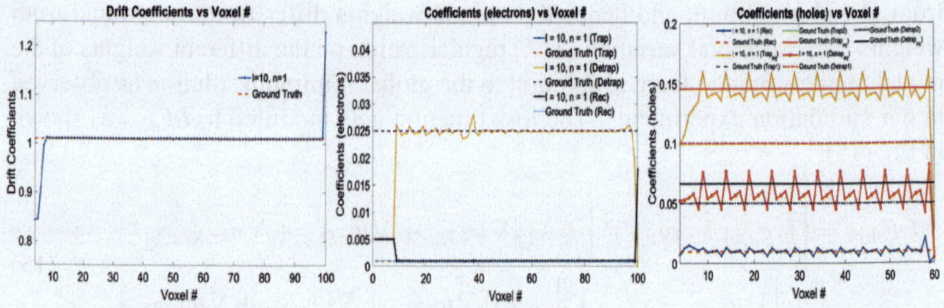

(a) Drift Coefficients (b) Electron Coefficients (c) Hole Coefficients

Fig. 7 (**a**) Drift coefficients (μ_e), (**b**) electron coefficients (w_{eT}, w_{eD}, w_{eRec}), and (**c**) hole coefficients ($w_{hT,\,1}, w_{hD,\,1}, w_{hT,\,2}, w_{hD,\,2}, w_{hRec}$) for e-h injection at voxel positions 9 with stride of 5 voxels until voxel 59 for *k*, *l*, *n* in Physical Model-3 as shown in the plot legend. *Rec* in the legend refers to recombination coefficients. Readers are suggested to enlarge the figure for closer view. Figure adapted with permission [33]

Similarly, in the ground truth simulation data, we considered fraction of charge getting detrapped from trap center 1 and 2 are 0.10 and 0.05, respectively. In the same manner, the detrapping 1 and 2 lifetimes are 94.9122 ns and 194.9573 ns, respectively. Considering Eqs. 19, and equivalent trapping probability as 0.1459, we compute the equivalent detrapping lifetime $\tau_{dt,\,eq}$ as 145.9 ns. The fraction of holes remaining after detrapping from the equivalent trapping level would be 0.9338. Thus, the fraction of charges getting detrapped from the equivalent trapping center is $1 - 0.9338 = 0.0662$. For the trained weights as shown in Fig. 13, the error values of drift coefficients (μ_e), trapping (w_{eT}), detrapping (w_{eD}), and recombination coefficients (w_{eRec}) for electrons are 1.32×10^{-5}, 0.0412, 0.0316, and 0.0277, respectively, which are computed for voxels 9 to 99. Similarly, the error values of the equivalent trapping ($w_{hT,\,eq}$), equivalent detrapping ($w_{hD,\,eq}$), and recombination coefficients ($w_{h,\,Rec}$) for holes are 0.0954, 0.1957, and 0.3378, respectively, which are computed for voxels 1 to 59. The arithmetic mean of the error of these material properties is 0.1042.

3.3 Identifying Defects Without a Priori Information

In the previous sections of this chapter, all the defects of the RTSD are considered while developing the physics-based machine learning model for characterizing the properties of the RTSD and CdZnTe in particular. However, the presence of defects in RTSD—trapping, detrapping, and recombination for electrons and holes are often unknown. In this section, we demonstrate the capabilities of physics-inspired machine learning (PI-ML) models of RTSDs with all defects in identifying only defects which are present in the material. Three cases are considered: (i)

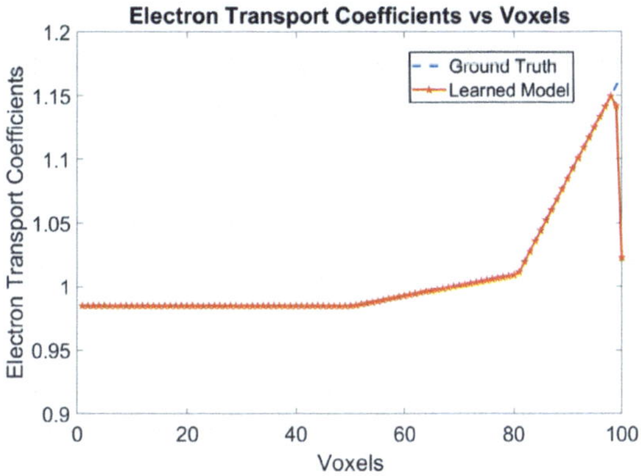

Fig. 8 Electron transport coefficients using Physical Model-3 for e-h injections at voxels 9, 14, 19, 24, 29, 34, 39, 44, 49, 54, and 59. Figure adapted with permission [34]

only trapping centers with no recombination, (ii) no trapping centers and only recombination, and (iii) no trapping centers and no recombination.

The PI-ML models in general can be designed considering either all the charge transport trainable weights or a reduced set of weights, as described in the previous sections of this chapter. In this section, we demonstrate our approach based on the Physical Model-3 and show that our model learns only the material defects which are present in the data from the classical model. Thus, the dependency of the PI-ML model of the RTSD on the precise properties of the material is eliminated. One can develop these PI-ML models just based on the general transport equations. The model possesses the capability to identify the defects present in the material from these generalized models and training data.

Numerical Experiments

(a) ***No Defects and No Recombination.***

The ground truth data has been generated using the classical model with no trapping centers for electrons and holes, along with no recombination of electrons and holes in the bulk of the crystal. The Physical Model-3 is trained with this ground truth data, and the trained weights are analyzed. Figure 8 shows the variation of electron transport coefficients over the voxels. The root of normalized mean squared error (RNMSE) error for the electron transport coefficients is 0.0014. The e-h injections are for every five voxels from Voxel 9 until Voxel 59.

The electron and hole trapping coefficients are shown in Fig. 9a, b, respectively. The model is correctly able to identify the trapping coefficients of 0 for both electrons and holes. For holes, both the trapping probabilities are learned as 0 from voxels 1 to 59, while for electrons, only one trapping probability is learned as 0 from voxels 9 up to 99, which is the correct ground truth parameter of the detector as well. The RNMSE values for the electron and hole trapping coefficients are 9.6517×10^{-4} and 1.6113×10^{-8}, respectively. We show only the trapping coefficients for electrons and holes and not the detrapping coefficients, since, if the trapping coefficients for electrons and holes are 0, then the detrapping coefficients do not matter.

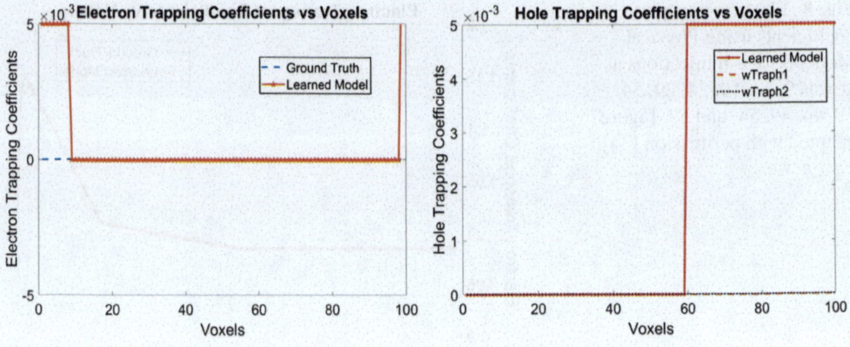

(a) (b)

Fig. 9 (**a**) Electron trapping coefficients and (**b**) hole trapping coefficients using Physical Model-3 for e-h injections at voxels 9, 14, 19, 24, 29, 34, 39, 44, 49, 54, and 59. Figure adapted with permission [34]

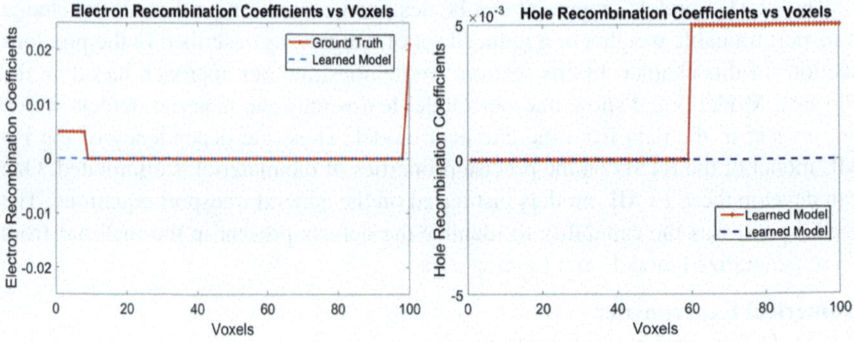

(a) (b)

Fig. 10 (**a**) Electron recombination coefficients and (**b**) hole recombination coefficients using Physical Model-3 for e-h injections at voxels 9, 14, 19, 24, 29, 34, 39, 44, 49, 54, and 59. Figure adapted with permission [34]

The electron and hole recombination coefficients are shown in Fig. 10a, b, respectively. The model is correctly able to identify the recombination coefficients of 0 for both electrons and holes. For holes, the recombination probabilities are learned as 0 from voxels 1 to 59, while for electrons, the recombination probabilities are learned as 0 from voxels 9 up to 99, which are the correct ground truth parameters of the detector. The RNMSE values for the electron and hole recombination coefficients are 9.6538×10^{-4} and 1.2609×10^{-7}, respectively. The mean RNMSE value for the trained coefficients (transport of electrons, trapping of electrons, and holes along with recombination of electrons and holes) is 6.6×10^{-4}, which is a small value.

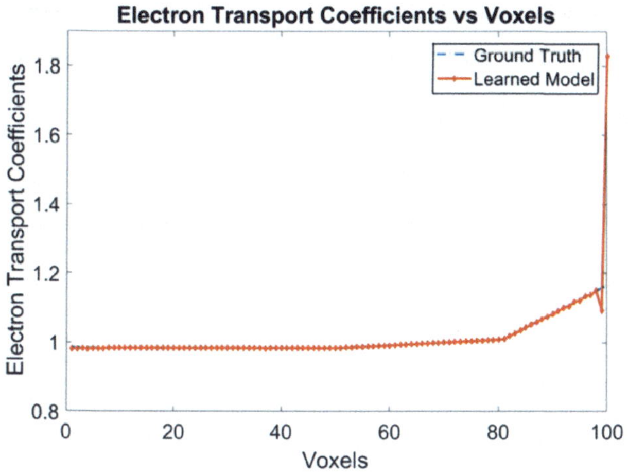

Fig. 11 Electron transport coefficients for e-h injections at voxels 9, 14, 19, 24, 29, 34, 39, 44, 49, 54, and 59. Figure adapted with permission [34]

(b) ***Defects and No Recombination.***

The ground truth data have been generated using the classical model with one trapping center for electrons and two trapping centers for holes with no recombination of electrons and holes in the bulk of the crystal. The Physical Model-3 is trained with this ground truth data, and the trained weights are analyzed. Figure 11 shows the variation of electron transport coefficients over the voxels. The RNMSE value for the electron transport coefficients is 0.0058. The e-h injections are for every five voxels from voxel 9 until voxel 59.

Figure 12a, b shows the electron and hole trapping coefficients, respectively. It is seen that the model can learn the piecewise linearly uniform electron trapping coefficients from voxel 9 to voxel 99 as shown in Fig. 6a. Similarly, for holes, the model can learn the equivalent hole trapping coefficients, which are piecewise linearly uniform as well. The hole coefficients are learned from voxel 59 all the way toward the cathode (left end) up to voxel 1 as shown in Fig. 6b. The RNMSE values for the electron and hole trapping coefficients are 0.5845 and 0.0510, respectively.

Similarly, Fig. 13a, b shows the electron and hole detrapping coefficients, respectively. It is seen that the model can learn the piecewise linearly uniform electron detrapping coefficients from voxel 9 to voxel 99 as shown in Fig. 13a for most of the voxels. Around voxel 40, there is a single position where the electron detrapping coefficient is learned incorrectly, but for other voxels, the coefficients are learned correctly. Similarly, for holes, the model can learn the equivalent hole detrapping coefficients which are piecewise linearly uniform as well. The hole coefficients are learned from voxel 59 all the way toward the cathode (left end) up to voxel 1 as shown in Fig. 13b. The RNMSE values for the electron and hole detrapping coefficients are 1.0498 and 0.0728, respectively. We observe that

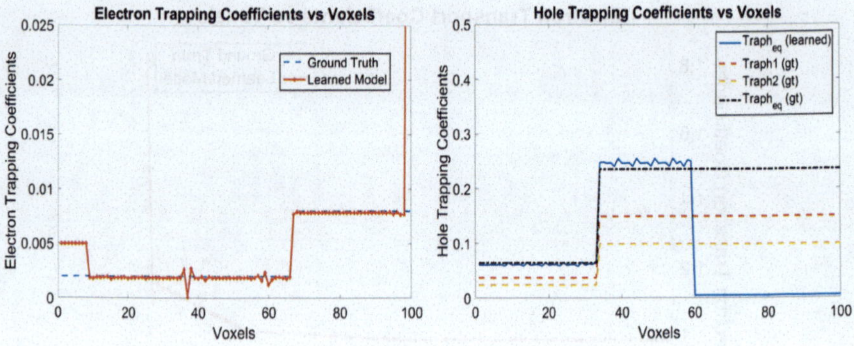

(a) **(b)**

Fig. 12 **(a)** Electron trapping coefficients and **(b)** hole trapping coefficients using Physical Model-3 for e-h injections at voxels 9, 14, 19, 24, 29, 34, 39, 44, 49, 54, and 59. Figure adapted with permission [34]

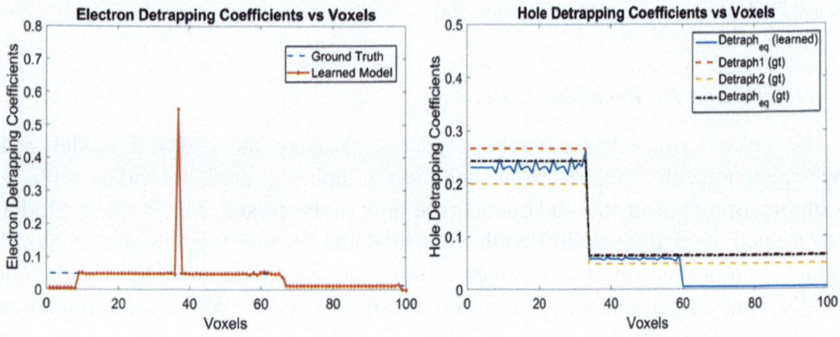

(a) **(b)**

Fig. 13 **(a)** Electron detrapping coefficients and **(b)** hole detrapping coefficients using Physical Model-3 for e-h injections at voxels 9, 14, 19, 24, 29, 34, 39, 44, 49, 54, and 59. Figure adapted with permission [34]

for multiple hole trapping centers, the hole trapping and detrapping coefficients are learned as the equivalent trapping and detrapping coefficients—which are the characteristics of Physical Model-3 [33].

The electron and hole recombination coefficients are shown in Fig. 14a, b, respectively. The model is correctly able to identify the recombination coefficients of 0 for both electrons and holes. For holes, the recombination probabilities are learned as 0 from voxels 1 to 59, while for electrons, the recombination probabilities are learned as 0 from voxels 9 up to 99, which are the correct ground truth parameters of the detector. The RNMSE values for the electron and hole recombination coefficients are 0.0026 and 8.6241×10^{-5}, respectively. The mean RNMSE value for the trained coefficients (transport of electrons, trapping, and detrapping coefficients of electrons and holes along with recombination of electrons

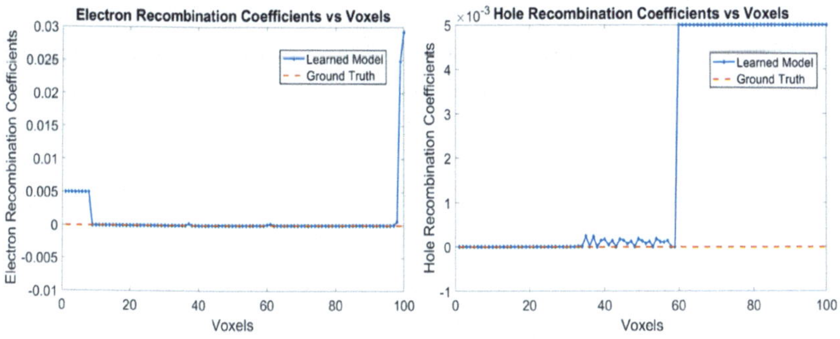

(a) **(b)**

Fig. 14 **(a)** Electron recombination coefficients and **(b)** hole recombination coefficients for e-h injections at voxels 9, 14, 19, 24, 29, 34, 39, 44, 49, 54, and 59. Figure adapted with permission [34]

Fig. 15 Electron recombination coefficients for e-h injections at Voxels 9, 14, 19, 24, 29, 34, 39, 44, 49, 54 and 59. Figure adapted with permission [34]

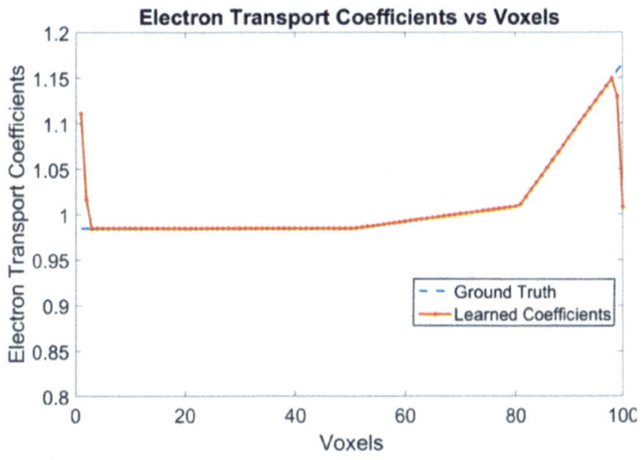

and holes) is 0.2524. This RNMSE value is attributed to the fact that one of the trained electron trapping and detrapping coefficients has a significantly high value near voxel 40 and voxel 100, which makes the RNMSE value high.

(c) ***No Defects and with Recombination.***

The ground truth data have been generated using the classical model with no electron and hole trapping centers but with just recombination for both electrons and holes. The Physical Model-3 is trained with this ground truth data. Figure 15 shows the variation of electron transport coefficients over the voxels, with the e-h injections for every five voxels from voxel 9 to voxel 59. The RNMSE value for electron transport coefficients is 0.0025.

Figure 16a, b shows the electron and hole trapping coefficients, respectively, for the voxels trained by the multiple e-h injections. Both these trapping coefficients converge to 0, which is the ground truth parameter in the data. The RNMSE values for the electron and hole trapping coefficients are 0.0018 and 0.0011, respectively.

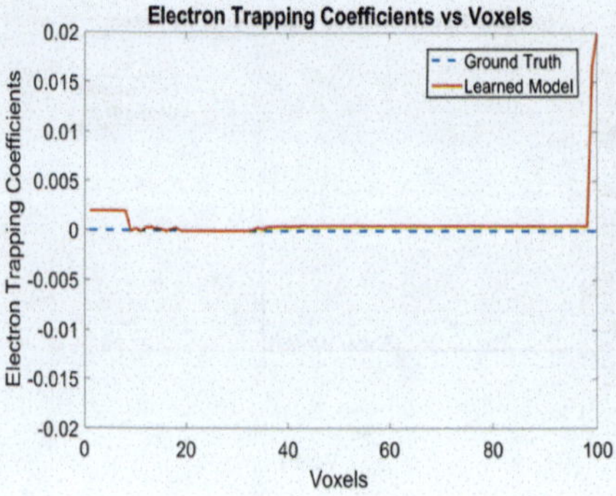

(a)

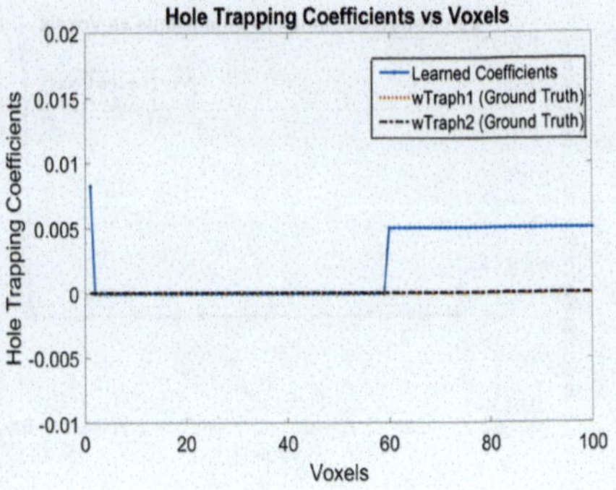

(b)

Fig. 16 (**a**) Electron trapping coefficients and (**b**) hole trapping coefficients for e-h injections at voxels 9, 14, 19, 24, 29, 34, 39, 44, 49, 54, and 59. Figure adapted with permission [34]

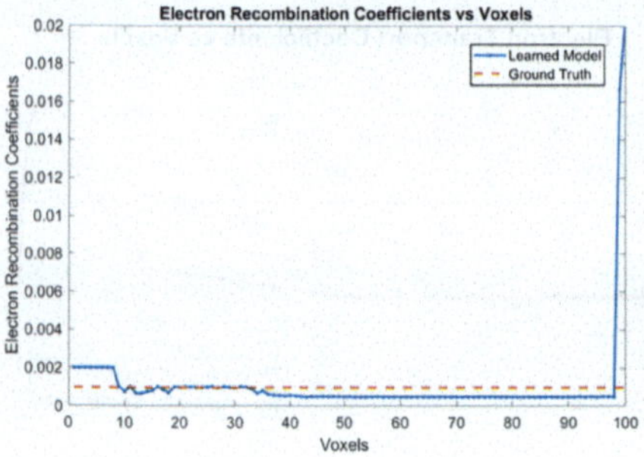

(a)

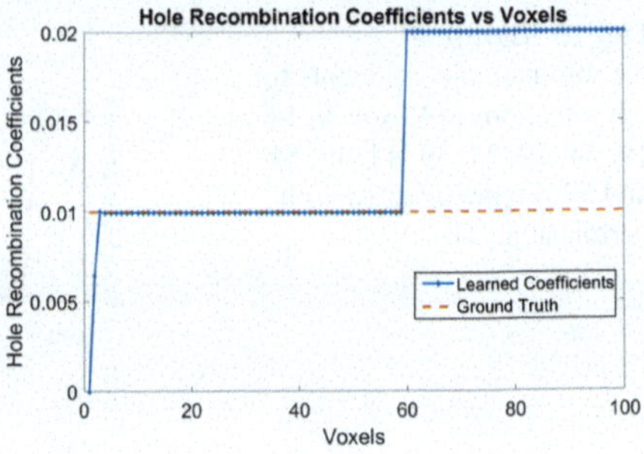

(b)

Fig. 17 (**a**) Electron recombination coefficients and (**b**) hole recombination coefficients for e-h injections at voxels 9, 14, 19, 24, 29, 34, 39, 44, 49, 54, and 59. Figure adapted with permission [34]

Similarly, Fig. 17a, b shows the electron and hole recombination coefficients, respectively, for the voxels trained by the multiple e-h injections. The hole recombination coefficients converge to the ground truth parameters better than the electron recombination coefficients. For the electron recombination coefficients, the learned coefficients converge better for voxels 9 to 34, but the rest of the coefficients converge to a value which is close to the ground truth recombination coefficients. The RNMSE values of the electron and hole recombination coefficients are 1.6937 and 0.1383, respectively. The mean RNMSE value for the trained coefficients (transport of electrons, trapping of electrons and holes along with recombination of electrons and holes) is 0.3675. The mean RNMSE value is significantly contributed due to the high value of the learned electron recombination coefficients near the anode (Voxel 100).

4 Machine Learning-Based Full Model of Detector in 3D

In Sect. 3, we introduced a novel physics-based machine learning (PBML) model to determine the material properties such as transport, trapping, detrapping, and recombination coefficients of electrons and holes at spatially discretized locations in the RTSD in 1D. In this Section, we extend the model to characterize the RTSD over a discretized sub-pixelated 3D volume. Our novel physics-based machine learning approach is the first to characterize a full 3D charge transport model of the RTSD. In this work, we first discretize the RTSD between pixelated electrodes spatially into three dimensions—x, y, and z. The resulting discretization is termed as voxels in 3D space. In each voxel, the different physics-based charge transport properties such as drift, trapping, detrapping, and recombination of charges are modeled as trainable model weights. The drift of the charges in 3D considers second-order nonlinear motion which is observed in practice with the RTSDs. Based on the electron-hole pair injections as input to the PBML model, and signals at the electrodes, free and trapped charges (electrons and holes) as outputs of the model, the PBML model determines the trainable weights by backpropagating the loss function. The trained weights of the model represent one-to-one relation to that of the actual physical charge transport properties in a voxelized detector.

4.1 Machine Learning-Based Physical Model in 3D

In RTSDs, electron and hole transport properties play a significant role in selecting detectors for end applications. When high-energy X-rays or γ photons hit the CdZnTe detector, it undergoes photoelectric effect, Compton scattering, or pair production and deposit energy in the detector creating electron-hole pairs in these processes. The Shockley-Ramo theorem states that the moving charge within a detector induces charge on the electrodes of the detector which are extracted [56]. The weighting potential when a point charge is at a particular location determines the charge (or current) induced in the electrodes. In our study, we consider a single large electrode (cathode) and a pixelated anode (11×11) as shown in Fig. 18. The weighting potential of the 121 anode pixels is not identical, and there are some variations in the weighting potential of central pixels, edge pixels, and corner pixels. Moreover, the weighting potential under one pixel also shows slight lateral variation [61].

Additionally, the variation of voltage along the volume of the detector is uniform in most of the region between anodes and the cathode. For those electrons and holes under one pixel, the charges will drift perpendicularly to their corresponding electrodes. However, the charges in between the anode pixels will have different drift behavior [61]. The nonuniformity of the electric field is shown in Fig. 19. In most of the region, the electric field is perpendicular to the cathode and anode pixels. However, near the anode, the electric field bends toward the nearest anode.

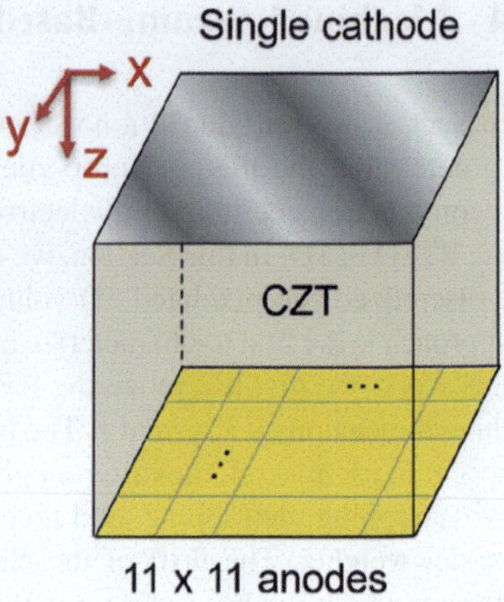

Fig. 18 Pixelated CdZnTe detector. The detector configuration consists of a single cathode and 11 × 11 pixelated anodes. Figure adapted with permission [35]

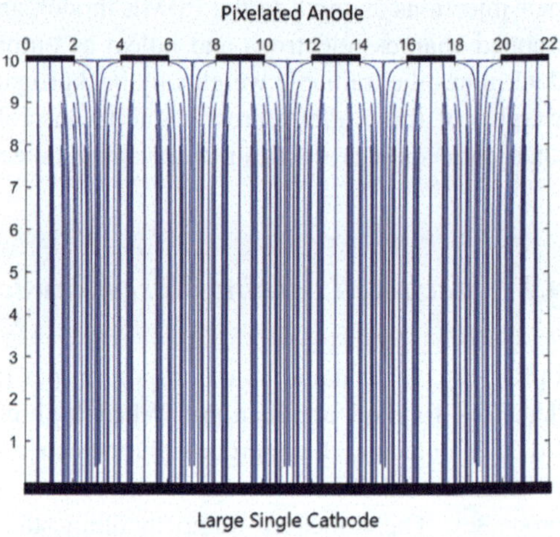

Fig. 19 Variation of electric field between the anode pixels and cathode (for GaAs crystal). The electric field becomes nonuniform near the anode pixels. Figure adapted with permission [35]

The nonuniform electric field in between the anode pixels is shown in Fig. 19. In the region of uniform electric field, the motion of electrons and holes will drift perpendicular to the electrodes. The nonuniform electric field under the gap will bend the trajectory of electrons closer to the nearest anode. The electrons in the gap between the anode pixels will be collected to the closest anode pixel.

3D Learning Model

The configuration of CdZnTe detector in our PBML model consists of a CdZnTe crystal with a single planar cathode on one end and several anode pixels (11 × 11) on the other end as shown in Fig. 20 (left). Out of the 121 anode pixels, we just use the cuboidal volume of the crystal under the central anode pixel and the cathode on the other end and subdivide this volume into voxels as shown in Fig. 20. In the x-direction (OX), we divide into M divisions in the y-direction (OY); we subdivide into N divisions and in z-direction (OZ); and we subdivide into P divisions. Thus,

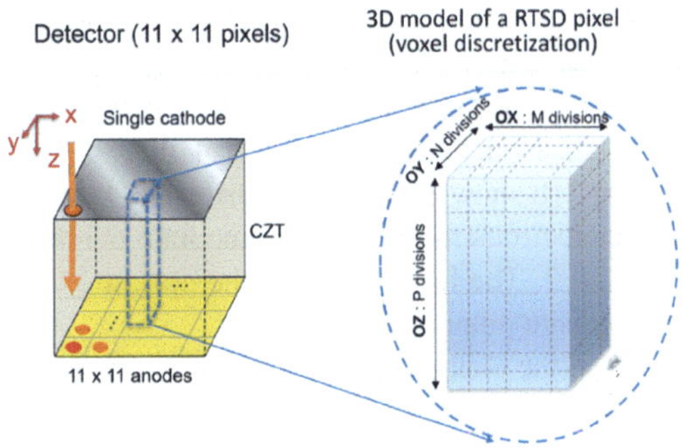

Fig. 20 CdZnTe with 121 pixelated anodes at one end and a single planar cathode on the opposite end (left). A single pixel is further discretized into subpixels (right). Figure adapted with permission [35]

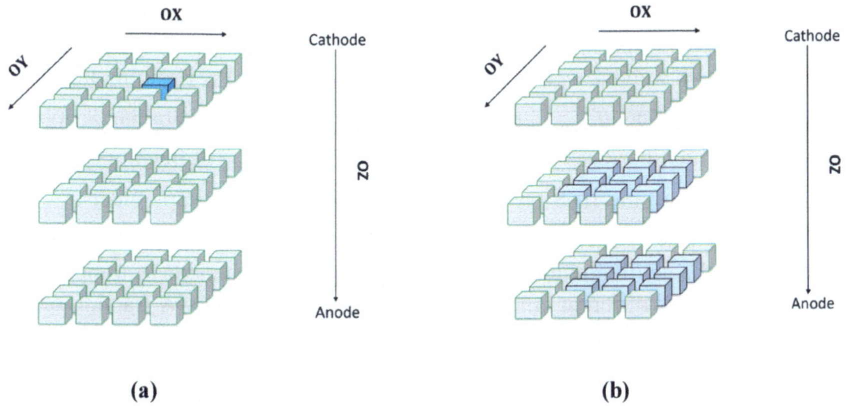

Fig. 21 (**a**) Electrons at a particular voxel position at time $t = 0$ and (**b**) electrons drift in the neighboring voxels at time $t = t_1$. Figure adapted with permission [35]

the total voxels are $M \times N \times P$. Since the single anode pixel is subdivided in x- and y-direction into M and N divisions, we consider the number of subpixels as $M \times N$.

The material properties of the CdZnTe for electrons and holes are considered at each voxel d. These are the drift coefficients for electrons and holes, electron trapping lifetime, electron detrapping lifetime, hole trapping 1 lifetime, hole detrapping lifetime, hole trapping 2 lifetime, hole detrapping 2 lifetime, electron lifetime, and hole lifetime, respectively. Each of the material properties is thus converted into $M \times N \times P$ discrete values.

The motion of charges is taken into account by considering nonuniform electric field between the electrodes (cathode and anode) as shown in Fig. 20. The charge transfer thus occurs not only in z-direction (OZ) which is perpendicular to the electrodes of opposite polarity but also in x- and y-direction (OX) and (OY) as well. Figure 21a shows electrons at a particular voxel at time $t = 0$. Due to the nonuniform electric field, the electrons move in 3D direction. However, the motion in z-direction (OZ) is much significant compared to the motion in the (OX) and (OY) directions, since the electric field is much stronger in the z-direction. When electron moves in z-direction, we assume the motion of the electrons in the x-y plane to be constrained in the immediate neighboring locations. For a voxel position of (i, j, k), the electron drifts to immediate neighboring locations of $(i+1, j)$, $(i-1, j)$, $(i, j+1)$, $(i, j-1)$, $(i+1, j+1)$, $(i-1, j-1)$, $(i+1, j-1)$, and $(i-1, j+1)$ in the x-y plane. Figure 21b shows the electron drift in 3D during a time step. From the z position of k, the electrons move in the immediate neighboring voxels having z position $k+1$ and $k+2$. The holes move in the opposite direction from anode to cathode due to its opposite polarity.

Once the charge, for example, holes drift from neighboring voxels to a voxel (i, j, k) at time t, it is added to the existing holes in that voxel from the previous time $t = t - 1$ to form net hole charges. Eq. (17) shows the phenomena of the drift of holes during the time t, forming total hole charge $q_{h,\text{int}}^{t,(i,j,k)}$ A fraction of this net hole charge then combines with the intrinsic electron concentration in bulk of the material at voxel (i, j, k) with weights $w_{hRec,\,(i,j,k)}$ as shown in Eq. (18). The resulting free holes after recombination are $q_{h,\text{int1}}^{t,(i,j,k)}$ as shown in Eq. (19). The concentration of holes in the r th trap center increases based on the fraction $w_{hT,\,r,\,(i,j,k)}$ of free holes $q_{h,\text{int1}}^{t,(i,j,k)}$ getting trapped and the fraction $w_{hD,\,r,\,(i,j,k)}$ of holes in that trap center $\tilde{q}_{h,r}^{t,(i,j,k)}$ which is detrapped in Eq. (20). We consider R trapping centers for holes. $q_{h,mob}^{t,(i,j,k)}$ in Eq. (21) is the net free/mobile holes considering the phenomena of trapping and detrapping in the R different hole trapping centers of the RTSD. A fraction of this free holes $q_{h,mob}^{t,(i,j,k)}$ drifts with weight $w_{h,\,(i,j,k)}$ to the neighboring voxels as shown in Eq. (22). Similar equations are also valid for electrons which now move from anode to cathode due to opposite polarity.

For illustrative purposes, Fig. 22 shows the operation in a voxel (4, 7, 3) at time t. At any time $t-1$, the charge in that voxel is $q_h^{t-1,(4,7,3)}$ for holes and $q_e^{t-1,(4,7,3)}$ for electrons. The electrons from adjacent voxels in 3D drift at time t and are added onto the existing charge. Some of the net electrons gets recombined in the bulk of the crystal with the intrinsic hole concentration in the bulk of the material based on the electron lifetime probability $w_{e,\,(4,7,3)}$. The electrons then get trapped with probability $w_{eT,\,(4,7,3)}$ to the trap level in the detector and detrap electrons back as excess electron concentration over bulk with a probability $w_{eD,\,(4,7,3)}$. Following the 3D electric field, a fraction of the electron charges may be left behind in the voxel (4,7,3), while the remaining fraction drifts to the neighboring voxels. Exactly same operations are repeated for holes, with the holes drifting from anode to cathode.

$$q_{h,\text{int}}^{t,(i,j,k)} = q_h^{t-1,(i,j,k)} + \Big(q_{h,o}^{t,(i+1,j,k)} + q_{h,o}^{t,(i,j+1,k)} + q_{h,o}^{t,(i,j,k+1)} + q_{h,o}^{t,(i+1,j,k+1)} + q_{h,o}^{t,(i,j+1,k+1)} + \cdots \Big), \tag{17}$$

$$q_{h,Rec}^{t,(i,j,k)} = w_{hRec,(i,j,k)} \times q_{h,\text{int}}^{t,(i,j,k)}, \tag{18}$$

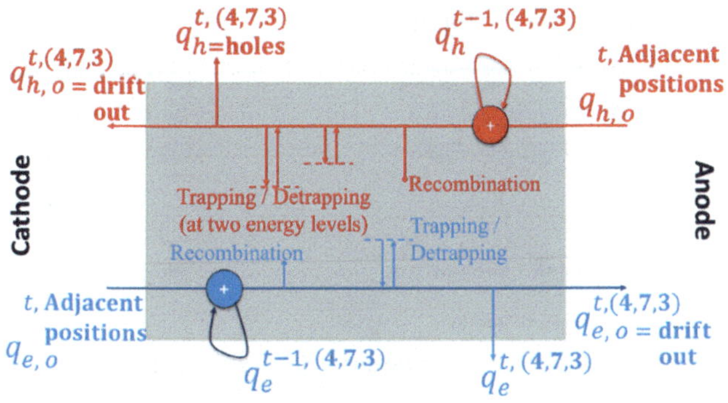

Fig. 22 Operation in voxel (4,7,3) at time t. Figure adapted with permission [35]

$$q_{h,\text{int1}}^{t,(i,j,k)} = q_{h,\text{int}}^{t,(i,j,k)} - q_{h,Rec}^{t,(i,j,k)}, \tag{19}$$

$$\tilde{q}_{h,r}^{t,(i,j,k)} = \tilde{q}_{h,r}^{t,(i,j,k)} + \left(w_{hT,r,(i,j,k)} \times q_{h,\text{int1}}^{t,(i,j,k)}\right) - \left(w_{hD,r,(i,j,k)} \times \tilde{q}_{h,r}^{t,(i,j,k)}\right), \tag{20}$$

$$\begin{aligned} q_{h,mob}^{t,(i,j,k)} &= q_{h,\text{int1}}^{t,(i,j,k)} \times \left(1 - w_{hRec,(i,j,k)} - \sum\nolimits_{r=1}^{R} w_{hT,r,(i,j,k)}\right) \\ &+ \sum\nolimits_{r=1}^{R} \left(\tilde{q}_{h,r}^{t,(i,j,k)} \times w_{hD,r,(i,j,k)}\right), \end{aligned} \tag{21}$$

$$q_{h,o}^{t,(i,j,k)} = w_{h,(i,j,k)} \times q_{h,mob}^{t,(i,j,k)}, \tag{22}$$

The mathematical operations in each voxel in this 3D PBML model are also described in simplified voxelized 1D models in the previous section. In our PBML model of charge transport, the voxels are connected in a bidirectional manner in 3D to form a complicated PBML model. A simplified representation of the model is shown in Fig. 23. The left of the model has a planar cathode, and the right consists of sub-pixelated anodes. The RTSD between the cathode and an anode pixel is subdivided into 3D voxels as shown with blue squares in Fig. 23. The input electron-hole charge pair injections can be in any of these voxels which depends on the position of interaction of high energy photons with the RTSD. The electrons drift toward the anode through the neighboring voxels in the z-direction (of Figs. 22 and 23) as shown by blue lines in the figure. On the other hand, the red lines show the drift of holes from one voxel to the neighboring voxels toward the cathode in the negative z-direction (of Figs. 22 and 23). The bidirectional connection shown by dotted lines indicates the movement of electrons and holes in the lateral direction (OX and OY) which basically represents lateral (second order) drift of charges in x-y plane.

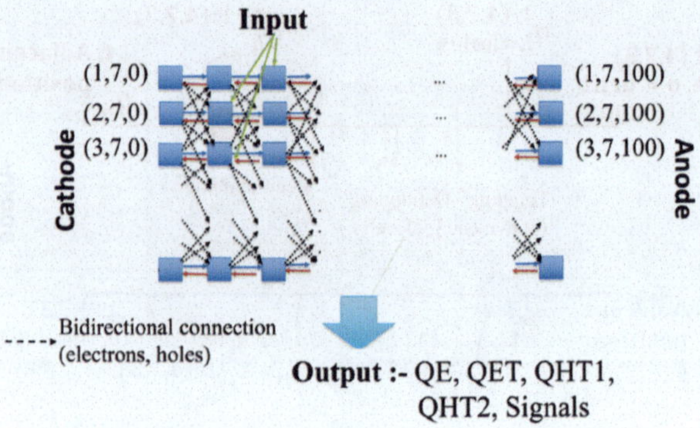

Fig. 23 Interconnection among voxels in the 3D model. Each blue squares represents a voxel shown in Fig. 22. Figure adapted with permission [35]

The model is trained with the input-output pairs of data. The input data is the positions of injected electron-hole pair, and the output data are the signals obtained at the electrodes along with the electrons and holes (free and trapped) in each of the voxels over time. The weighting potential as defined by the electrode and detector configuration is nonuniform over the sub-pixels. During training the model, the loss function is computed as the sum of the squared errors between the signals at the electrodes and charges in the voxels compared to the ground truth along with the $error^2_{voltage}$. The overall loss function for training this model is shown in Eq. (7). The loss function is shown for Z trapping centers for electrons and R trapping centers for holes for a general model with several trapping centers for electrons and holes. However, we perform simulation experiments considering CZT detector with two trapping centers for holes and one trapping center for electrons [14, 31]. In the loss function, the errors due to the signals and voltage are grouped together; free and trapped electron charges are grouped together; and free and trapped charges due to holes are grouped together with weighting terms k, l, and n, respectively. Clearly, as k, l, and/or n are varied, the errors due to those terms vary. The higher the value of these parameters, the lower the errors associated with those terms. In these error terms, the subscript gt for a particular parameter X (for instance, X is *signal* or q_e) refers to the ground truth data for that parameter X generated in MATLAB using the classical equations, and the subscript L for the same parameter X refers to the data generated by the PBML model.

$$\begin{aligned} LF = & \, k\left[\left(signal_{gt} - signal_L\right)^2 + error^2_{voltage}\right] \\ & + l\left[\left(q_{e,gt} - q_{e,L}\right)^2 + \sum\nolimits_{z=1}^{z}\left(q_{e,z,gt} - q_{e,z,L}\right)^2\right] \\ & + n\left[\left(q_{h,gt} - q_{h,L}\right)^2 + \sum\nolimits_{r=1}^{R}\left(q_{ht,r,gt} - q_{ht,r,L}\right)^2\right], \end{aligned} \tag{23}$$

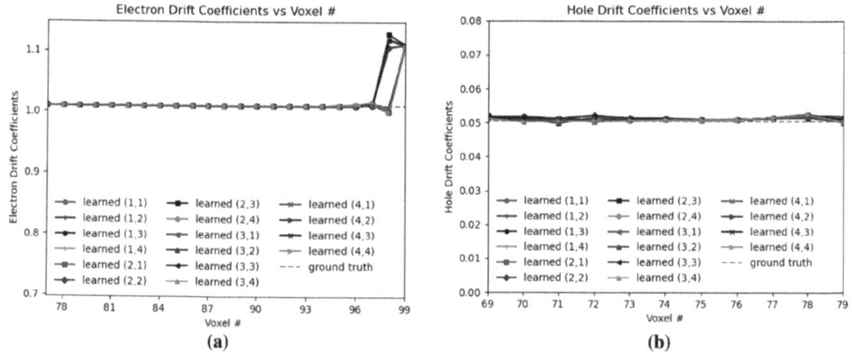

Fig. 24 Electron and hole drift coefficients in 3D physical model for e-h injections at OZ positions 77 and 79 for all sub-pixels. (**a**) Electron drift coefficients and (**b**) hole drift coefficients. Figure adapted with permission [35]

Numerical Experiments for All 3D Sub-pixels

The experimental results have been shown in 3D for all sub-pixels within a pixel. The central pixel volume as shown in Fig. 20 (right) has been divided into 4 × 4 sub-pixels. The simulation data has been generated with anode pixel divided into 12 × 12 sub-pixels in OX and OY, as mentioned earlier in this Section. The data (signals, free and trapped charges, electrode weighting potentials) for the 4 × 4 sub-pixels has been converted from 12 × 12 sub-pixels by simply taking the arithmetic mean of the corresponding data (signals, free and trapped charges, electrode weighting potentials) as 3 × 3 sub-pixels in OX and OY in a nonoverlapping manner.

Figure 24a, b shows the variation of electron and hole drift coefficients for e-h injection at OZ positions of 77 and 79 for all the sub-pixels. The electron drift coefficients are learned from OZ position 77 to 99, and the hole drift coefficients are learned from OZ position 69 to 79. For the electron drift coefficients, $Err(w_{eDrift})$ is 0.0117, while for the hole drift coefficients, $Err(w_{hDrift})$ is 0.0152. We observe mismatch in the learned electron drift coefficient near the anode compared to the ground truth and a slight difference in the learned coefficients of voxel 78 and other voxels for the hole drift coefficient compared to the ground truth.

Figure 25a, b shows the electron trapping and detrapping coefficients for e-h injection at OZ positions of 77 and 79 for all the sub-pixels. The electron coefficients are learned from the leftmost point of e–h injection up to position 99 in OZ direction. For the electron trapping and detrapping coefficients, the errors *Err*( ) are of the similar order, $Err(w_{eT})$ is 0.0696, and $Err(w_{eD})$ is 0.0560. Similarly, Fig. 26a, b shows the hole trapping 1 and detrapping 1 coefficients. For hole trapping 1 and detrapping 1 coefficients, the errors are of the similar order, $Err(w_{hT1})$ is 0.0228, and $Err(w_{hD1})$ is 0.0195, respectively. The hole coefficients are learned from OZ position 69 to position 79 for all the sub-pixels. Hole Trapping 2 and Detrapping 2 coefficients are shown in Fig. 27a, b, respectively. For hole trapping 2 and detrapping 2 coefficients, $Err(w_{hT2})$ is 0.0115, and $Err(w_{hD2})$ is 0.0047,

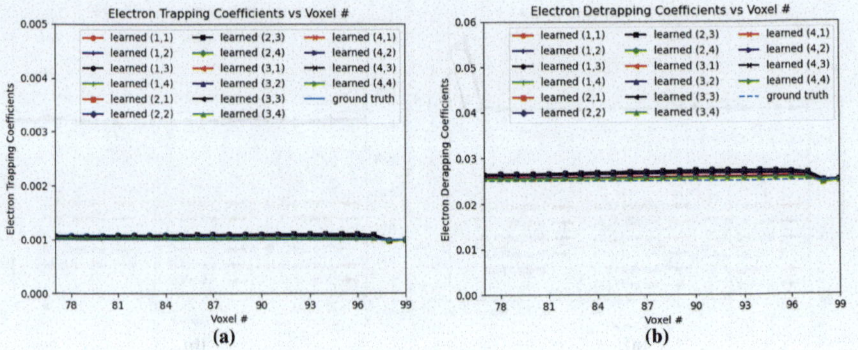

Fig. 25 Electron trapping and detrapping coefficients in 3D physical model for e-h injections at OZ positions 77 and 79 for all sub-pixels. (**a**) Electron trapping coefficients and (**b**) electron detrapping coefficients. Figure adapted with permission [35]

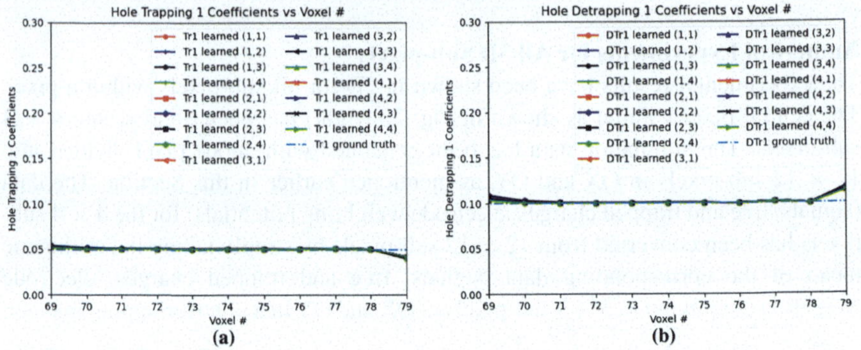

Fig. 26 Hole trapping 1 and detrapping 1 coefficients in 3D physical model for e-h injections at OZ positions 77 and 79 for all sub-pixels. (**a**) Hole trapping 1 coefficients and (**b**) hole detrapping 1 coefficients. Figure adapted with permission [35]

respectively. It is clear from Fig. 27 that $Err(w_{hT2})$ convergence to the ground truth is inferior to $Err(w_{hD2})$. For both the hole trapping 1 and 2 along with hole detrapping 1 and 2 coefficients, the trained weights converge to the ground truth values.

Figure 28a, b shows the electron and hole recombination coefficients, respectively. For the electron recombination coefficients, the learned coefficients are close to the ground truth at positions close to the injection OZ, while the errors increased as it went further away from the positions of injection toward the anode. For electron recombination coefficients, $Err(w_{eRec})$ is 0.0347 for OZ position 77 to 92. For injection positions 77 and 79, the electron recombination coefficients do not converge beyond voxel 92 due to decrease in gradients in those voxels. For hole recombination coefficients as well, similar effect can be observed, where the learned coefficients from OZ positions of 69 to 79 deviate slightly from the ground

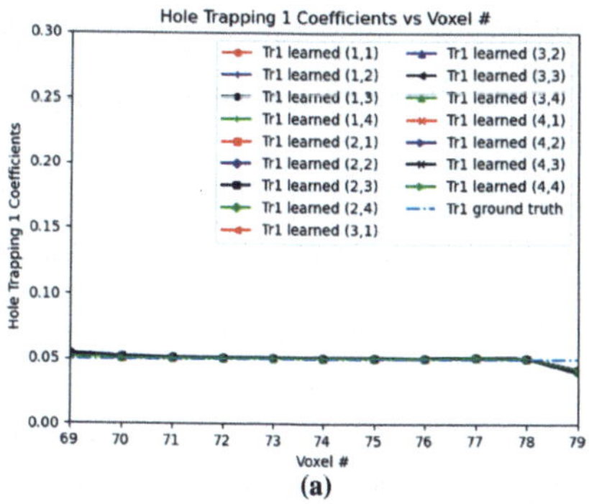

Fig. 27 Hole trapping 2 and detrapping 2 coefficients in 3D physical model for e-h injections at OZ positions 77 and 79 for all sub-pixels. (**a**) Hole trapping 2 coefficients and (**b**) hole detrapping 2 coefficients. Figure adapted with permission [35]

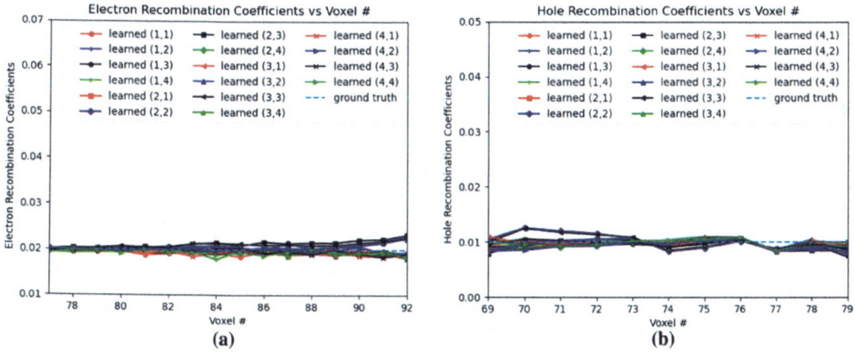

Fig. 28 Electron and hole recombination coefficients in 3D physical model for e-h injections at OZ positions 77 and 79 for all sub-pixels. (**a**) Electron recombination coefficients and (**b**) hole recombination coefficients. Figure adapted with permission [35]

truth values at OZ positions farther away from the OZ injection positions of 77 and 79. For the hole recombination coefficients, $Err(w_{hRec})$ is 0.0873. It is apparent from Fig. 28 that the error $Err(w_{hRec})$ in the hole recombination coefficient with respect to the ground truth is inferior compared to that of the electron recombination coefficients $Err(w_{eRec})$. For all of these learned material properties using the PBML approach, the mean $Err()$ is 0.0333.

5 Discussion and Conclusion

We demonstrate how a novel combination of machine learning along with physics knowledge of charge transport can be used to characterize the room-temperature semiconductor detector (RTSD) at a much finer scale spatially. This is vital for developing more sophisticated algorithms and its implementation in large-scale medical, industrial, and defense applications.

We consider pixelated configuration of RTSD in demonstrating our physics-based machine learning approach, but this method can be extended to other electrode configurations as well. The RTSD crystal is virtually discretized into small volumes or voxels where the physical laws governing the electron and hole charge motion in the RTSD hold true. A learning-based model has been developed by incorporating these physical laws—transport, trapping, detrapping, and recombination of electron and hole charges for each of these voxels. The electron-hole charge pair is injected at one or multiple positions, and the outputs of the model are the signals at the electrodes, along with free and trapped charges in the crystal. The CZT detector in this model is considered to have one trapping center for electrons and two for holes. However, this physics-based machine learning approach can be extended to any number of trapping centers for electrons and holes. Once trained on the ground truth data, the learning model has the trained weights which are the material properties in terms of drift, trapping, detrapping, and recombination coefficients for both electrons and holes. We study the effects of prioritizing the different terms of the loss function and observe improvements in the convergence properties of the trainable weights of the model.

The data required to train the learning-based models must be generated in real world using several experimental setups and expertise. This is not repeatable and scalable for numerous RTSD crystals which are required for several applications. To address this issue, several reduced order models have been highlighted in this chapter, which requires fewer data for training the model. Additionally, often in the RTSD, some of the charge transport properties may be present, while some other properties may be absent. We show that a generalized physics-based machine learning modeling approach possesses the capability to directly identify those properties which are only present in the RTSD and the ones which are absent. Apart from the effectiveness of this novel physics-based machine learning approach in characterizing the RTSD in 1D, we show the effectiveness of this approach for 3D discretization of the RTSD as well. This characterization of the RTSD properties in 3D opens the identification of defects in further higher spatial resolution, aiding in development of advanced event-based imaging algorithms.

There are a few limitations of our physics-based machine learning approach in characterizing RTSDs. Our model has been developed by considering a forward physics model (classical model). However, there are other physical equations which describe the charge transport in RTSDs. Although our physics-based machine learning methodology for finding the material properties in micron scale is still valid for those forward models (classical models), the detailed implementation in that case needs to be carried out. Additionally, since this physics-based learning model has been based on equations used in a classical model, the physics-based learning model has the risks of "inverse crime" [62] when the same (or nearly the same) theoretical ingredients are employed to synthesize as well as invert data in an inverse problem. To avoid this pitfall, one way could be to use a relaxed theoretical model compared to the one exactly used as the forward model. Alternatively, a traditional machine learning or deep learning model can be used in conjunction with the physics-based learning model. In this work, the machine learning-based models have been developed without considering noise such as Poisson noise, Quantization noise, Sampling Noise, or other Statistical noise and actual experimental data from sensor hardware electronics. Developing models incorporating this information a priori is one of the future research directions. This physics-based machine learning also needs to be trained based on training data on GPUs. Although the training time is a few hours, this also needs to be accounted for while implementing this model. However, this drawback exists in mostly all deep learning and artificial intelligence models, but since our physics-based model uses far less training weights, the training time is faster than conventional deep learning models and faster than the classical approaches using non-deep learning methods or experimental approaches.

Overall, the chapter presents a novel approach of physics-based machine learning model for characterizing RTSDs. This work has been discussed in more details in [32–41, 63]. Our approach in characterization of the RTSD is the first novel approach in this field. Additionally, we have used most of the knowledge of the physical world in this model, some of which may vary from crystal to crystal. We have demonstrated several cases of this model—with few data, without a priori knowledge of the RTSD, 1D and 3D discretization of the RTSD. Development of learning-based models which use fewer physical laws, which can infer the physics from data, is another interesting approach which is within the scope of future work. Incorporating the sensor models to image reconstruction and reuse of imperfect RTSD sensors is one of the applications for the learning-based physical model. Our learning-based model can be extended to sub-pixel imaging and improve the quality of image reconstruction.

References

1. C. Scheiber, G.C. Giakos, Medical applications of CdTe and CdZnTe detectors. Nucl. Inst. Methods Phys. Res. A **458**(1–2), 12–25 (2001)
2. T.E. Schlesinger, J.E. Toney, H. Yoon, E.Y. Lee, B.A. Brunett, L. Franks, R.B. James, Cadmium zinc telluride and its use as a nuclear radiation detector material. Mater. Sci. Eng. R: Rep. **32**(4–5), 103–189 (2001)
3. A.P.H. Butler, N.G. Anderson, R. Tipples, N. Cook, R. Watts, J. Meyer, A.J. Bell, T.R. Melzer, P.H. Butler, Bio-medical X-ray imaging with spectroscopic pixel detectors. Nucl. Inst. Methods Phys. Res. A **591**(1), 141–146 (2008)
4. S.D. Sordo, L. Abbene, E. Caroli, A.M. Mancini, A. Zappettini, P. Ubertini, Progress in the development of CdTe and CdZnTe semiconductor radiation detectors for astrophysical and medical applications. Sensors **9**(05), 3491–3526 (2009)
5. P.M. Johns, J.C. Nino, Room temperature semiconductor detectors for nuclear security. J. Appl. Phys. **126**(4), 040902 (2019)
6. U.N. Roy, G.S. Camarda, Y. Cui, R. Gul, G. Yang, J. Zazvorka, V. Dedic, J. Franc, R.B. James, Evaluation of CdZnTeSe as a high-quality gamma-ray spectroscopic material with better compositional homogeneity and reduced defects. Sci. Rep. **9**(1), 7303 (2019)

7. N. Zhang, A. Yeckel, A. Burger, Y. Cui, K.G. Lynn, J.J. Derby, Anomalous segregation during electrodynamic gradient freeze growth of cadmium zinc telluride. J. Cryst. Growth **325**(1), 10–19 (2011)
8. A.E. Bolotnikov, G.S. Camarda, Y. Cui, G. Yang, A. Hossain, K. Kim, R.B. James, Characterization and evaluation of extended defects in CZT crystals for gamma-ray detectors. J. Cryst. Growth **379**, 46–56 (2013)
9. G.A. Carini, A.E. Bolotnikov, G.S. Camarda, R.B. James, High-resolution X-ray mapping of CdZnTe detectors. Nucl. Inst. Methods Phys. Res. A **579**(1), 120–124 (2007)
10. M. Amman, J.S. Lee, P.N. Luke, Electron trapping nonuniformity in high-pressure-Bridgman-grown CdZnTe. J. Appl. Phys. **92**(6), 3198–3206 (2002)
11. G.S. Camarda, A.E. Bolotnikov, Y. Cui, A. Hossain, S.A. Awadalla, J. Mackenzie, H. Chen, R.B. James, Polarization studies of cdznte detectors using synchrotron x-ray radiation. IEEE Trans. Nucl. Sci. **55**(6), 3725–3730 (2008)
12. U.N. Roy, G.S. Camarda, Y. Cui, R.B. James, Advances in CdZnTeSe for radiation detector applications. Radiat. Ther. **1**(2), 123–130 (2021)
13. M.C. Veale, P. Booker, S. Cross, M.D. Hart, L. Jowitt, J. Lipp, A. Schneider, P. Seller, R.M. Wheater, M.D. Wilson, C.C.T. Hansson, Characterization of the uniformity of high-flux CdZnTe material. Sensors **20**(10), 2747 (2020)
14. E.Y. Lee, R.B. James, R.W. Olsen, H. Hermon, Compensation and trapping in CdZnTe radiation detectors studied by thermoelectric emission spectroscopy, thermally stimulated conductivity, and current-voltage measurements. J. Electron. Mater. **28**(6), 766–773 (1999)
15. G.C. Tepper, R. Kessick, R.B. James, L. Van den Berg, Contactless measurements of charge traps and carrier lifetimes in detector-grade cadmium zinc telluride and mercuric iodide, in *Hard X-Ray, Gamma-Ray, and Neutron Detector Physics II*, vol. 4141, (SPIE, 2000, November), pp. 76–88
16. Z. Medunić, Ž. Pastuović, M. Jakšić, N. Skukan, Studying of trap levels by the use of focused ion beams. Nucl. Inst. Methods Phys. Res. B **231**(1–4), 486–490 (2005)
17. M. Pavlović, M. Jakšić, H. Zorc, Z. Medunić, Identification of deep trap levels from thermally stimulated current spectra of semi-insulating CdZnTe detector material. J. Appl. Phys. **104**(2), 023525 (2008)
18. M.L. Rodrigues, Z. He, High-flux experiments and simulations of pulse-mode 3d-position-sensitive CdZnTe pixelated detectors, in *2011 IEEE Nuclear Science Symposium Conference Record*, (IEEE, 2011, October), pp. 4677–4688
19. R.M. Blakney, H.P. Grunwald, Small-signal current transients in insulators with traps. Phys. Rev. **159**(3), 658 (1967)
20. M. Jung, J. Morel, P. Fougeres, M. Hage-Ali, P. Siffert, A new method for evaluation of transport properties in CdTe and CZT detectors. Nucl. Inst. Methods Phys. Res. A **428**(1), 45–57 (1999)
21. M. Prokesch, D.S. Bale, C. Szeles, Fast high-flux response of CdZnTe x-ray detectors by optical manipulation of deep level defect occupations. IEEE Trans. Nucl. Sci. **57**(4), 2397–2399 (2010)
22. G.C. Tepper, R. Kessick, C. Szeles, Investigation of the electronic properties of cadmium zinc telluride surfaces using pulsed laser microwave cavity perturbation, in *Hard X-Ray and Gamma-Ray Detector Physics III*, vol. 4507, (SPIE, 2001, December), pp. 79–89
23. Q. Zheng, F. Dierre, J. Franc, J. Crocco, H. Bensalah, V. Corregidor, E. Alves, E. Ruiz, O. Vela, J.M. Perez, E. Dieguez, Investigation of generation of defects due to metallization on CdZnTe detectors. J. Phys. D. Appl. Phys. **45**(17), 175102 (2012)
24. J. Willard, X. Jia, S. Xu, M. Steinbach, V. Kumar, Integrating scientific knowledge with machine learning for engineering and environmental systems. ACM Comput. Surv. **55**(4), 1–37 (2022)
25. G.E. Karniadakis, I.G. Kevrekidis, L. Lu, P. Perdikaris, S. Wang, L. Yang, Physics-informed machine learning. Nat. Rev. Phys. **3**(6), 422–440 (2021)
26. Y. Khoo, J. Lu, L. Ying, Solving for high-dimensional committor functions using artificial neural networks. Res. Math. Sci. **6**, 1–13 (2019)

27. J. Han, A. Jentzen, E, W., Solving high-dimensional partial differential equations using deep learning. Proc. Natl. Acad. Sci. **115**(34), 8505–8510 (2018)
28. T.W. Hughes, I.A. Williamson, M. Minkov, S. Fan, Wave physics as an analog recurrent neural network. Sci. Adv. **5**(12), eaay6946 (2019)
29. L. Lu, P. Jin, G. Pang, Z. Zhang, G.E. Karniadakis, Learning nonlinear operators via DeepONet based on the universal approximation theorem of operators. Nat. Mach. Intell. **3**(3), 218–229 (2021)
30. S. Markidis, The old and the new: Can physics-informed deep-learning replace traditional linear solvers? Front. Big Data, 92 (2021)
31. M. Rodrigues. High-flux experiments and simulations of pulse-mode position-sensitive CdZnTe pixelated detectors. PhD thesis, University of Michigan (2012)
32. S. Banerjee, M. Rodrigues, A.H. Vija, A.K. Katsaggelos, A learning-based physical model of charge transport in room-temperature semiconductor detectors. IEEE Trans. Nucl. Sci. **69**(1), 2–16 (2021)
33. S. Banerjee, M. Rodrigues, M. Ballester, A.H. Vija, A.K. Katsaggelos, Learning-based physical models of room-temperature semiconductor detectors with reduced data. Sci. Rep. **13**(1), 168 (2023)
34. S. Banerjee, M. Rodrigues, M. Ballester, A.H. Vija, A. Katsaggelos, Identifying defects without a priori knowledge in a room-temperature semiconductor detector using physics inspired machine learning model. Sensors **24**(1), 92 (2023)
35. S. Banerjee, M. Rodrigues, M. Ballester, A.H. Vija, A.K. Katsaggelos, A physics based machine learning model to characterize room temperature semiconductor detectors in 3D. Sci. Rep. **14**(1), 7803 (2024)
36. S. Banerjee, M. Rodrigues, M. Ballester, A.H. Vija, A.K. Katsaggelos, Machine learning approaches in room temperature semiconductor detectors, in *X-ray Photon Processing Detectors: Space, Industrial, and Medical Applications*, (Springer, Cham, 2023), pp. 67–94
37. M. Ballester, J. Kaspar, F. Massanes, S. Banerjee, A.H. Vija, A.K. Katsaggelos, Modeling and Simulation of Charge-Induced Signals in Photon-Counting CZT Detectors for Medical Imaging Applications (2024). arXiv preprint arXiv:2405.13168
38. S. Banerjee, M. Rodrigues, Siemens Medical Solutions USA Inc, Space time electron-hole charge transport network for solid-state material studies. U.S. Patent 11,480,608 (2022)
39. S. Banerjee, M. Rodrigues, Siemens Medical Solutions USA Inc, Inverse neural network for particle detection in a solid-state-devices. U.S. Patent Application 16/850,306 (2021)
40. A.H. Vija, M. Rodrigues, S. Banerjee, A. Katsaggelos, Northwestern University and Siemens Medical Solutions USA Inc, Enhancement of weak signal for machine training neural network representing a solid-state detector. U.S. Patent Application 17/302,718 (2022)
41. S. Banerjee, M. Rodrigues, A.H. Vija, A. Katsaggelos, Northwestern University and Siemens Medical Solutions USA Inc, Solid-state detector characterization by machine learning-based physical model with reduced defect levels. U.S. Patent Application 17/938,686 (2024)
42. E. Diéguez, Growth of $Cd_{0.9}Zn_{0.1}Te$ bulk crystals (2011)
43. W.T.R.W. Shockley, W.T. Read Jr., Statistics of the recombinations of holes and electrons. Phys. Rev. **87**(5), 835 (1952)
44. R.N. Hall, Electron-hole recombination in germanium. Phys. Rev. **87**(2), 387 (1952)
45. M.L. Rodrigues, Z. He, Properties and spectroscopic performance of semiconductor detectors under high-flux irradiation, in *Medical Applications of Radiation Detectors*, vol. 8143, (SPIE, 2011, September), pp. 51–61
46. T.H. Prettyman, Method for mapping charge pulses in semiconductor radiation detectors. Nucl. Inst. Methods Phys. Res. A **422**(1–3), 232–237 (1999)
47. J. Durst, P. Bartl, B. Kreisler, T. Michel, G. Anton, Monte Carlo simulation of pixelated photon counting X-ray detectors like the Medipix2 and the Medipix3 using high-Z sensor materials, in *2008 IEEE Nuclear Science Symposium Conference Record*, (IEEE, 2008, October), pp. 4990–4994
48. M. Picone, A. Glière, P. Massé, A three-dimensional model of cdznte gamma-ray spectrometer. Nucl. Inst. Methods Phys. Res. A **504**(1–3), 313–316 (2003)

49. S.M. Sze, Y. Li, K.K. Ng, *Physics of semiconductor devices* (Wiley, 2021)
50. M. Prokesch, CdZnTe for gamma and x-ray applications, in *Solid-State Radiation Detectors: Technology and Applications*, (2015), pp. 17–48
51. A. Musiienko, J. Pipek, P. Praus, M. Brynza, E. Belas, B. Dryzhakov, D. Mao-Hua, M. Ahmadi, R. Grill, Deciphering the effect of traps on electronic charge transport properties of methylammonium lead tribromide perovskite. Sci. Adv. **6**(37), eabb6393 (2020)
52. A. Musiienko, R. Grill, P. Moravec, G. Korcsmáros, M. Rejhon, J. Pekárek, H. Elhadidy, L. Šedivý, I. Vasylchenko, Dual-wavelength photo-Hall effect spectroscopy of deep levels in high resistive CdZnTe with negative differential photoconductivity. J. Appl. Phys. **123**(16), 161502 (2018)
53. W. Shockley, Currents to conductors induced by a moving point charge. J. Appl. Phys. **9**(10), 635–636 (1938)
54. S. Ramo, Currents induced by electron motion. Proc. IRE **27**(9), 584–585 (1939)
55. G.F. Knoll, *Radiation Detection and Measurement* (Wiley, 2010)
56. Z. He, Review of the Shockley–Ramo theorem and its application in semiconductor gamma-ray detectors. Nucl. Inst. Methods Phys. Res. A **463**(1–2), 250–267 (2001)
57. P.J. Werbos, Backpropagation through time: what it does and how to do it. Proc. IEEE **78**(10), 1550–1560 (1990)
58. D.E. Rumelhart, G.E. Hinton, R.J. Williams, *Learning Internal Representations by Error Propagation* (California University/La Jolla Institute for Cognitive Science, San Diego, 1985)
59. D.P. Kingma, J. Ba, Adam: A method for stochastic optimization (2014). arXiv preprint arXiv:1412.6980
60. M. Abadi, A. Agarwal, P. Barham, E. Brevdo, Z. Chen, C. Citro, G.S. Corrado, A. Davis, J. Dean, M. Devin, S. Ghemawat, Tensorflow: Large-scale machine learning on heterogeneous distributed systems (2016). arXiv preprint arXiv:1603.04467
61. Z. Chen, Y. Zhu, Z. He, Intrinsic photopeak efficiency measurement and simulation for pixelated CdZnTe detector. Nucl. Inst. Methods Phys. Res. A **980**, 164501 (2020)
62. A. Wirgin, The inverse crime (2004). arXiv preprint math-ph/0401050
63. S. Banerjee, Learning based intelligent imaging. PhD thesis, Northwestern University (2022)

Enhanced 3D X-Ray Tomography: Deep Learning–Based Advanced Algorithms for Fiber Instance Segmentation

Vijay Kumar Pediredla, Mengqi Fang, Yankai Cao, Andre Phillion, Mark Martinez, and Bhushan Gopaluni

1 Introduction

In the realm of the pulp and paper industry, the meticulous craft of producing high-quality paper handsheets relies heavily on both the inherent properties of the pulp fibers and the operational parameters of the pulping process [1]. Understanding the intricate interplay between the operational variables and their influence on the final product's fibrous structure is crucial for ensuring the production of paper products with desired characteristics. 3D X-ray tomography has emerged as a powerful tool for studying the microstructure of materials, including paper handsheets. This nondestructive imaging technique enables the visualization of internal structures with high resolution, providing valuable insights into fiber arrangement and distribution. Also, by correlating the fibrous structure with process conditions, operational optimization and improved product quality are achieved [2, 3]. However, achieving this requires thorough analysis and interpretation of fiber characteristics and operating parameters and necessitates adjusting process conditions.

In recent years, there has been a noticeable trend toward leveraging advanced image processing techniques, particularly utilizing 3D X-ray tomography coupled with deep learning algorithms, to analyze the microstructure of paper handsheets. The deep learning algorithms were built upon the principles of artificial neural

The original version of the chapter has been revised. A correction to this chapter can be found at https://doi.org/10.1007/978-3-031-75653-5_13

V. K. Pediredla (✉) · M. Fang · Y. Cao · B. Gopaluni
UBC, Vancouver, Canada
e-mail: vijay.pediredla@ubc.ca

A. Phillion · M. Martinez
Department of Materials Science and Engineering, McMaster University, Hamilton, ON, Canada

K. (K.) Iniewski, L. (K.) Cai (eds.), *Deep Learning for Advanced X-ray Detection and Imaging Applications*, https://doi.org/10.1007/978-3-031-75653-5_9

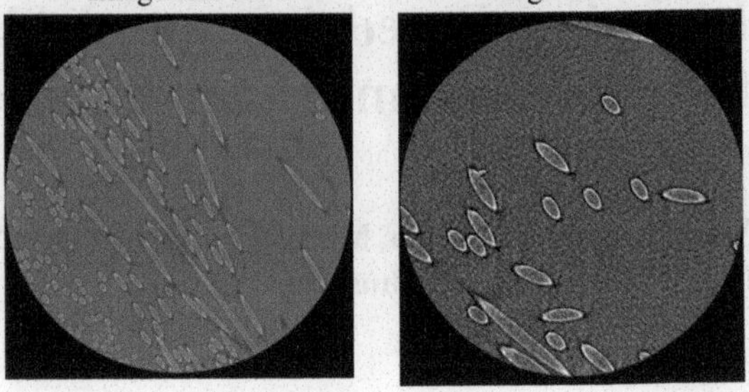

Fig. 1 Nylon fiber bundle tomographic data sets

networks and possess unparalleled capabilities in discerning intricate patterns and features within visual data. The goal is to capitalize on these capabilities to understand the complex structural properties of fibers embedded within paper handsheets. These algorithms have proven pivotal in the new era of precision and efficiency in image analysis, offering highly accurate models that have transformed various applications, including image segmentation. This work aims to harness the power of these cutting-edge algorithms to delve into the intricate properties of fibers within paper handsheets. Two datasets of Nylon bundle fiber tomographic images were utilized in this study, as illustrated in Fig. 1. Data set 1 is large, fully labeled, and annotated, making it threshold friendly. On the other hand, data set 2 is the small, target set, unlabeled, and thus is not threshold friendly, posing challenges for segmentation. The proposed method in this study is applied to data set 2 to demonstrate its effectiveness.

2 Background: Image Segmentation Methods

Image segmentation is a process that involves splitting an image into multiple segments to extract meaningful information. Over the years, advancements in image segmentation techniques have brought about significant changes in various fields, from medical imaging to autonomous driving. These techniques enable the automated analysis and comprehension of visual data by accurately delineating regions of interest through various software tools. Object detection identifies a bounding box corresponding to a class in the image, whereas image segmentation constructs a pixel-wise representation corresponding to an object in the image. This distinction allows for a much deeper understanding of image segmentation, addressing issues with greater precision and leading to more significant outcomes. Additionally, with the developments in deep learning algorithms and computational capabilities, convolutional neural networks (CNNs) are becoming the foundation for modern segmentation methods, offering improved accuracy and efficiency [4]. Image segmentation is divided broadly into two categories, one is semantic segmentation, and the other is instance segmentation.

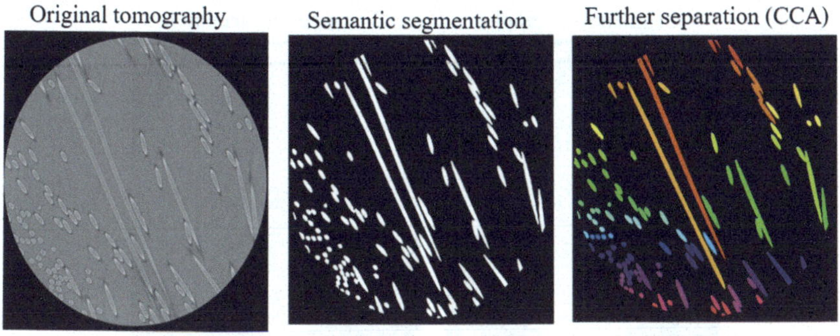

Fig. 2 Semantic segmentation of the image data set 1

2.1 *Semantic Segmentation*

The semantic segmentation method aims to allocate semantic labels to each pixel in an image, thus providing a pixel-wise classification of objects and regions. This approach is critical in tasks such as scene understanding, where the ability to distinguish between different object categories is paramount. State-of-the-art semantic segmentation models, such as U-Net [5] and DeepLab [6], leverage deep learning architectures to achieve remarkable accuracy. For instance, the U-Net architecture demonstrates exceptional performance in segmenting biomedical images [5]. Similarly, the DeepLab model showcases the effectiveness of combining CNNs with conditional random fields (CRFs) for semantic segmentation tasks [6]. To further refine the segmentation results, connected components analysis (CCA) with labeling is employed to detect individual panels in images and convert segment features to fusion decisions [7]. Integrating semantic segmentation with CCA is applied to the Nylon bundle image data set 1, as shown in Fig. 2.

2.2 *Instance Segmentation*

Instance segmentation extends semantic segmentation by not only classifying pixels but also describing individual objects within each class. This fine-grained segmentation is particularly useful in scenarios where precise object boundaries need to be identified, such as object detection and tracking. Mask R-CNN [8], a popular instance segmentation framework, integrates region-based convolutional neural networks with mask prediction branches. The Mask R-CNN introduces this architecture, demonstrating its efficacy in accurately segmenting objects in images with complex backgrounds. To further enhance the performance, the Feature Pyramid Network (FPN) [9] is employed, utilizing a top-down network path incorporating lateral connections. This approach facilitates the extraction of features

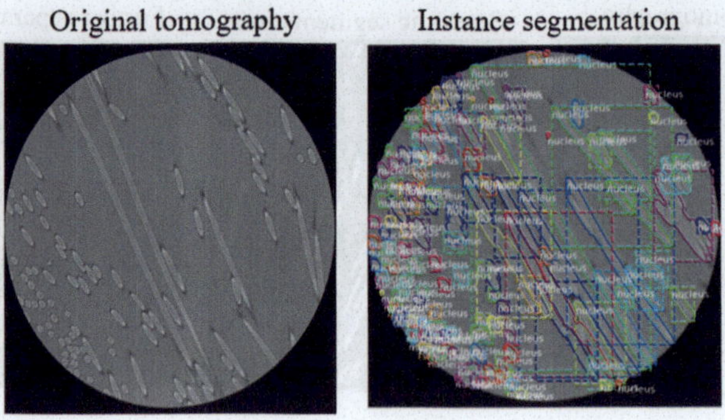

Fig. 3 Instance segmentation of the image data set 1

in a staged manner. Additionally, datasets like Microsoft's Common Objects in Context (COCO) [10] capture instances with intricate spatial layouts, further enriching the potential for advancements in image processing methodologies. Instance segmentation is applied to the Nylon bundle image data set 1, as shown in Fig. 3.

In summary, the selection between semantic and instance segmentation hinges on the unique demands of the application [11]. Semantic segmentation offers a comprehensive view of the scene by categorizing pixels into classes, while instance segmentation provides a more detailed depiction of individual objects. When prioritizing object-level analysis, instance segmentation proves advantageous; however, for broader image comprehension tasks, semantic segmentation may suffice. Ultimately, choosing the segmentation approach should be guided by the specific objectives and constraints of the application at hand.

3 Image Segmentation Algorithms for Fiber Analysis

In the context of pulp fiber analysis, image segmentation plays a crucial role in characterizing nylon fiber bundles and paper pulp. By accurately understanding individual fibers within bundles, segmentation algorithms facilitate quantitative analysis of fiber properties such as length, width, coarseness, and orientation. This information is vital for understanding the structural properties of paper pulp and optimizing the manufacturing process [1, 2]. In [12], the significance of image segmentation in analyzing fiber bundle scale for non-crimp fabric-reinforced composites is elaborated. A graph-based segmentation method [13] is developed and validated on a synthetic fiber data set to identify the individual fibers accurately. A novel three-step algorithm is introduced to isolate individual papermaking fibers in μCT images of paper handsheets [14]. Also, watershed transformation algorithms are utilized for segmenting fiber networks in paper handsheets, emphasizing the utility of these approaches in handling complex fiber arrangements. Moreover, there is a discernible trend toward leveraging 3D imaging modalities, as demonstrated in [15], utilizing X-ray tomography to capture the intricate three-dimensional structure of fibers, enabling more comprehensive analysis.

Researchers are increasingly focusing on hybrid segmentation approaches, combining traditional image processing methods with innovative techniques to enhance segmentation accuracy and robustness [16]. Advancements in computational hardware and software have further enabled the processing of large-scale 3D image datasets, facilitating the development of robust segmentation algorithms capable of handling volumetric data efficiently. In conclusion, these advancements in image segmentation methodologies are poised to revolutionize our understanding of fiber materials and their applications across various industries.

In pulp fiber analysis, semantic segmentation is often preferred over instance segmentation due to several reasons. Firstly, semantic segmentation simplifies the segmentation process by categorizing regions of interest without distinguishing between individual fiber instances, making it suitable for analyzing homogeneous fiber distributions in paper handsheets. Secondly, semantic segmentation algorithms are computationally efficient, making them preferable for processing large-scale image datasets. Since semantic segmentation does not require delineating individual fiber instances, it achieves faster processing speeds while maintaining satisfactory segmentation accuracy. Moreover, semantic segmentation is well-suited for applications where the primary goal is to analyze overall fiber characteristics and distributions rather than identifying individual fibers. Overall, the choice between semantic and instance segmentation depends on the specific requirements of the analysis task, with semantic segmentation offering a streamlined and computationally efficient approach for analyzing overall fiber characteristics in pulp and paper fiber materials.

3.1 Deep Learning Algorithms for Image Segmentation

In recent years, deep learning has significantly advanced semantic segmentation by enabling the creation of highly accurate and efficient algorithms. Various deep-learning architectures have been explored in image segmentation. A fully convolutional network (FCN) [17] was introduced to achieve state-of-the-art performance by converting traditional classification networks into fully convolutional ones. Subsequent studies have focused on improving FCNs, including the integration of skip connections in U-Net [18] and the implementation of atrous spatial pyramid pooling [19].

However, in the domain of paper pulp fibers, there is limited literature specifically addressing the semantic segmentation of paper pulp fibers using deep learning techniques. Notably, one recent study utilized deep learning to automate wood species detection and classification in microscopic images [20]. Another study applied supervised machine learning techniques, including Lasso regression, support vector machine (SVM), feed-forward neural networks (FFNN), and recurrent neural networks (RNN), applied to fiber data from pulp testing micrographs [21]. By leveraging the advantages of each of those algorithms, this method achieved a maximum accuracy of 81%, offering the potential for consistent and efficient online quality control in the pulp and paper industry. While these endeavors mark significant progress in the field of paper pulp fiber segmentation, there remains ample opportunity for further exploration to develop specialized deep learning algorithms that address the specific challenges posed by the complex and heterogeneous nature of paper pulp fiber images. As an effort toward that, a hybrid deep-learning-based fiber segmentation model is proposed and validated in this study.

3.2 Vision Transformers in Image Recognition

The emergence of vision transformers (ViTs) represents a paradigm shift in image recognition tasks. ViTs have demonstrated superior performance on large-scale image datasets [22]. By leveraging self-attention mechanisms, transformers capture long-range dependencies in images, enabling more effective feature extraction and segmentation. The establishment of TextileNet [23], a fashion textile dataset, has provided a standardized framework for training and evaluating deep learning models, including CNNs and ViTs. This framework aids in textile material identification and categorization. Incorporating ViTs into the analysis of paper handsheets presents an exciting opportunity for enhancing segmentation accuracy and efficiency. Notably, these algorithms exhibit versatility across diverse environments, showcasing their ability to effectively depict complex real-world scenarios.

3.3 Challenges and Considerations

While deep learning algorithms have demonstrated their potential for automatic and accurate instance segmentation across various real-world scenarios, their application in fiber segmentation faces several challenges. These include the scarcity of labeled samples, the limited availability of 3D tomographic data, and the inherent difficulties in achieving high accuracy with deep learning-based instance segmentation methods. Moreover, the dense occurrence of fibers, their arbitrary orientations, and varied locations further complicate the training of deep neural networks like Mask-RCNN, often resulting in lower segmentation accuracy compared to traditional approaches such as watershed segmentation. Additionally, errors in 2D tomogram instance segmentation can propagate and affect the accuracy of 3D reconstruction, posing additional challenges. To address the challenges above, the study introduces a hybrid 3D image segmentation algorithm [3]. This technique focuses on utilizing nylon-fiber bundles as data sets for the initial investigation.

4 Deep Learning-Based Hybrid 3D Fiber Segmentation Algorithm

Our primary goal is to accurately segment and label the individual nylon fibers within the 3D images. A novel hybrid 3D fiber segmentation algorithm that involves deep learning-assisted semantic segmentation that generates 2D images from 3D images for fiber extraction is proposed in this study. A series of four sequential image processing blocks is illustrated in Fig. 4. These blocks consist of (1) U-net-aided semantic segmentation for fiber extraction, (2) elliptical contour estimation with watershed segmentation for fiber separation, (3) Gaussian modeling for 3D reconstruction, and (4) Gaussian probability density function-based modeling for 3D object refining. Notably, supervised learning is employed only in the initial step, while the subsequent steps do not rely on labeled datasets, thus adopting unsupervised approaches. The upcoming subsections provide a comprehensive explanation of the dataset utilized, and each of these four sequential function blocks.

4.1 Data Preprocessing and Selection of Dataset

The two Nylon bundle fiber datasets utilized in the study are 2D tomogram slices produced by slicing along their z-axis and differentiated based on their threshold compatibility. Dataset 1, being large, fully labeled, and annotated, is considered threshold-friendly, whereas dataset 2, smaller in size and lacking labeling, presents challenges for threshold-based segmentation. The proposed methodology is specifically applied to dataset 2 to demonstrate its effectiveness under these challenging conditions. The Nylon fiber bundle dataset 2 has a dimension of $1013 \times 964 \times 500$ voxels with a resolution of 1 μm. To facilitate model training, dataset 2 is divided into training and validation sets with a ratio of 0.85 and 0.15, respectively. A subset of images from the training set is manually annotated to create ground truth pixel-wise labels. The labeled training dataset is augmented by rotating and distorting the images, as shown in Fig. 5a. Subsequently, scaling operations expand dataset 2 to enhance model robustness and generalization, which also aligns the intensity distribution of the image dataset 1, as shown in Fig. 5b. The analysis of the X-ray 3D image volume from dataset 2 is conducted using the developed approach for

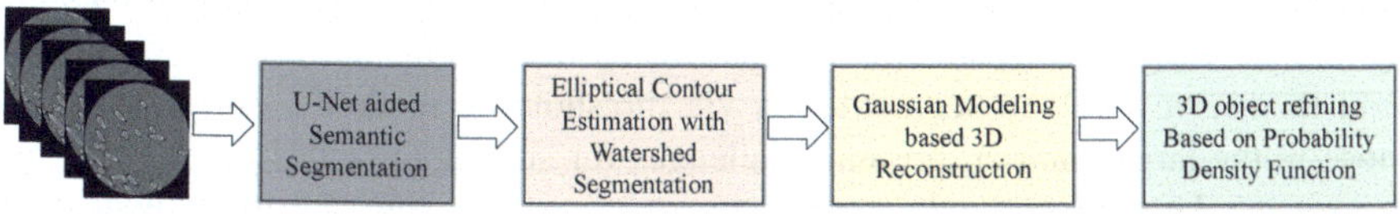

Fig. 4 The block diagram of the proposed deep-learning-based fiber segmentation architecture

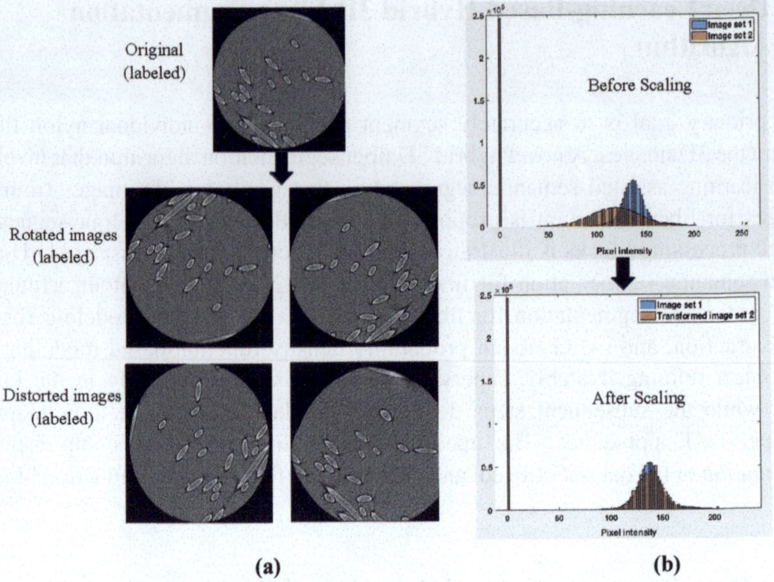

Fig. 5 Image augmentation and image scaling intensity distributions of dataset 2

performance evaluation. Each step outlined in the proposed approach is applied to the dataset as described in the subsequent sections.

4.2 U-Net for Fiber Extraction

U-Net is a CNN commonly employed for segmenting biomedical images, sharing similarities with 2D nylon fiber tomograms. Comprising encoder and decoder components, U-Net efficiently identifies image features through down-sampling and recovers spatial information via up-sampling, making it ideal for semantic segmentation tasks. In our case, segmenting the compressed nylon fiber bundle necessitates training the U-Net model on fully annotated compressed samples. However, as only fully labeled uncompressed samples are available, partially annotated images from the compressed bundle are incorporated into the training dataset. These images undergo manual labeling, with the remaining uncompressed nylon fiber images designated for the test dataset. While some nylon fiber specimens lend themselves well to segmentation via thresholding methods, others, such as those in the target dataset, present challenges for such approaches due to varying experimental conditions and sampling procedures. Two steps are implemented to prepare an effective dataset for training the U-Net model, as shown in Fig. 6. Firstly, the dataset is carefully selected through the naïve thresholding method and scaled based on the mean and standard deviation of the image grayscale intensity in the

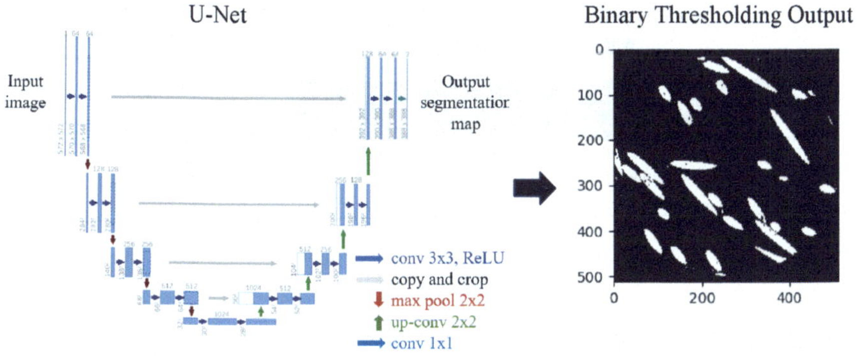

Fig. 6 U-Net model and its binary thresholding output

target dataset, ensuring consistency in distribution. This process generates a new scaled dataset. Subsequently, a subset of images in the target dataset is manually annotated and undergoes image augmentation, resulting in a new dataset. These annotated images are combined with the scaled dataset to create an integrated dataset for U-Net training. This U-Net model has a 512 × 512 input size with four down layers and four upper layers. Each layer uses a 3 × 3 kernel and ReLU activation. Upon completion of training, the U-Net model efficiently labels the entire target dataset in less than a second of processing time. Following 20 epoch iterations, the U-Net model demonstrates thorough training, achieving high accuracy rates of 97.22% for training and 97.50% for validation. Additionally, the model exhibits low loss values of 0.0262 for training and 0.0313 for validation. These metrics indicate that the U-Net model has effectively learned the features of the dataset during training, resulting in accurate predictions on both the training and validation sets.

4.3 *Watershed Instance Segmentation*

Elliptical contour estimation is a vital component in image processing, contributing to various applications such as object detection, recognition, and segmentation. This block presents the concept of elliptical contour estimation and its integration with the watershed algorithm. The U-Net model processes grayscale 2D image slices to produce categorical outputs, effectively distinguishing fibers from background areas. Subsequently, binarized images are generated, initiating the instance segmentation process to separate fibers from each other. Conventional marker-controlled watershed algorithms face challenges in separating elliptical fused objects. To overcome this, the proposed methodology employs the elliptical contour estimation algorithm, providing improved separation between closely connected elliptical fibers. However, overlapping areas between estimated elliptical contours may occur.

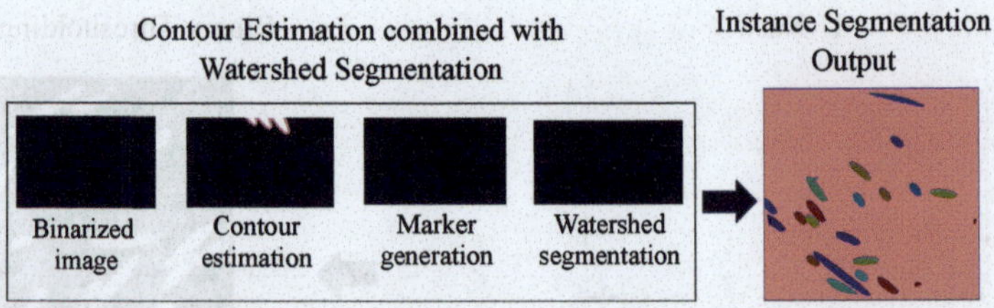

Fig. 7 Proposed elliptical contour estimation with water segmentation and its output

To achieve smoother instance segmentation, a three-step approach is developed in this block, starting with applying the elliptical contour estimation algorithm to generate multiple contours, reducing the size of contours until they are separate objects, and, finally, utilizing obtained markers to apply marker-controlled watershed segmentation, as illustrated in Fig. 7. The performance is demonstrated by manually counting the number of fibers across 500 slices and comparing them with the instance numbers provided by watershed segmentation. The developed elliptical contour-based watershed segmentation yields reliable marker locations, enhancing the accuracy compared to non-elliptical contour-based approaches. Analysis of all 500 slices reveals that the proposed approach achieves over 90% accuracy for 77.6% of the slices, with only 6.2% exhibiting accuracy below 85%. Despite potential manual annotation errors, such as recognizing circular and oval shapes, the proposed method yields satisfactory accuracy. Additional steps are developed in the following subsections to tackle these challenges.

4.4 3D Fiber Reconstruction

In the 3D reconstruction step, the segmented 2D slices containing individual nylon fibers are merged into 3D space by aligning and linking labeled cross sections belonging to the same fiber. This process involves stitching together the segmented 2D image slices to create a comprehensive 3D image, considering the spatial and geometric layout of the nylon fiber bundle, as shown in Fig. 8. To achieve this, object-tracking algorithms are utilized to track the location of fibers between adjacent 2D slices based on individual IDs. While the centroid matching algorithm has been traditionally employed for its simplicity and processing speed, it tends to lose accuracy in densely distributed nylon fiber scenarios or when instance segmentation biases exist from previous watershed segmentation steps.

A novel fiber tracking approach utilizing a Gaussian probability distribution model is proposed to address the limitations of centroid matching. This model compresses each individual fiber object into a Gaussian distribution on each 2D slice. Instead of tracing centroid movement using Euclidean distance, KL divergence is utilized as a matching criterion to measure the similarity between Gaussian

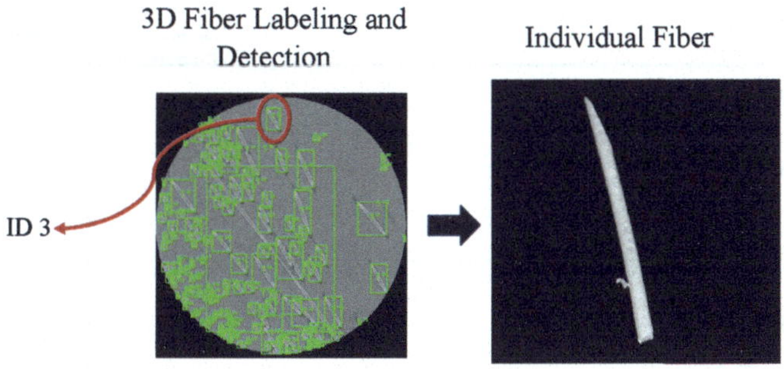

Fig. 8 Gaussian modeling-based 3D fiber reconstruction

distributions. By computing the KL divergence criterion between the current object distribution and suspected objects in previous slices, fibers with minimal divergence below a predefined threshold are connected, indicating they belong to the same fiber. This approach enhances accuracy in tracking densely distributed or biased instances, improving the overall fidelity of 3D reconstruction. The binarized 3D image volume is resliced from the X direction, resulting in 1013 slices. The watershed algorithm is reapplied to each slice, for instance, segmentation. Each object on a slice is represented by a multivariate Gaussian distribution, and its KL divergence with adjacent objects in the previous six slices is computed. If the minimal KL divergence is less than 1, the object in the previous slice with the minimal divergence is chosen as a potential connection candidate; otherwise, the current object is assigned a new label. For example, ID 3 is highlighted in Fig. 8 and is identified as the individual fiber using this proposed methodology.

4.5 3D Object Refining

The Gaussian distribution and KL divergence metrics encapsulate crucial information regarding the segmented nylon fibers. Once the 3D reconstruction process is finalized, individual fibers within the 3D images become identifiable. However, due to inaccuracies in instance segmentation, some extracted 3D fibers may erroneously include attached parts of neighboring fibers. An outlier detection and replacement mechanism are used to address this issue as shown in Fig. 9. This approach involves computing and storing the cross-sectional area of each fiber in every slice, facilitating the detection of abnormal increases in fiber cross-section areas indicative of segmentation errors. Outliers are identified based on predetermined detection thresholds, and erroneous slices are corrected through linear interpolation to remove redundant fiber components introduced by faulty segmentation, as shown in Fig. 10.

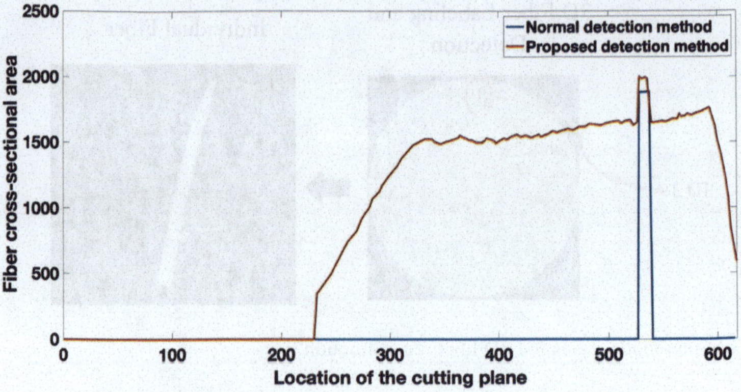

Fig. 9 Fiber cross-section detection

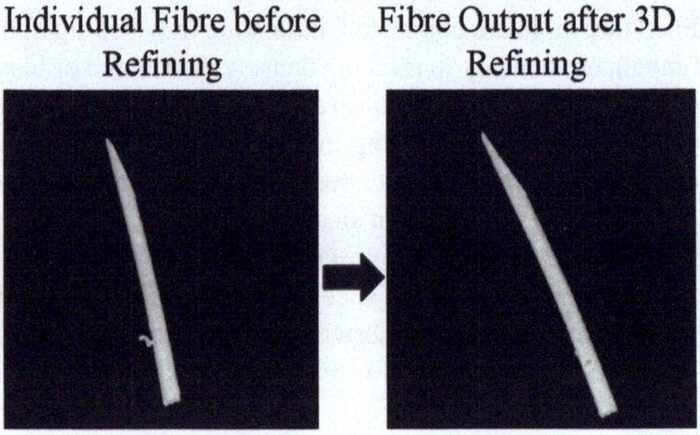

Fig. 10 3D Object refining output

5 Applications and Future Directions in Fiber Segmentation

The proposed hybrid 3D fiber segmentation methodology holds promise for various applications beyond fiber segmentation in 3D X-ray tomographic images. Firstly, it can be extended to other materials science fields where accurate segmentation of fibrous structures is crucial, such as composite materials or geological samples. Additionally, this methodology could find applications in medical imaging, particularly in fields like histopathology or radiology, where the identification and analysis of intricate structures are essential for diagnosis and treatment planning. Moreover, in industrial settings, such as quality control in manufacturing processes or material characterization in research and development, precisely segmenting and analyzing fibrous materials could improve product quality and process efficiency.

In terms of future directions, integrating deep learning-based image segmentation techniques opens avenues for further advancements. Continued research in developing more efficient and accurate deep learning models tailored to the unique characteristics of pulp fibers could significantly enhance the segmentation process. Additionally, exploring novel architectures, such as attention mechanisms or graph neural networks, could further improve the performance of segmentation algorithms, particularly in handling complex fiber arrangements or challenging imaging conditions. Furthermore, advancements in computational hardware and algorithms could enable real-time or near-real-time segmentation of large-scale 3D X-ray tomographic datasets, facilitating rapid analysis and decision-making in various industrial and scientific applications.

6 Conclusions

This chapter delves into the integration of deep learning techniques within 3D fiber segmentation algorithms, aiming to demonstrate their effectiveness. Through an extensive review of existing literature, the study showcases methodologies and techniques employed in leveraging deep learning for 3D fiber segmentation. It elucidates how these methods are applied across different scenarios to enhance segmentation accuracy and efficiency. Additionally, a novel hybrid 3D fiber segmentation algorithm is introduced, which utilizes deep learning-assisted semantic segmentation to generate 2D images from 3D data for fiber extraction. The algorithm combines an image processing method, elliptical contour estimation, with the marker-controlled watershed technique to approximate the boundaries of fibers within an image with elliptical shapes. Individual fibers are identified using a 3D reconstruction modeling technique that combines KL divergence measurement and the Gaussian method. The proposed method is validated using nylon fiber image data sets, which share similarities with wood fibers in polymer composition. This validation demonstrates the efficacy of the approach in handling various properties of paper handsheets.

References

1. R. Wimmer, G. Downes, R. Evans, G. Rasmussen, J. French, Direct effects of wood characteristics on pulp and handsheet properties of eucalyptus globulus. Holzforschung **56**(3), 244–252 (2002)
2. P. Kibleur, J. Aelterman, M.N. Boone, J. Van den Bulcke, J. Van Acker, Deep learning segmentation of wood fiber bundles in fiberboards. Compos. Sci. Technol. **221**, 109287 (2022)
3. M. Fang, A. Sibellas, J. Drummond, Y. Cao, A. Phillion, M. Martinez, V. Pediredla, B. Gopaluni, Deep learning-based hybrid reconstruction algorithm for fibre instance segmentation from 3D x-ray tomographic images. Can. J. Chem. Eng (2023)
4. S. Minaee, Y. Boykov, F. Porikli, A. Plaza, N. Kehtarnavaz, D. Terzopoulos, Image segmentation using deep learning: A survey. IEEE Trans. Pattern Anal. Mach. Intell. **44**(7), 3523–3542 (2021)
5. O. Ronneberger, P. Fischer, T. Brox, *U-net: Convolutional networks for biomedical image segmentation*, in medical image computing and computer-assisted intervention–MICCAI 2015, 18th international conference, Munich, Germany, October 5–9, 2015, Proceedings, Part III 18. (Springer, 2015), pp. 234–241

6. L.C. Chen, G. Papandreou, I. Kokkinos, K. Murphy, A.L. Yuille, Deeplab: Semantic image segmentation with deep convolutional nets, atrous convolution, and fully connected crfs. IEEE Trans. Pattern Anal. Mach. Intell. **40**(4), 834–848 (2017)
7. A. Thyagharajan, B. Ummenhofer, P. Laddha, O.J. Omer, S. Subramoney, *Segment-fusion: Hierarchical context fusion for robust 3D semantic segmentation*, in Proceedings of the IEEE/CVF conference on computer vision and pattern recognition (2022), pp. 1236–1245
8. K. He, G. Gkioxari, P. Dollár, R. Girshick, *Mask r-cnn*, in Proceedings of the IEEE international conference on computer vision (2017), pp. 2961–2969
9. T.Y. Lin, P. Dollár, R. Girshick, K. He, B. Hariharan, S. Belongie, *Feature pyramid networks for object detection*, In Proceedings of the IEEE conference on computer vision and pattern recognition (2017), pp. 2117–2125
10. T.Y. Lin, M. Maire, S. Belongie, J. Hays, P. Perona, D. Ramanan, P. Dollar, C.L. Zitnick, *Microsoft coco: Common objects in context*, in Computer vision–ECCV 2014: 13th European conference, Zurich, Switzerland, September 6–12, 2014, Proceedings, Part V 13 (Springer, 2014), pp. 740–755
11. H. Yu, Z. Yang, L. Tan, Y. Wang, W. Sun, M. Sun, Y. Tang, Methods and datasets on semantic segmentation: A review. Neurocomputing **304**, 82–103 (2018)
12. R.M. Auenhammer, L.P. Mikkelsen, L.E. Asp, B.J. Blinzler, Automated X-ray computer tomography segmentation method for finite element analysis of non-crimp fabric reinforced composites. Compos. Struct. **256**, 113136 (2021)
13. E.L. Wernersson, A. Brun, C.L. Luengo Hendriks, *Segmentation of wood fibres in 3D CT images using graph cuts*, in image analysis and processing–ICIAP 2009: 15th international conference Vietri sul Mare, Italy, September 8–11, 2009 Proceedings 15 (Springer, Berlin/Heidelberg, 2009), pp. 92–102
14. Y. Sharma, Automated fibre segmentation in micro-CT images of paper, Doctoral dissertation, University of British Columbia, 2014
15. M. Hu, A. Olsson, S. Hall, T. Seifert, Fibre directions at a branch-stem junction in Norway spruce: A microscale investigation using X-ray computed tomography. Wood Sci. Technol. **56**(1), 147–169 (2022)
16. A.V. Patel, T. Hou, J.D.B. Rodriguez, T.K. Dey, D.P. Birnie III, Topological filtering for 3D microstructure segmentation. Comput. Mater. Sci. **202**, 110920 (2022)
17. J. Long, E. Shelhamer, T. Darrell, *Fully convolutional networks for semantic segmentation*, in Proceedings of the IEEE conference on computer vision and pattern recognition (2015), pp. 3431–3440
18. O. Ronneberger, P. Fischer, T. Brox, U-net: Convolutional networks for biomedical image segmentation. In Medical image computing and computer-assisted intervention–MICCAI 2015: 18th international conference, Munich, Germany, October 5–9, 2015, proceedings, part III 18 (Springer, (2015), pp. 234–241
19. J. Chen, C. Wang, Y. Tong, AtICNet: Semantic segmentation with atrous spatial pyramid pooling in image cascade network. EURASIP J. Wirel. Commun. Netw. **2019**, 1–7 (2019)
20. L. Nieradzik, J. Sieburg-Rockel, S. Helmling, J. Keuper, T. Weibel, A. Olbrich, H. Stephani, Automating wood species detection and classification in microscopic images of fibrous materials with deep learning (2023). arXiv preprint arXiv:2307.09588
21. S.B. Lindström, R. Amjad, E. Gåhlin, L. Andersson, M. Kaarto, K. Liubytska, et al., Pulp particle classification based on optical fiber analysis and machine learning techniques. Fibers **12**(1), 2 (2023)
22. S. Khan, M. Naseer, M. Hayat, S.W. Zamir, F.S. Khan, M. Shah, Transformers in vision: A survey. ACM Comput. Surv. **54**(10s), 1–41 (2022)
23. S. Zhong, M. Ribul, Y. Cho, M. Obrist, TextileNet: A material taxonomy-based fashion textile dataset (2023) arXiv preprint arXiv:2301.06160

Machine Learning–Based Image Processing in Radiotherapy

Shinichiro Mori and Yasukuni Mori

1 Introduction

Recent advancements in radiotherapy are closely related to imaging technologies. The most conformal radiation treatment cannot be implemented without the latest imaging and image-guidance technologies. Radiotherapy involves treating the patient by irradiating a radiation beam from outside the body rather than directly visualizing and treating the tumor inside the body as in surgical procedures. Therefore, medical imaging in radiotherapy is crucial for enhancing image quality and emphasizing the image information needed by clinical staff to better understand the internal conditions of the body.

The clinical workflow in radiotherapy involves several key steps: patient consultation, planning image acquisition, treatment planning, and treatment beam delivery. The following topics are within the broad range of image processing techniques used in radiotherapy: image analysis and understanding (image segmentation, object detection, and localization); image quality enhancement and restoration (image denoising, image enhancement, image restoration); and image generation and transformation (image synthesis). Deep learning has significantly enhanced artificial intelligence performance beyond conventional image processing and machine learning, and several deep learning techniques have been developed for this workflow. Computer-aided detection/diagnosis (CAD) is useful for decision support during patient consultation [1]. Acquiring computed tomography (CT) images is essential for radiotherapy to calculate dose distribution accurately. CT image artifact reduc-

S. Mori (✉)
Institute for Quantum Medical Science, National Institutes for Quantum Science and Technology, Inage-ku, Chiba, Japan
e-mail: mori.shinichiro@qst.go.jp

Y. Mori
Graduate School of InformaticsChiba University, Inage-ku, Chiba, Japan

K. (K.) Iniewski, L. (K.) Cai (eds.), *Deep Learning for Advanced X-ray Detection and Imaging Applications*, https://doi.org/10.1007/978-3-031-75653-5_10

tion techniques prevent impairments in quantitative analysis and degradation in dose calculation [2]. Synthetic CT images generated from magnetic resonance imaging (MRI) are particularly useful in MR-guided radiotherapy [3]. During the treatment planning process, multiple image modalities are used for contouring volumes of interest (VOIs), including target volumes and organs at risk. This is because the information from different image modalities varies. Deformable image registration is employed to register VOIs on MR images and/or PET images for planning of CT images [4]. A major challenge in manual segmentation is that it is time-consuming and its accuracy depends on the user's skill. Deep learning-based segmentation has addressed these issues effectively [5]. Additionally, dose prediction helps reduce treatment planning time by exploring optimization parameters before performing numerical dose calculations [6].

Treatment beam delivery includes patient setup and treatment beam irradiation. Patient setup is the process of registering the actual patient position to that determined in the treatment plan [7]. The accuracy of this procedure could have impact on treatment results. For treatment beam irradiation, if the treatment beam is turned on when the tumor position is outside the beam field due to respiration, sufficient treatment results cannot be achieved. Recent motion management techniques capture the tumor position in real time, ensuring accurate delivery of the treatment beam [8, 9]. Multiple imaging modalities are generally used in radiotherapy. As an example of a proton therapy system, the treatment room features a CT on rails, a robotic couch, and a rotating gantry equipped with a pair of X-ray imaging systems (Fig. 1). The robotic couch, with six degrees of freedom, positions the patient at the treatment isocenter. The rotating gantry houses an X-ray fluoroscopic imaging system for patient positioning and respiratory gating. All components, including the flat panel detector and X-ray tubes, are integrated into the rotating gantry.

Several articles on image-guided radiation therapy (IGRT) have been introduced [10–14], and we do not propose to review them. In this chapter, we highlight the use of deep learning-based image processing in radiotherapy and explore the technologies that will be critical for future advancements in this field.

2 Image Quality Enhancement

2.1 Overview

Image quality enhancement, including contrast adjustment, sharpening, color correction, and other techniques, improves the visual quality and perceptibility of images. This helps clinical staff understand the rich information within the patient's body more effectively. For example, enhanced image quality facilitates easier visual inspection during the patient setup process by allowing more accurate comparisons between the reference image and the current image acquired in the treatment room.

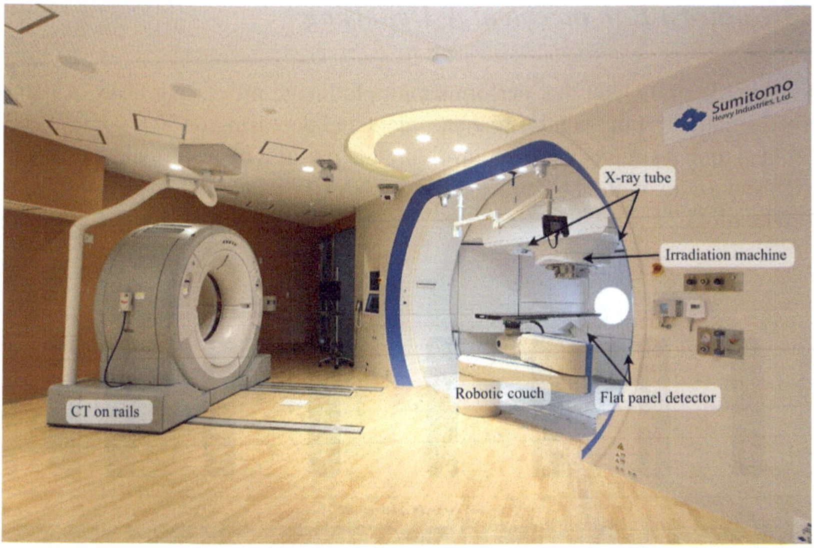

Fig. 1 Photo of a proton therapy room: The treatment room is equipped with a pair of X-ray imaging systems and a CT on rails to visualize patient anatomy. (Source: Sumitomo Heavy Industries, Ltd.)

Additionally, it improves tumor detection accuracy in medical images for tumor tracking.

Multiple image processing techniques, rather than simple image processing with filtering, are sometimes required to achieve the image quality necessary for radiotherapy. By using deep neural networks (DNNs), it is possible to accomplish image processing tasks that conventional filtering methods could not handle. However, performing complex image processing requires a large number of layers. Increasing the number of layers leads to longer computation times; therefore, it is necessary to adjust the network structure in a well-balanced manner, especially for real-time processing. Real-time image processing in commercial imaging systems has been made possible using specialized hardware like the graphics processing unit (GPU) and field-programmable gate array (FPGA). However, developing applications for them demands more advanced programming skills and longer development times compared to CPU-based programming. Although programming with the GPU is simpler than with the FPGA, it still requires more expertise than CPU programming.

Since development of image processing programming by DNNs is easier than development of GPU/FPGA-based image processing, as several DNN frameworks are freely available, image quality enhancement techniques are increasingly incorporating methods that use DNNs. A typical technique is image denoising [15, 16], which has demonstrated better performance than such currently available methods as block-matching and 3D and nonlocally centralized sparse representation.

2.2 Contrast Enhancement + Denoising

One example of a DNN that performs multiple image processing tasks is a single DNN that handles both contrast enhancement and denoising on X-ray images [17]. Mori [17] took a simple approach involving the use of a convolutional autoencoder (CAE) and a convolutional neural network (CNN) (Fig. 2a). The CAE comprised an "encoder block" and a "decoder block." The encoder block was designed to extract

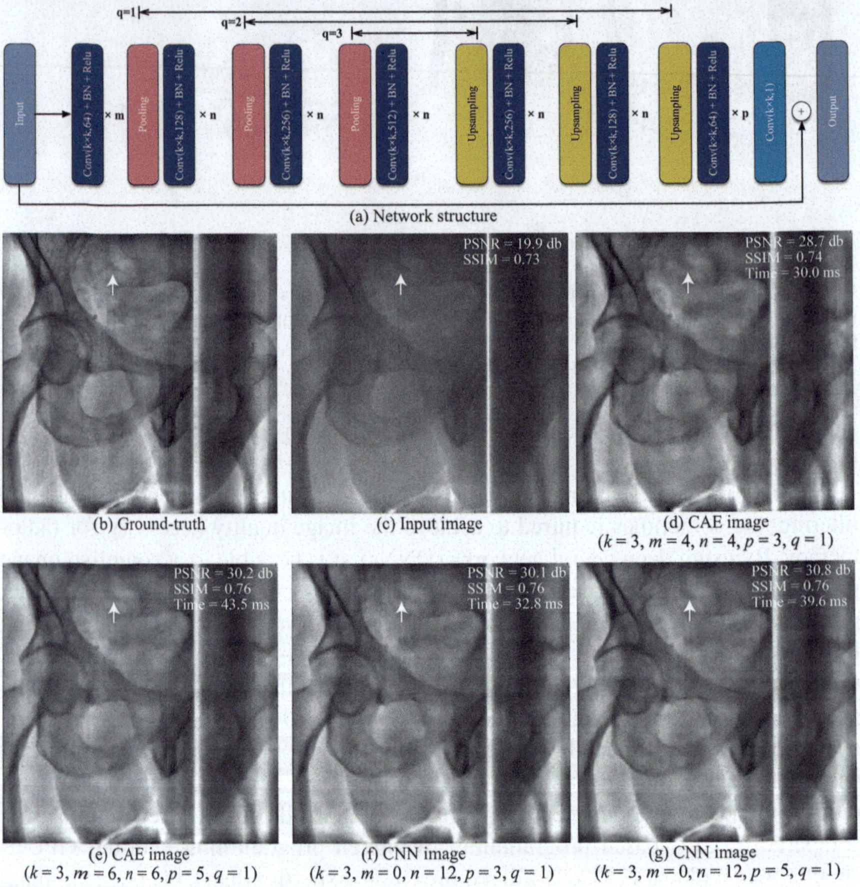

Fig. 2 (**a**) The network structure including convolution (Conv (the kernel size, the number of channels)), BN, ReLU, pooling, and upsampling layers. The stride size was set at two pixels in both the pooling and upsampling layers. (**b**) Ground-truth image. (**c**) Input original X-ray image. (**d**) CAE image predicted with the network consisting of a pair of pooling and upsampling layers and four convolutional layers, and (**e**) six convolutional layers. (**f**) CNN image predicted with the network consisting of a pair of pooling and upsampling layers and 12 convolutional layers, with an additional 3 convolutional layers, and (**g**) 5 convolutional layers added before and after the pooling and upsampling layers. Reproduced from [17] with permission of the copyright holder

features from the input data while reducing its spatial dimensions. In contrast, the decoder block reconstructed these features back into the spatial format of the input data. Additionally, the last convolutional layer with a kernel size of 1×1 (or 3×3) and a single channel was employed to produce a single grayscale image suitable for clinical purposes. The incorporation of shortcut connections helped mitigate the issue of vanishing gradients within the deep architecture. As an example of the CAE, the parameter q (Fig. 2a), which represents the number of pooling-upsampling layer pairs, was selected to be greater than 2. The CNN included one or multiple sets of convolution, batch normalization (BN), and rectified linear unit (ReLU) layers, with the last layer being a convolutional layer producing a single feature map. The input X-ray image was added after the last layer. The ground-truth image was averaged over ten continuously acquired images to reduce image noise, and then it was enhanced using contrast-limited adaptive histogram equalization (CLAHE) to improve image contrast (Fig. 2b). The input X-ray image, offered for comparison, had lower contrast, and it was noisy (Fig. 2c).

The predicted images using CAE (Fig. 2d–e) and CNN (Fig. 2f–g) demonstrated improved quality, closely matching the ground-truth image. A small tissue density was visible in the ground-truth image (marked with an arrow in Fig. 2b), but it was not visible in the input X-ray image (the arrow in Fig. 2c marks where the density should have been seen). Consequently, the output images also did not visualize this detail. The peak signal-to-noise ratio (PSNR) values increased with the number of convolutional layers; however, the computation time also increased. The network for image processing, from CLAHE to image denoising, was concluded to be effective in reducing the radiation dose to the patient.

3 Object Detection and Localization

3.1 Overview

In the thoracic and abdominal areas, respiratory-correlated imaging, such as 4D-CT, or respiratory-gated treatment can be performed by monitoring patient respiration. Commercial external tracking systems are widely used in both particle and photon beam therapies. These systems come in two types: One uses artificial markers on the patient's abdomen tracked by infrared cameras, which requires precise marker placement to correlate with tumor movement; the other uses cameras to monitor the patient's upper body surface, which eliminates the need for markers and provides motion information over a larger area. These methods assume that external motion is, however, well correlated with the actual internal tumor motion, which is questionable.

Internal organ movement includes respiratory motion (cycles of a few seconds), high-frequency cardiac motion, and peristaltic motion (cycles of several tens of seconds). Respiratory patterns can vary in phase shift, drift, amplitude, and period. Accurate real-time tumor tracking requires internal respiratory monitoring, achieved through fiducial marker tracking and markerless tracking. Tracking motion with particle beams is relatively straightforward by modifying the scanning pattern to directly follow the tumor motion [18]. A significant uncertainty in fiducial marker tracking is that markers not implanted in the tumor may not accurately represent its position. In lung cancer cases, tumor motion is greatest in the craniocaudal direction and more pronounced in the lower and peripheral lung regions, though this can vary in diseased lungs [19]. One study found fiducial marker displacement was over 10 mm in about 33% of lung cancer cases [20].

A major challenge in IGRT is real-time tumor tracking without fiducial markers. Detecting 3D tumor positions from 2D digital fluoroscopy is difficult. Integrating CAD into radiotherapy requires real-time accuracy. Diagnostic images lack real-time geometric data, and online images lack prior information like tumor and critical organ contours. Computer vision and machine learning enhance detection and recognition, thereby improving treatment accuracy. "Supervised learning IGRT," combining IGRT and machine learning with patient-specific data, is a promising approach.

3.2 Markerless Tumor Tracking

One method for markerless tracking using imaging patterns involves multiple-template matching with fluoroscopic images [21]. This technique requires preparing template images before irradiation, which may slow down treatment throughput. To address this, several alternative methods have been proposed. Li et al. [22] suggested generating template images from treatment planning 4D-CT images. This approach mitigates the quality differences between digitally reconstructed radiography (DRR) and fluoroscopic images by using subtraction templates from temporally averaged images. Other techniques aim to enhance treatment throughput by employing image pattern features with the mean shift or level set methods. However, these approaches require calculating the tumor position on the first frame, which is challenging due to motion artifacts. Another approach involves a DNN-based markerless tracking algorithm that uses DRR images as training data [8, 23]. Currently, the DNN training process takes several hours, making it impractical to integrate DNNs trained with X-ray fluoroscopic images into clinical applications. Additionally, there are insufficient fluoroscopic images available for use as training data. DRR images, however, are generated by projecting CT data along the X-ray imaging beam path, allowing for the generation of a sufficient number of images with variable positions. This enables the training of DNNs with DRR images before treatment.

An example of this technique [8] is introduced below. Hirai et al. [8] reported that the advantage of this approach was to express the tumor position with a Gaussian morphology (target probability map (TPM)) to stabilize position detection and shorten computation time by inputting subimages defined by regions of interests (ROIs) on the DRR image into different channels. This algorithm had six steps: The first three were learning stages conducted before treatment, and the last three were tracking stages during beam irradiation (Fig. 3).

Step 1: Define ROI and TPM positions on DRR images.
Rectangular ROIs were set on 4D-DRR images from 4D-CT planning data, placed where the greatest displacement was observed. Due to low contrast in fluoroscopic images, ROIs were positioned to capture sufficient tumor displacement. TPM positions were defined at the tumor centers for each respiratory phase on the DRR images, with TPM sizes matching the ROI sizes (yellow square in Fig. 3a).

Step 2: Prepare training data.
Training data were generated from 4D-DRR images and reference tumor positions. To augment the data, DRR images were generated with randomly shifted CT positions simulating patient setup errors, and corresponding tumor positions were projected onto the DRRs. These images and positions were further rotated and translated (Fig. 3b).

Step 3: Carry out network training.
The DNN was trained to produce a TPM by importing subimages within the ROIs and minimizing the loss function (Fig. 3c).

Step 4: Clipping ROI images on the fluoroscopic images.
Fluoroscopic image pixel sizes, acquired during treatment, were resized to match DRR image sizes. Pixel values were normalized similarly to the training DRR images. Subimages were clipped from ROIs at positions defined in Step 1 on the resized fluoroscopic images (Fig. 3d).

Step 5: Generating a TPM.
Subimages from ROIs were assigned to respective channels, and a TPM was generated by importing these subimages into the DNN (Fig. 3e).

Step 6: Calculating tumor position.
The tumor position on the fluoroscopic image was derived through positional weighted averaging of the TPM. Tumor positions were calculated on two fluoroscopic images from fluoroscopic imaging systems, and the 3D tumor position was derived from the 2D positions (Fig. 3f).

Hirai et al. [8] found respiratory-induced organ motion was greater in the superior-inferior (SI) direction than in the lateral direction. ROI positions included large displacement regions, leading to image variation mainly in the SI direction, complicating lateral position resolution. This was corrected using a linear regression model between SI and lateral tumor positions from planning 4D-CT data.

Tracking positional accuracy was achieved with Euclidean distances of 1.90 ± 0.65 mm for five lung cases and 1.37 ± 0.81 mm for five liver cases. The greatest tracking error occurred in the SI direction, with tumor tracking accuracy declining around the 120th frame (Fig. 4a). To illustrate this, Hirai et al. visualized the TPM and overlaid it on the fluoroscopic image (Fig. 4b, c). At T1, the tumor tracking closely matched the reference, and the TPM displayed a Gaussian morphology (Fig. 4b). However, by frame 2, the tracking positional error was approximately 5 mm, as the TPM lost its Gaussian shape and higher values

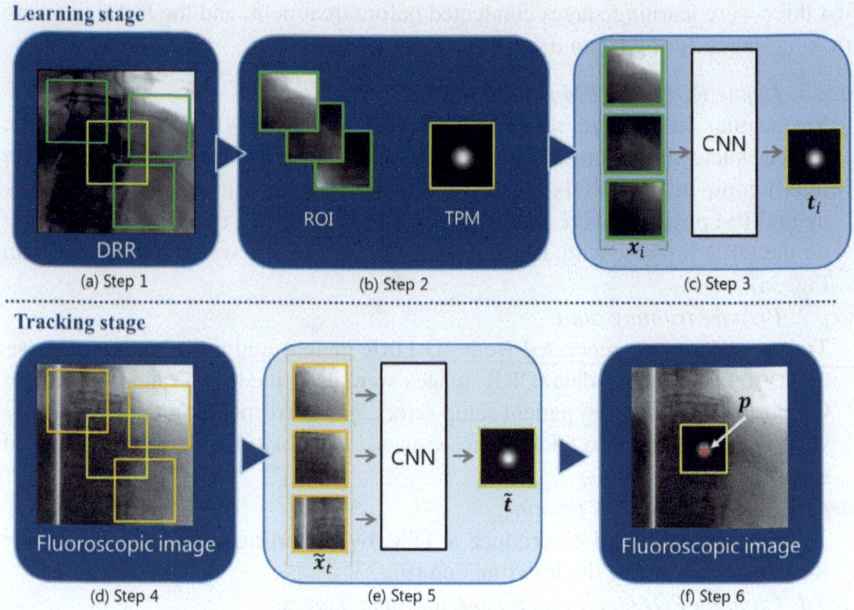

Fig. 3 (**a**) Step 1: Define ROIs on the DRR images. (**b**) Step 2: Prepare training data from 4D-DRR subimages and their TPMs using the ground-truth tumor positions. (**c**) Step 3: Optimize the parameters of the tumor tracking DNN. (**d**) Step 4: Clip the subimages of ROIs from the X-ray fluoroscopic images. (**e**) Step 5: Generate the TPM using the DNN. (**f**) Step 6: Calculate the tumor position on the X-ray fluoroscopic images from the TPM. Reproduced from [8] with permission of the copyright holder

were observed on the right side. The average computation time for tumor tracking using a pair of fluoroscopic images across all ten patients was 39.8 ± 3.7 ms per frame.

From this brief introduction to one DNN-based markerless tumor tracking method, both fiducial marker tracking and markerless tracking were seen to have individual benefits and drawbacks. The choice of tracking method depends on the patient's condition and the treatment strategy, and it is the responsibility of the medical staff to make this decision.

4 Image Synthesis

4.1 Overview

Medical image synthesis plays a crucial role in clinical applications such as supporting medical diagnosis, planning treatment, segmenting an image, and providing image super-resolution. The acquisition of images from multiple modalities is

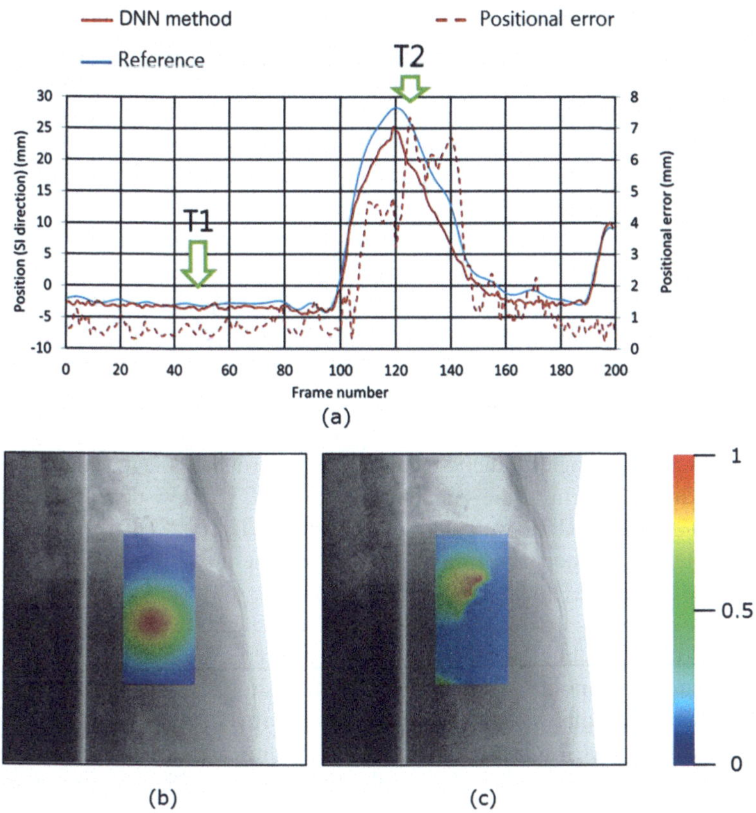

Fig. 4 A representative liver case. (**a**) Tumor positions along the SI direction for both the reference and DNN method (left axis), with positional error plotted (right axis). TPM overlaid on the X-ray fluoroscopic image at (**b**) time = T1 (50th frame) and (**c**) time = T2 (130th frame) for the fluoroscopic image. Reproduced from [8] with permission of the copyright holder

particularly essential for diagnosis and treatment. For instance, images obtained from different modalities like MRI and CT provide distinct types of information. MRI is well-suited for revealing detailed structures of soft tissues, making it effective for observing the brain, muscles, and tumors. On the other hand, CT excels in evaluating dense tissues such as bones, blood vessels, and lungs, and it is widely used in diagnosing trauma, hemorrhage, and pulmonary diseases. By combining multiple modalities, it is possible to integrate these different pieces of information for a more comprehensive diagnosis.

AI technologies, especially DNN have advanced as effective methods for generating and integrating images from multiple modalities. These technologies enable rapid and efficient medical image synthesis, offering a new means to improve the quality of health care. From the perspective of image generation methods, DNN-based approaches have become mainstream in recent years. In medical image synthesis using DNN, adversarial generative models [24] and diffusion models [16] are commonly employed. Generative adversarial networks (GANs), proposed by Goodfellow et al. [25], involve two neural networks, a generator and a discriminator, competing with each other to generate realistic images. The generator produces fake images that look real, while the discriminator determines whether the images are real or fake. Through this competition, the generator gradually acquires the ability to produce realistic images.

Diffusion models generate data sequentially. These models use a "diffusion process" that adds noise to the original data incrementally and a "reverse diffusion process" that removes the noise to reconstruct the original data. Diffusion models are powerful generative models capable of high-quality, realistic data generation and are applied in various fields, including medical imaging.

From the viewpoint of image modality conversion, image generation can be classified into intra-modality image synthesis and intermodality image synthesis. The intra-modality image synthetic technique transforms an image acquired in one modality to another form of the same modality, e.g., CT to low-dose CT, MRI T1-weighted images to MRI T2-weighted images, and PET [26–28]. The intermodality image synthetic technique transforms images from one modality to another: MR to CT, CT to MR, and PET to CT [29, 30].

These technologies are not only expected to have significant applications in the medical field but also hold promise for further future developments. For example, the construction of automatic diagnostic systems using generated medical images and real-time image analysis could significantly improve the speed and accuracy of medical examinations. Moreover, these advancements are expected to contribute to personalized medicine tailored to individual patients. The progress in image generation technologies will likely lead to more advanced and efficient image processing in medical settings.

4.2 Generation of MR Images for Conversion to CT Images

CT and MRI are widely used techniques for visualizing anatomical structures of the human body. Utilizing both imaging modalities is essential in the implementation of radiation therapy. CT images not only visualize the anatomical structures of patients but also facilitate dose calculation within a patient's body by utilizing the relative electron density computed from CT values. Conversely, CT images lack soft tissue contrast, making it preferable to use MR images, which offer superior tissue contrast, for accurate segmentation of tumors and normal organs.

In recent years, MRI-only radiotherapy has gained attention by leveraging the superior visualization capabilities of MRI and replacing CT with MRI for conducting radiation therapy. This approach offers several advantages, including simplification of the treatment process, reduction in treatment costs, minimization of patient radiation exposure, and elimination of registration errors between CT and MR images. However, MR images do not contain electron density information, rendering them inadequate for dose calculation on their own. Consequently, a method has been developed to synthesize pseudo-CT images from MR images and perform dose calculations based on the virtual electron density information.

Approaches that learn the correlation between CT and MR images using the DNN have been proposed to generate synthetic CT images. Notably, GANs, which involve a generator producing pseudo-images and a discriminator determining the authenticity of these images, have shown significant success in image domain transformation tasks. GANs employ unsupervised learning, enabling them to learn features without providing transformed MRI data of CT images, making them highly compatible with tasks such as MRI to CT conversion, where obtaining exact CT image pairs for MRI is challenging.

An example of conversion of MRI to a CT image using a model that incorporates perceptual loss proposed by Johnson et al. [31] into simple GANs is introduced in this section. Perceptual loss is a loss function based on high-level features extracted from a pretrained network. By using this loss function, the similarity of images can be appropriately measured, which cannot be captured by pixel-level loss, enabling the conversion of high-quality images between different modalities.

Perceptual loss captures the perceptual and semantic information of images, calculated using high-level feature maps extracted from a pretrained image classification model. It consists of feature reconstruction loss, ensuring that the output image and the content image have similar high-level features, and style reconstruction loss, ensuring that the output image and the style image share similar styles (colors, textures, common patterns).

Feature reconstruction loss enhances the capability of an output image of the model to share the same feature representation as the content image. Rather than matching pixel values directly, it aims to match higher-order features. This loss is calculated as the Euclidean distance between the feature maps of the output image $\hat{y}$ generated by the generator for input image y and the content image y_c:

$$L^j_{feature} = \frac{1}{C_j H_j W_j} \left| \Phi_j(\hat{y}) - \Phi_j(y_c) \right|^2_2.$$

Here $\Phi_j(\cdot)$ represents the feature map output of the jth layer of the pretrained network Φ. C_j, H_j, and W_j denote the number of channels, height, and width of the feature map at that layer, respectively. The final feature reconstruction loss is calculated as the sum of $L^j_{feature}$ over multiple layers as follows:

$$L_{feature} = \sum_{j \in J_f} L^j_{feature},$$

where J_f is the set of layer indices of the feature maps used.

The main objective of using feature reconstruction loss is to train the network to capture higher-order perceptual and semantic information of the content image rather than pixel-level loss. Consequently, the output image appears visually like the content image, even if individual pixel values do not match perfectly. Therefore, by considering the input image y as the MR image and the output image $\hat{y}$ as the generated synthetic CT image, and treating y as the content image ($y_c = y$), it is expected that by minimizing the feature reconstruction loss between y and $\hat{y}$ during training, $\hat{y}$ will be transformed to preserve the visual similarities of y, such as its shape, which cannot be captured by pixel-level loss.

Style reconstruction loss is a loss function used to ensure that the style of an image generated by the generator matches that of a style image. Style refers to visual features such as color, texture, and patterns. This loss is calculated using the feature maps from intermediate layer of a pretrained DNN and captures the correlations between features using Gram matrices. The Gram matrix $G_j(x)$ for input x at the jth layer of the network is calculated based on the inner products of its feature map. When the shape of the feature map at jth layer of the network for input x is $C_j \times H_j \times W_j$, the Gram matrix is $C_j \times C_j$, and its (c, c') element is given by:

$$G_j(x)_{c,c'} = \frac{1}{C_j H_j W_j} \sum_{h=1}^{H_j} \sum_{w=1}^{W_j} \Phi_j(x)_{h,w,c} \Phi_j(x)_{h,w,c'},$$

where $\Phi_j(x)$ represents the C_j-dimensional feature vectors at each point of the feature map with shape $H_j \times W_j$. Style reconstruction loss is calculated as the squared Frobenius norm between the Gram matrices of the generated image $\hat{y}$ and the style image y_s at each layer, as follows:

$$L_{style} = \sum_{j \in J_s} \left| G_j(\hat{y}) - G_j(y_s) \right|_F^2,$$

where J_s represents the set of layer indices of the pretrained DNN whose feature maps are used.

Style reconstruction loss aims not merely for pixel-level matching but for ensuring the consistency of the overall style of the image. This means adjusting the image such that its overall color distribution and texture patterns match. By specifying an actual CT image (which does not need to be paired with the MR image) as the style image, it is expected that the synthetic generated CT image $\hat{y}$ will be transformed to match the style of a true CT image.

An example of modality conversion from MRI to CT using this method is presented; Fig. 5 is a schematic diagram of the network. The model was a very simple one that incorporated two perceptual losses into the conventional GAN setup. The final loss was computed as the weighted average of each loss as follows:

$$L = \lambda_a L_{adv} + \lambda_f L_{feature} + \lambda_s L_{style},$$

where L_{adv} was the loss calculated by the GANs. The generator utilized a U-net base structure incorporating a residual network, while the discriminator employed a patchGAN. The pretrained network used for calculating the perceptual loss was

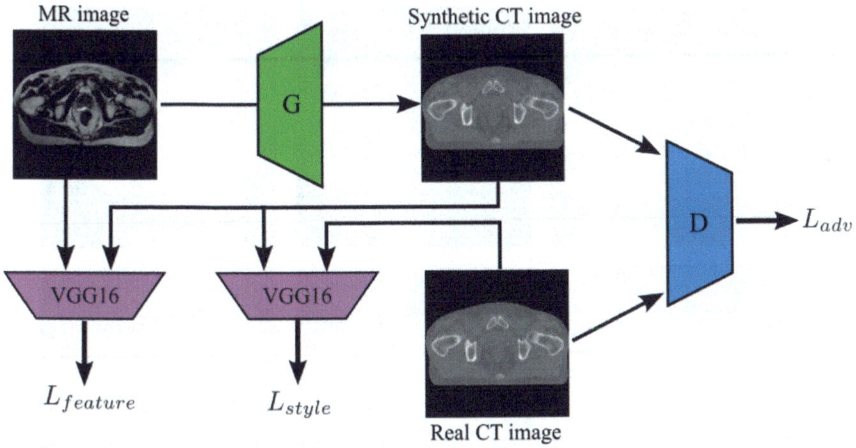

Fig. 5 GAN-based network structure with perceptual loss. $L_{feature}$ was calculated between the MR images and the synthetic CT images, while L_{style} was calculated between the true CT images and the synthetic CT images. L_{adv} was a loss function based on the standard GAN. The values of the hyperparameters λ_a, λ_f, and λ_s were set to 0.4, 0.1, and 0.1, respectively

VGG16. For the feature reconstruction loss, feature maps from the 2nd, 4th, 7th, 10th, and 13th layers were used, and for the style reconstruction loss, feature maps from the 2nd, 4th, 7th, and 10th layers were used.

Figure 6 shows an example of a CT image generated from an MR image using this method. The body contours and bone shapes were appropriately converted from the MR image to the CT image. Additionally, the grayscale color and texture of the synthetic CT image could be considered as relatively well-generated. This could be attributed to the feature reconstruction loss and style reconstruction loss.

4.3 Generation of DRR Images for Conversion to X-Ray Images

In image synthesis using CT imaging, where a fixed pixel value is assigned to each material (Hounsfield unit), generating an X-ray image from a DRR image presents challenges [32]. The pixel values in an X-ray image can fluctuate depending on factors like object size and imaging conditions (e.g., kV, mAs), and typically, X-ray images exhibit lower contrast and higher spatial resolution compared to DRR images. Nonetheless, this technique can be advantageous for the patient positional verification process, which has been improved through the use of automatic image registration software. Despite these advancements, clinical staff still require additional time to visually inspect and compare reference DRR images with X-ray images acquired in the treatment room due to significant differences

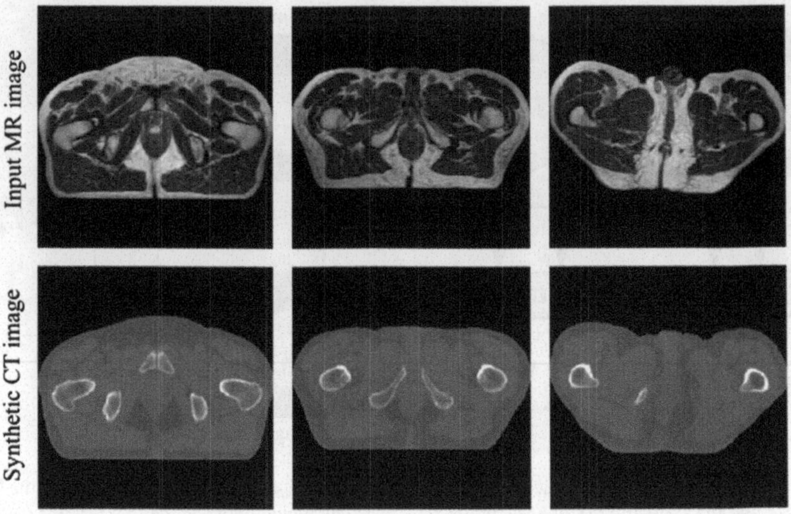

Fig. 6 Examples of CT images generated from MR images using the proposed method. The top row displays the original MR images, while the bottom row shows the corresponding generated CT images

in image quality. Ultimately, this technique could enhance throughput, especially when images from two different modalities need visual comparison. The machine learning-based tumor tracking, as introduced in the previous section, was trained using DRR images to detect tumors on X-ray images. If synthetic X-ray images were used for the training process, tumor detection accuracy could be improved.

One example of this technique utilizes a U-net architecture with shortcut connections, as shown in Fig. 7 [33]. L1 loss and L2 loss, including mean absolute error (MAE) and mean squared error (MSE), have been conventionally employed as loss functions in image synthesis. These loss functions, however, only assess distortion, often resulting in predicted images that are blurry and resemble DRR images. To accurately capture the detailed structures and comprehensive information present in the original X-ray image, it is essential to evaluate image quality by considering both distortion and perceptual quality [34].

Figure 8 displays the ground-truth X-ray image, input DRR image, and synthetic X-ray image [32]. In Fig. 8a, the treatment machine cover edge and patient's earlobe were visible in the vertical direction, indicated by red and blue arrows, respectively; however, these features were not visualized in the input DRR image (Fig. 8b). Visual inspection showed the quality of the synthetic X-ray image to closely resemble that of the ground-truth X-ray image. The impact of the pixel value variations on between difference medical imaging modalities would be minimized by applying an attention mechanism to the network structure [35].

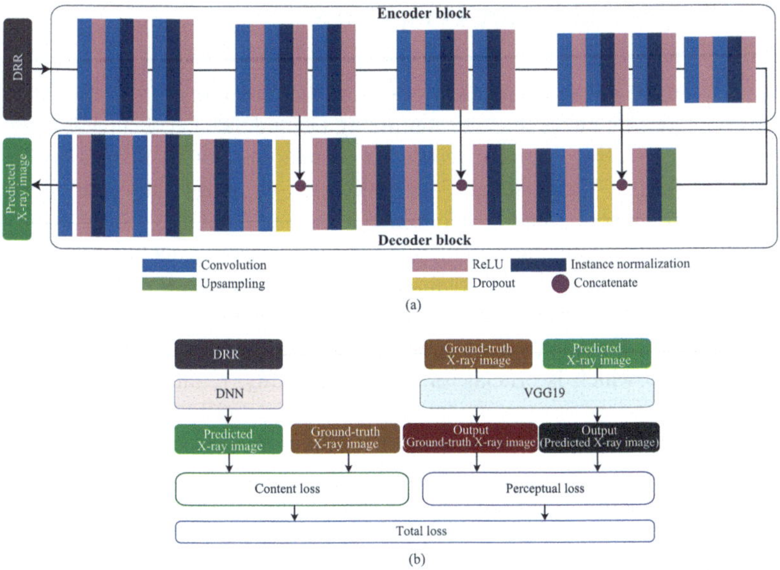

Fig. 7 (**a**) The network structure for the DNN designed to generate X-ray images. The structure encompasses an input layer receiving the DRR image data, convolutional layers for feature extraction, and deep layers with shortcut connections to prevent information loss. (**b**) The loss calculation combines content loss and perceptual loss. Content loss minimizes pixel-wise differences between the DRR and X-ray images, focusing on precise pixel values. Perceptual loss maintains textural and structural similarities by comparing features from pretrained models. Reproduced from [32] with permission of the copyright holder

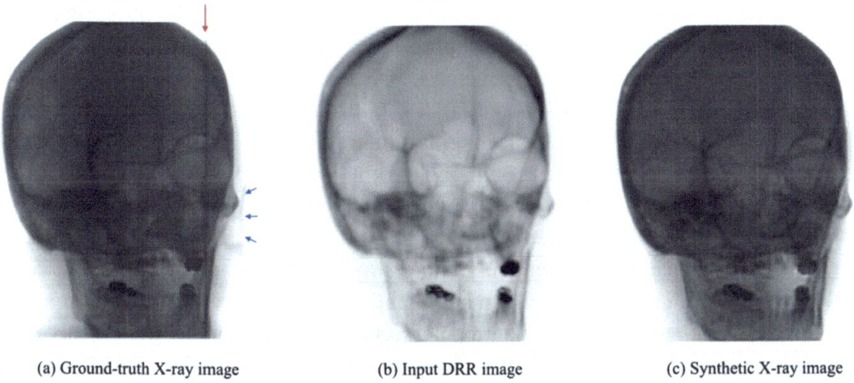

Fig. 8 (**a**) Ground-truth X-ray image. The red arrow indicates the edge of the treatment machine cover, and the blue arrows point to the earlobe. (**b**) Input DRR image. (**c**) Synthetic X-ray image produced by the DNN. Reproduced from [32] with permission of the copyright holder

5 Challenges and Future Perspective

In this chapter, we discussed applications of medical imaging in radiation therapy, particularly focusing on recent studies utilizing DNN. DNN excels in automatically learning and extracting intricate features and patterns within images with high precision, making it highly suitable for tasks such as image classification, object detection, segmentation, and generation. Consequently, DNN holds significant promise for diverse applications in medical image processing, including diagnosing diseases, detecting lesions, and planning surgeries. However, integrating these technologies into the medical field presents several challenges.

One major challenge is ensuring the generalization performance of machine learning models, which is crucial for their practicality, reliability, and efficiency. Robustness of a model to new data determines its utility and dependability. Typically, machine learning assumes that training and test data are drawn from the same distribution. However, medical imaging data often come from different environments or facilities and can vary over time. This makes it unrealistic to expect training and test environments to follow the same distribution, posing a significant challenge known as domain shift. For example, when training data and ground-truth image data are acquired at different times and on different days, they may include interfractional and intrafractional anatomical changes and misalignment. Consequently, it is uncertain whether synthetic images trained on such data can achieve sufficient image quality.

Another important challenge is the interpretability and transparency of DNN models. For healthcare professionals and patients to understand and trust AI decisions, it is essential to present the results of deep learning models, which tend to be black box in nature, in an understandable manner. Enhancing the interpretability of AI can foster trust and facilitate its appropriate use in clinical settings.

As discussed in this chapter, several DNN technologies have been integrated into radiotherapy. Radiology, encompassing both diagnostic radiology and radiotherapy, heavily relies on medical imaging, making it more feasible that radiology incorporates DNN technologies compared to other medical fields. Integrating DNN into radiology is not expected to replace medical staff but rather to increase throughput, reduce healthcare disparities, and allow medical staff to spend more time on other critical tasks, ultimately enhancing the overall quality of health care. Transitioning from research to clinical applications is a long process, and extending these innovations to other treatment centers takes even longer. However, this path is essential and undoubtedly worth the effort, especially when undertaken with colleagues from both domestic and international institutions.

References

1. H.R. Roth et al., Improving computer-aided detection using convolutional neural networks and random view aggregation. IEEE Trans. Med. Imaging **35**, 1170–1181 (2016)

2. S. Mori, R. Hirai, Y. Sakata, Using a deep neural network for four-dimensional CT artifact reduction in image-guided radiotherapy. Phys. Med. **65**, 67–75 (2019)
3. H. Xiao, MR-based synthetic CT generation using a deep convolutional neural network method. Med. Phys. **44**, 1408–1419 (2017)
4. B.D. de Vos et al., A deep learning framework for unsupervised affine and deformable image registration. Med. Image Anal. **52**, 128–143 (2019)
5. B. Ibragimov, L. Xing, Segmentation of organs-at-risks in head and neck CT images using convolutional neural networks. Med. Phys. **44**, 547–557 (2017)
6. Z. Liu et al., A deep learning method for prediction of three-dimensional dose distribution of helical tomotherapy. Med. Phys. **46**, 1972–1983 (2019)
7. S. Miao, Z.J. Wang, Y. Zheng, R. Liao, in *2016 IEEE 13th International Symposium on Biomedical Imaging (ISBI)*. (2016), pp. 1430–1434
8. R. Hirai, Y. Sakata, A. Tanizawa, S. Mori, Real-time tumor tracking using fluoroscopic imaging with deep neural network analysis. Phys. Med. **59**, 22–29 (2019)
9. Y. Sakata, R. Hirai, K. Kobuna, A. Tanizawa, S. Mori, A machine learning-based real-time tumor tracking system for fluoroscopic gating of lung radiotherapy. Phys. Med. Biol. **65**, 085014 (2020)
10. M. Guckenberger et al., Precision required for dose-escalated treatment of spinal metastases and implications for image-guided radiation therapy (IGRT). Radiother. Oncol. **84**, 56–63 (2007)
11. D. Verellen et al., Innovations in image-guided radiotherapy. Nat. Rev. Cancer **7**, 949–960 (2007)
12. G.T. Chen, G.C. Sharp, S. Mori, A review of image-guided radiotherapy. Radiol. Phys. Technol. **2**, 1–12 (2009)
13. S. Mori, S. Zenklusen, A.C. Knopf, Current status and future prospects of multi-dimensional image-guided particle therapy. Radiol. Phys. Technol. **6**, 249–272 (2013)
14. M. Mueller et al., The markerless lung target tracking AAPM grand challenge (MATCH) results. Med. Phys. **49**, 1161–1180 (2022)
15. P. Vincent, H. Larochelle, I. Lajoie, Y. Bengio, P.A. Manzagol, Stacked denoising autoencoders: Learning useful representations in a deep network with a local denoising criterion. J. Mach. Learn. Res. **11**, 3371–3408 (2010)
16. S. Pan et al., 2D medical image synthesis using transformer-based denoising diffusion probabilistic model. Phys. Med. Biol. **68** (2023)
17. S. Mori, Deep architecture neural network-based real-time image processing for image-guided radiotherapy. Phys. Med. **40**, 79–87 (2017)
18. S.S. Korreman, T. Juhler-Nottrup, A.L. Boyer, Respiratory gated beam delivery cannot facilitate margin reduction, unless combined with respiratory correlated image guidance. Radiother. Oncol. **86**, 61–68 (2008)
19. Y. Seppenwoolde et al., Precise and real-time measurement of 3D tumor motion in lung due to breathing and heartbeat, measured during radiotherapy. Int. J. Radiat. Oncol. Biol. Phys. **53**, 822–834 (2002)
20. H. Shirato et al., Speed and amplitude of lung tumor motion precisely detected in four-dimensional setup and in real-time tumor-tracking radiotherapy. Int. J. Radiat. Oncol. Biol. Phys. **64**, 1229–1236 (2006)
21. Y. Cui, J.G. Dy, G.C. Sharp, B. Alexander, S.B. Jiang, Multiple template-based fluoroscopic tracking of lung tumor mass without implanted fiducial markers. Phys. Med. Biol. **52**, 6229–6242 (2007)
22. R. Li, J.H. Lewis, L.I. Cervino, S.B. Jiang, A feasibility study of markerless fluoroscopic gating for lung cancer radiotherapy using 4DCT templates. Phys. Med. Biol. **54**, N489–N500 (2009)
23. W. Takahashi, S. Oshikawa, S. Mori, Real-time markerless tumour tracking with patient-specific deep learning using a personalised data generation strategy: Proof of concept by phantom study. Br. J. Radiol. **93**, 20190420 (2020)
24. X. Dong et al., Automatic multiorgan segmentation in thorax CT images using U-net-GAN. Med. Phys. **46**, 2157–2168 (2019)

25. I.J. Goodfellow et al., Generative adversarial networks. arXiv **1406**, 2661 (2014)
26. E. Kang, J. Min, J.C. Ye, A deep convolutional neural network using directional wavelets for low-dose X-ray CT reconstruction. Med. Phys. **44**, e360–e375 (2017)
27. E. Gong, J.M. Pauly, M. Wintermark, G. Zaharchuk, Deep learning enables reduced gadolinium dose for contrast-enhanced brain MRI. J. Magn. Reson. Imaging **48**, 330–340 (2018)
28. X. Dong et al., Deep learning-based attenuation correction in the absence of structural information for whole-body positron emission tomography imaging. Phys. Med. Biol. **65**, 055011 (2020)
29. H. Arabi et al., Comparative study of algorithms for synthetic CT generation from MRI: Consequences for MRI-guided radiation planning in the pelvic region. Med. Phys. **45**, 5218–5233 (2018)
30. Y. Lei et al., CT prostate segmentation based on synthetic MRI-aided deep attention fully convolution network. Med. Phys. **47**, 530–540 (2020)
31. J. Johnson, A. Alahi, L. Fei-Fei, Perceptual Losses for Real-Time Style Transfer and Super-Resolution (2016) (10.48550/arXiv.1603.08155)
32. S. Mori et al., Deep neural network-based synthetic image digital fluoroscopy using digitally reconstructed tomography. Phys. Eng. Sci. Med. **46**, 1227–1237 (2023)
33. V. Badrinarayanan, A. Kendall, R. Cipolla, SegNet: A deep convolutional encoder-decoder architecture for image segmentation. IEEE Trans. Pattern Anal. Mach. Intell. **39**, 2481–2495 (2017)
34. B. Yochai, M. Tomer, *The perception-distortion tradeoff*. in Proceedings of the IEEE Conference on Computer Vision and Pattern Recognition (CVPR), pp. 6228–6237 (2018)
35. A. Vaswani et al., Attention Is All You Need (2017) (10.48550/arXiv.1706.03762)

Deep Learning–Based Image Reconstruction of Coded-Aperture Imaging in Nuclear Security Applications

Qi Liu, Yihang Wang, Xinhong Bai, Yi Cheng, Xiangguo Tuo, Qingxian Zhang, and Jianbin Zhou

1 Introduction

Nuclear power plays a key role in the world's low-carbon energy structure, 411 nuclear power reactors were operational, and 58 reactors were under construction [1]. The increasing number of nuclear facilities inevitably produces radioactive waste that may be obtained by terrorists or criminal organizations for non-peaceful use. As reported by the International Atomic Energy Agency (IAEA) Incident and Trafficking Database (ITDB), 4075 incidents of nuclear and other radioactive materials out of regulatory control were confirmed between 1993 and 2022. Of these incidents, 344 were connected with trafficking or malicious use including high-enriched uranium, plutonium, and plutonium beryllium neutron sources [2]. Therefore, accurate localization of radioactive sources is crucial in fields of nuclear safety and security, decommissioning and decontamination of nuclear facilities, and public health [3, 4].

In order to obtain the spatial distribution of radioactive materials, various gamma-ray imaging techniques have been developed and successfully applied to the above applications, including pinhole camera [5–7], coded-aperture cameras [8–11],

Q. Liu (✉) · Y. Wang · X. Bai · X. Tuo
School of Physics and Electronic Engineering, Sichuan University of Science and Engineering, Zigong, China

Y. Cheng
School of Physics and Electronic Engineering, Sichuan University of Science and Engineering, Zigong, China

College of Nuclear Technology and Automation Engineering, Chengdu University of Technology, Chengdu, China

Q. Zhang · J. Zhou
College of Nuclear Technology and Automation Engineering, Chengdu University of Technology, Chengdu, China

K. (K.) Iniewski, L. (K.) Cai (eds.), *Deep Learning for Advanced X-ray Detection and Imaging Applications*, https://doi.org/10.1007/978-3-031-75653-5_11

Compton cameras [12–14], rotational modulation collimator (RMC) systems [15, 16], and time-encoded imaging systems [17–20].

Coded-aperture and Compton imaging systems have received the most development and are best suited for field use because of their excellent performance and portability. Compton cameras can provide a 4π field of view (FOV), but the angular resolution is typically worse than 10°, and the performance is degraded for low-energy photons (<250 keV) because of their dependence on the Compton scattering kinematics. Coded-aperture cameras rely on the spatial modulation technique and their angular resolution is typically better than 6° due to the mechanical collimation using a heavy metal with a specific pattern. However, the FOV of such systems is limited, and the sources out of the fully coded field of view (FCFOV) are likely to reconstruct false hotspots due to the partial encoding effect that will be described in the Sect. 2. A false localization of radioactive sources may lead personnel to accidentally receive excessive exposure to radiation and cause harm to health in nuclear security applications.

In this chapter, we will briefly introduce the theory of the coded-aperture imaging technique in Sect. 2.1. The principle of partial encoding effect is described, and its influence is experimentally studied with a gamma source and a real coded-aperture camera. A brief review of existing methods to address the above issue is presented in Sect. 2.2. The existing methods of image reconstruction are summarized in Sect. 2.3. A deep learning method based on a hybrid network structure of convolutional neural network (CNN) and Swin Transformer to reconstruct radioactive source distribution directly using the project data is proposed in Sect. 3. Both simulation and experimental results of different source conditions are presented in Sect. 4.

2 Background

Coded-aperture imaging is derived from the pinhole imaging technique that can directly project the radiation scene within the FOV of the imaging system into a position-sensitive detector through a single hole of the mask, while the detector is surrounded by thick shielding to decrease background noise and keep sufficient signal-to-noise ratio (SNR). The angular resolution of a pinhole camera is inversely proportional to the size of the hole. A better angular resolution can be obtained by narrowing the hole size, but the portion of incident photons will significantly decrease at the same time. Therefore, the trade-off between the spatial resolution and SNR must be considered during the mask design phase. In order to address the above issue, the coded-aperture imaging technique replaces the single hole with multi-holes to increase the SNR while maintaining the same angular resolution as the single hole case.

2.1 Theory of Coded-Aperture Imaging

Coded-aperture imaging technique for radiation detection was firstly proposed in X-ray astronomical applications by Ables and Dicke independently [21, 22]. Coded-aperture cameras are comprised of a position-sensitive detector that can be represented by a matrix $\boldsymbol{D}$, and a coded mask $\boldsymbol{A}$ containing opaque elements filled with heavy metal such as tungsten and transparent elements that permit the incident photons to pass through. In the early works, these holes were randomly opened on the mask and a point source can project multiple shadows on the position-sensitive detector. The counting distribution becomes more complicated than the single hole case, and a source position corresponds to a specific pattern rather than a small area of the detector. Therefore, the image formation is divided into an encoding and reconstruction process, respectively.

During the encoding phase, the photons emitted from the radioactive sources $\boldsymbol{\lambda}$ within the FOV of imaging systems were modulated by mask $\boldsymbol{A}$ and detected by the detector $\boldsymbol{P}$, which can be represented by:

$$\boldsymbol{P} = \boldsymbol{\lambda} * \boldsymbol{A} + \boldsymbol{B} \tag{1}$$

where the symbol $_*$ indicates the correlation operator and $\boldsymbol{B}$ represents all the events that are not modulated by the mask.

The reconstructed distribution of radioactive sources $\hat{\boldsymbol{\lambda}}$ can be calculated through a cross-correlation between the project data $\boldsymbol{P}$ and a decoding matrix $\boldsymbol{G}$ as follows:

$$\hat{\boldsymbol{\lambda}} = \boldsymbol{P} * \boldsymbol{G} = \boldsymbol{\lambda} * \boldsymbol{A} * \boldsymbol{G} + \boldsymbol{B} * \boldsymbol{G} \tag{2}$$

The radioactive distribution $\hat{\boldsymbol{\lambda}}$ will be perfectly reconstructed if $\boldsymbol{A} \, _* \, \boldsymbol{G}$ is the Kronecker delta function, except for the last noise term $\boldsymbol{B} \, _* \, \boldsymbol{G}$ including high-energy cosmic ray, electronics noise, as well as the photons that are not modulated by the mask. This is the so-called cross-correlation (CC) reconstruction method. However, $\boldsymbol{A} \, _* \, \boldsymbol{G}$ is not a delta function for the random mask pattern in general, and the reconstructed images appear an inherent noise. Since the random distribution of the transparent elements from the mask, the unique optimization is required for each specific application, and such masks are well suited for the applications that required a special shape such as the rotating cylindrical mask of a time-encoded imaging system. The above issues were addressed, and a number of portable coded-aperture cameras have been developed, since the uniformly redundant arrays (URA) were proposed by Fenimore and Canon [23]. URA-based coded masks typically have around ~50% open fraction and satisfy the condition that $\boldsymbol{A} \, _* \, \boldsymbol{G}$ is an ideal delta function without sidelobes. However, the shape of a URA mask must be rectangular, and the difference between the vertical and horizontal dimensions must equal two. Modified uniformly redundant arrays (MURAs) which are derived from quadratic residues rather than pseudo-noise (PN) sequences were proposed to address the above issue. The geometric configurations of MURAs can be linear, hexagonal, and square. The square masks can be constructed by a 2 × 2 mosaic of the basic pattern, and the detector size only equal to the basic pattern is required to fully cover the projection pattern. The only limitation is that the mask's dimension in each direction must be a prime p, so that the mask has a total of p^2 elements. Besides the MURAs, the pseudo-noise product (PNP) arrays can also provide a perfectly square geometry, but with only a 25% open fraction. The above coded apertures have fixed open fractions due to their Cartesian product structure, and imaging objects may have various sizes in different applications, extended sources requiring a relatively low open fraction of the coded aperture than point sources [24]. Recently, Singer aperture has been developed and provides flexibility in the construction of coded aperture with an open fraction of around $\frac{1}{p}$. A two-dimensional aperture can be obtained either by taking a one-dimensional Singer set and wrapping it onto a two-dimensional aperture or by taking the Cartesian product of a pair of one-dimensional Singer aperture directly [25–27].

Angular resolution $\delta\theta$ and FOV were determined by the system geometry such as the mask pattern, the element size of the mask and position-sensitive detector, and the mask-to-detector distance (MDD) and were given by:

$$\delta\theta = \arctan\left(\frac{w_m}{\text{MDD}}\right) \tag{3}$$

$$\text{FOV} = 2\arctan\left(\frac{m-d}{2\text{MDD}}\right) \tag{4}$$

where w_m is the base hole size in the mask, m and d are the sizes of the coded mask, and position-sensitive detector, respectively. Variable zooms can be implemented by changing the MDD of coded-aperture cameras. A better angular resolution can be obtained by increasing the MDD, once the hardware dimensions of a coded-aperture camera have been determined. However, the cost is that the FOV is further reduced. The FOV of a coded-aperture camera is typically smaller than 50° due to the two-plane geometric configuration composed of the mask and detector. The limited FOV not only affects the detectable area in field applications, but the reconstructed results may be interfered with by the partial encoding effect that will be described in Sect. 2.2.

2.2 *Partial Encoding Effect*

The modern coded-aperture imaging systems typically use the mask with a 2 × 2 mosaic of the basic pattern and expand the array dimensions from $p \times p$ to $(2p - 1) \times (2p - 1)$. Compared to the box geometry that the mask and detector have the same size, this 2 × 2 geometric arrangement is constructed, so that each source pixel within the FCFOV of the imaging system can cast a unique shadow pattern on the position-sensitive detector while providing a relatively larger FCFOV.

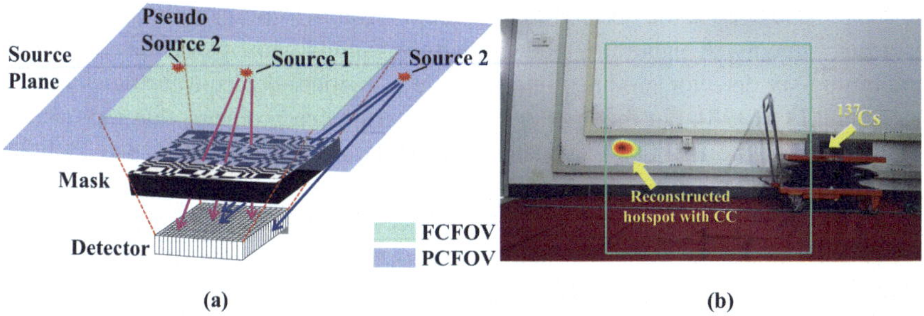

Fig. 1 (**a**) Schematic diagram of the partial encoding effect and (**b**) the reconstructed radiation image using the cross-correlation (CC) method superimposed on the optical image. The delimitation of the FCFOV was represented by the green square

However, the cyclic permutation feature of the mask poses difficulties with image reconstruction when radioactive sources are located out of the FCFOV of the coded-aperture camera.

Coded-aperture cameras are designed specifically for the localization of radioactive sources within the FCFOV such as the "Source 1" shown in Fig. 1a. All the detected photons emitted from the "Source 1" are modulated by the mask, so that the hotspot can be accurately reconstructed by using the cross-correlation method. In medical applications, the imaging process satisfies this condition, because the radioactive materials are placed in front of the imaging systems. However, the approximate locations of radioactive sources are often unknown in nuclear security applications, and there is a possibility that the radioactive sources are located at the partially coded field-of-view (PCFOV) of the imaging systems such as the "Source 2" shown in Fig. 1a. The emitted photons are partially modulated by the mask, while the rest directly interacts with the detector. The resulting shadow cast on the detector is partially identical to the shadow pattern of a source within the FCFOV due to the cyclic permutation feature of the MURA-based masks, which results in false hotspots (also called ghosts) in reconstructed images, as shown in Fig. 1a. This phenomenon is called the partial encoding effect and is experimentally illustrated in Fig. 1b. A ^{137}Cs source with an activity of 37 MBq was placed on a mobile platform with a 2 m source-to-mask distance (SMD). The position of the ^{137}Cs source is out of the FCFOV and on the right side of the imaging system. The result of the reconstructed radiation image using the cross-correlation method superimposes the optical image caught by the optical camera is shown in Fig. 1b. A false reconstructed hotspot was presented on the opposite side of the actual source position due to the partial encoding effect. Even worse, if the position of a radioactive source is near the boundary between the FCFOV and PCFOV of the imaging system, false hotspots will appear on both sides of the reconstructed images. This issue not only interferes with the accurate localization of radioactive sources but also affects the operator's judgment of the actual source number in the radiological scene. A false source localization poses a threat to public safety and personnel's health in nuclear security applications.

There have been some researches to address the above common issues of coded-aperture cameras. A panorama coded-aperture gamma camera was developed by combining six detector modules into a single hexagonal device to achieve a 360° horizontal FOV [28], but this method exists interference problems that radioactive sources within the FCFOV of a detector module are likely to be in the PCFOV of its adjacent detector module. Although the thick lead shields were placed between the detector modules, the partial encoding effect cannot be eliminated, because the contributions of the false reconstructed hotspots (ghosts) were derived from the photons that were modulated by the mask rather than the photons that directly interacted with the detector. Furthermore, only the horizontal direction was considered, and this approach was not suited for the vertical and diagonal directions. A multimode imager combining Compton with coded-aperture imaging techniques was designed based on the active mask configuration for nuclear threat detection [29]. The imager consists of 96, 1 cm^3 CdZnTe (CZT) detector elements arranged in a two-plane geometric configuration in which the front plane uses a random mask pattern with 32 detector elements to serve simultaneously as a coded mask in coded-aperture modality and a Compton scatter layer in Compton imaging modality, and the back plane is filled with 64 detector elements. However, an FCFOV is only available for sources that are directly on axis, because its mask size is the same as the detector plane in coded-aperture modality, which makes the accurate localization for the off-sources difficult. To address the problems of imaging artifacts and the coded-aperture modality suffers from a limited FOV, a spherical active coded-aperture imaging system was proposed by rearranging the detector elements into a 14-cm diameter sphere with 192 available detector locations rather than a planar mask-detector configuration, resulting in a 4π FOV for both imaging modalities [30]. The random mask pattern was chosen and optimized based on the "Great Deluge algorithm" to determine the optimal number and configuration of detector elements on the sphere, and the resulting optimized mask contains a total of 104 detector elements. But the coded-aperture modality can only be used in the imaging energy range of 50 ~ 400 keV, and the angular resolution is approximately 10° [31]. A hybrid gamma camera combining both the coded-aperture and Compton imaging modalities was developed based on a single Timpex3 chip hybridized with a 1 mm thick cadmium telluride semiconductor, which aims to take advantage of the high resolution of the coded-aperture imaging and the wide FOV of the Compton imaging technique [32]. However, the problem of the partial encoding effect still exists for the coded-aperture modality, and an additional measurement using the Compton imaging modality is required to identify the ghosts within the reconstructed image. Furthermore, the two advantages cannot be utilized simultaneously, because only one modality can be used at a time.

2.3 Image Reconstruction

The image reconstruction of coded aperture cameras can be performed with the cross-correlation between the known decoding matrix and the pattern recorded at the detector, as discussed in Sect. 2.1. This reconstruction method is mathematically well defined and suited for applications requiring high computational efficiency. It is assumed in Eqs. (1) and (2) that the mask is infinitely thin and perfectly opaque. But in security applications, the imaging energy range is broad, and a thick mask is needed to stop the high-energy photons, projection data will suffer from a strong vignetting effect for low-energy photons. In addition, the noise from photon scattering, transmission, and statistical fluctuation cannot be considered during the image reconstruction, which can cause a degradation in image quality. As discussed in Sect. 2.2, the imaging region of the cross-correlation method is limited to the FCFOV, because the false hotspots caused by partial encoding effect cannot be distinguished.

An iterative reconstruction algorithm such as the maximum likelihood expectation maximization (MLEM) algorithm can be used to estimate the object image with the highest likelihood for the measured projection data, which assumes that the photon counts follow a Poisson process [33]. A system matrix s_{ij} incorporating all the physical processes of the photon transport is required, and the algorithm can be formulated as

$$\lambda_j^{n+1} = \frac{\lambda_j^n}{\sum_i^I s_{ij}} \sum_i^I \frac{s_{ij} p_i}{\sum_k^J s_{ik} \lambda_k^n} \tag{5}$$

where I and J are the numbers of detector and image pixels, respectively; s_{ij} is the probability of photons emitted from image pixel j being detected at detector pixel i; p_i is the measured projection at detector pixel i; and λ_j^n is the estimate of the source distribution at image pixel j after n iterations. A system matrix considering the PCFOV can address the partial encoding effect [34]. But it suffers from the issues of computational time consumption and local convergence. In addition, the choice of the source pixel size also presents a trade-off between the angular resolution and reconstruction time, and it is a waste of computing resources if no radioactive source is located at the PCFOV of the imaging system, but the system matrix containing both the FCFOV and PCFOV were used.

Recently, deep learning techniques have been proposed for image reconstruction in coded aperture imaging [35–40]. A reconstruction method based on convolutional neural network (CNN) was developed and compared with the cross-correlation method. One million training data were simulated with the number of one to three 37 MBq ^{137}Cs sources, and the method was evaluated at different simulation conditions, because the output image and input projection data have the same size, so that the neural network including only two convolutional layers and fully connected layers is enough. The results show that the proposed method can suppress noise well and accurately reconstruct single source, poor performance observed with multiple sources. A similar model structure with more convolutional and fully connected layers is proposed to overcome the limitations of reconstruction performance for extended sources, because the size of output image is much larger than the input project data. Online training with a total of 200 million synthetic data with different shape, position, and intensity of sources is used. Both simulation and experimental tests show promising results on the problem of extended sources, but only low-energy sources were considered in this study, and the physical processes of high-energy photons such as transmission and scattering cannot be included in the synthetic data. In addition, it is impractical to obtain such a large amount of data using Monte Carlo simulations or experiments.

3 Methodology

In this section, we propose a deep learning reconstruction method based on a hybrid network structure of convolutional neural network (CNN) and Swin Transformer to reconstruct radioactive source distribution directly using the project data. Only 0.6 million training data were computed using Monte Carlo simulations with four different energies. This method can accurately reconstruct the actual source position rather than a single hot pixel when point sources lie between pixels. Not only is the CNN network with skip connections used to extract the compact feature maps from the input project data, but the transformer structure is also employed to capture long-term dependencies within the projection data.

3.1 *Configuration of the Coded-Aperture Camera*

The detector and mask of the coded-aperture camera adopted in this work are shown in Fig. 2. The camera is equipped with a MURA rank-19 mask based on a 2 × 2 cyclic replication of the basic pattern with an 8-mm thick tungsten-copper alloy

(a) **(b)**

Fig. 2 Photographs of (**a**) the detector module containing a 19 × 19 pixelated BGO scintillation crystal array coupled to a Hamamatsu H8500 PSPMT and (**b**) the coded mask

material, resulting in 37 × 37 pixels with dimensions of 88.8 × 88.8 × 8 mm^3. With consideration of the mask parameters, the detector module was composed of a 19 × 19 pixelated bismuth germanate (BGO) scintillation crystal with a pixel size of 2.4 × 2.4 × 10 mm^3 coupled to a Hamamatsu H8500 position-sensitive photomultiplier tube (PSPMT).

Compared to the semiconductor detector such as the cadmium telluride (CdTe) and cadmium zinc telluride (CdZnTe), BGO crystal has several advantages such as low-cost, relatively high density (7130 kg/m^3), no intrinsic radioactivity, nonhygroscopic, and high absorption efficiency, making it well suited for the gamma-ray imaging applications in the energy range from 50 keV to 1500 keV. The 8 × 8 multi-anode outputs of the H8500 were connected to a resistive network with the discretized positioning circuit, reducing the 64-pixel outputs to four output signals. Compared to the approach required that each channel has its amplifier to read independently, this signal processing method can effectively simplify the readout circuits. After digitizing the four signals with a 40 MSPS sampling frequency and 12-bit resolution, the incident position and energy can be extracted from the input waveform.

3.2 Data Collection

As a data-driven non-model framework, a deep neural network requires an advanced network architecture to enhance the expressive ability and adequate training data to improve the generalization performance. Data collection is the first process and plays a key role in deep learning applications. In general, training samples can be obtained with the data synthesis, Monte Carlo simulations, and data acquisition using a real imaging system with experimental measurements. Data synthesis is an efficient approach to obtain a large amount of training samples by using the simple geometrical model of the imaging system or analytical calculation of the detector response such as the linear attenuation. However, it neglects the physical processes during the data acquisition such as the inter-crystal scattering and crystal penetration, which leads to obvious discrepancies between the distributions of the real data and synthetic data. The most accurate way of acquiring training samples would be derived directly from the experimental measurements using the real imaging system and radioactive sources, all the physical effects such as the electronic noise, scattering and transmission of the mask, and scattering both within the detector and in the surrounding materials can be included. The experimental setup can be tedious due to the requirement of the precise positioning device to move the radioactive sources to different positions, which increase the cost and complexity of the measurements. The distribution of the projection data will also change if the hardware modules are replaced with a new one, which leads to discrepancies between the training data and the real data. Furthermore, the counting rate in nuclear security applications may be much lower than that of the medical applications, resulting in a long acquisition time to satisfy the counting statistics of the projection data, which makes the experimental measurements unpractical. An alternative to the experimental method is to simulate the imaging process using the Monte Carlo simulations. A variety of software packages such as the MCNP, Geant4, and GATE have been developed and based on the repeated random sampling of probability models of the gamma-ray transport through the matter, which offers the possibility of simulating the physical processes of a gamma camera very precisely. In this chapter, the simulation strategy was used to obtain the training samples, because it can achieve a good balance between the precision and the acquisition time of the training data, only at the cost of the computing resources.

The counting distribution of the shadow cast on the detector has different response characteristics when radioactive sources are located at different positions. The method introduced in this chapter was developed for nuclear security applications, so that only the far-field geometry was considered. All radioactive sources were simulated on a source plane with a 2-m source-to-mask distance to satisfy the far-field condition so that the detected photons can be considered as a parallel incident. A 59-mm mask-to-detector distance was set to provide a $4618 \times 4618\ \text{mm}^2$ PCFOV and a $1544 \times 1544\ \text{mm}^2$ FCFOV, respectively. The position information of the radioactive sources contained in the projection data is proportion to the ratio of the photons modulated by the mask. With the above considerations, the imaging region of $3000 \times 3000\ \text{mm}^2$ is considered in this study.

In each simulation, a single point radioactive source is simulated, and the position is uniformly sampled in the imaging region. Considering the trade-off between computational time and statistical fluctuation, a total of 10^6 photons were emitted from each source position, resulting in around 75×10^3 average counts. The computing time of each simulation for the given conditions was around 70 seconds on a workstation (OS: Ubuntu 22.04.2, RAM: 96 GB, CPU: Intel(R) Xeon(R) Gold 6132 2.6 GHz), resulting in a total of 486 days with a single-core computer to complete the data acquisition. Sources with the energy of 122 keV, 356 keV, 662 keV, and 1173 keV were simulated, and the dataset number of each energy ($\boldsymbol{p}_{112\ \text{keV}}$, $\boldsymbol{p}_{356\ \text{keV}}$, $\boldsymbol{p}_{662\ \text{keV}}$, $\boldsymbol{p}_{1173\ \text{keV}}$) is 0.15 million, resulting in a total of 0.6 million simulated projection data. Different data augmentation methods were used to obtain enough data for model training. Firstly, different single-source projection data were summed to obtain the projection data with four sources rather than use these single-source data to train model directly, and the method of projection data is represented as follows:

$$\begin{aligned}
\boldsymbol{p}_{1,t} &= \boldsymbol{p}_{122\ \text{keV},t} + \boldsymbol{p}_{356\ \text{keV},t} + \boldsymbol{p}_{662\ \text{keV},t} + \boldsymbol{p}_{1173\ \text{keV},t} \\
\boldsymbol{p}_{2,t} &= \boldsymbol{p}_{122\ \text{keV},n-t} + \boldsymbol{p}_{356\ \text{keV},t} + \boldsymbol{p}_{662\ \text{keV},t} + \boldsymbol{p}_{1173\ \text{keV},t} \\
\boldsymbol{p}_{3,t} &= \boldsymbol{p}_{122\ \text{keV},t} + \boldsymbol{p}_{356\ \text{keV},n-t} + \boldsymbol{p}_{662\ \text{keV},t} + \boldsymbol{p}_{1173\ \text{keV},t} \\
\boldsymbol{p}_{4,t} &= \boldsymbol{p}_{122\ \text{keV},t} + \boldsymbol{p}_{356\ \text{keV},t} + \boldsymbol{p}_{662\ \text{keV},n-t} + \boldsymbol{p}_{1173\ \text{keV},t} \\
\boldsymbol{p}_{5,t} &= \boldsymbol{p}_{122\ \text{keV},t} + \boldsymbol{p}_{356\ \text{keV},t} + \boldsymbol{p}_{662\ \text{keV},t} + \boldsymbol{p}_{1173\ \text{keV},n-t}
\end{aligned} \tag{6}$$

where $t = 0, 1, 2, \ldots\ldots, n$, and n is 0.15 million in this work.

In addition, the cyclic permutation feature of the MURA mask was utilized, and all the simulated projection data were rotated 180° to generate a symmetrical data around the center of imaging region. Same summation operation as the Eq. (6) was applied to the symmetrical data, resulting in a total of 1.5 million projection data with four sources randomly split to 90% into the training dataset, 5% into the validation dataset, and 5% into the test dataset, respectively. These three datasets were kept separate throughout the study, and the normalization was applied for each projection data.

Only point-source condition was considered in this study; the ground truth of each data is generated according to the distance between the actual source position and the center points of its three nearest pixels rather than simply using 1 to indicate the presence of a source and 0 to indicate no source, because the source position between pixels won't be distinguished. A larger number indicates the source is closer to the center of the pixel.

3.3 Deep Neural Network Architecture

Deep learning technique has achieved great successes in several computer vision tasks, because the advanced deep neural networks were used and trained with big data. CNN and its variants have achieved state of the art in image classification, detection, segmentation, synthesis, and reconstruction but lack the ability to extract long-term dependencies in images. Inspired by the great performance of transformers in the natural language processing field, vision transformers (ViTs) are proposed by dividing the input images or feature maps into patches and forming sequences to calculate self-attention, thereby capturing long-term dependencies well within the input image [41]. A deep learning method based on a hybrid network structure of CNN and Swin Transformer is proposed in this chapter to reconstruct radioactive source distribution directly using the project data [42]. The network is comprised of the module of shallow feature extraction, deep feature extraction, and image reconstruction, and an overview of the architecture is represented in Fig. 3.

Considering the trade-off between training computational cost and model performance, the feature maps of the input projection data are extracted and downsampling to the dimension of $14 \times 14 \times 256$ using the shallow feature extraction module. The lost structural information increases with the increasement of the number of convolutional layer and need to be compensated by increasing the number of channels. Zero padding is used to prevent significant information loss to affect the performance of the subsequent image reconstruction.

The obtained feature maps are then input to multiple cascade CNN and Swin Transformer blocks to implement the deep feature extraction, because image processing tasks have greatly benefited from very deep models. The cascade CNN is composed of a stack of ten identical blocks. Each block has two convolutional layers, a convolutional block attention module (CBAM) and a rectified linear unit (ReLU) activation function [43]. CBAM is a simple yet effective attention module for feed-forward convolutional neural networks, which incorporates both spatial and channel attention mechanisms to capture more discriminative features and enhance the representation of input data [44]. In addition, skip connection also known as

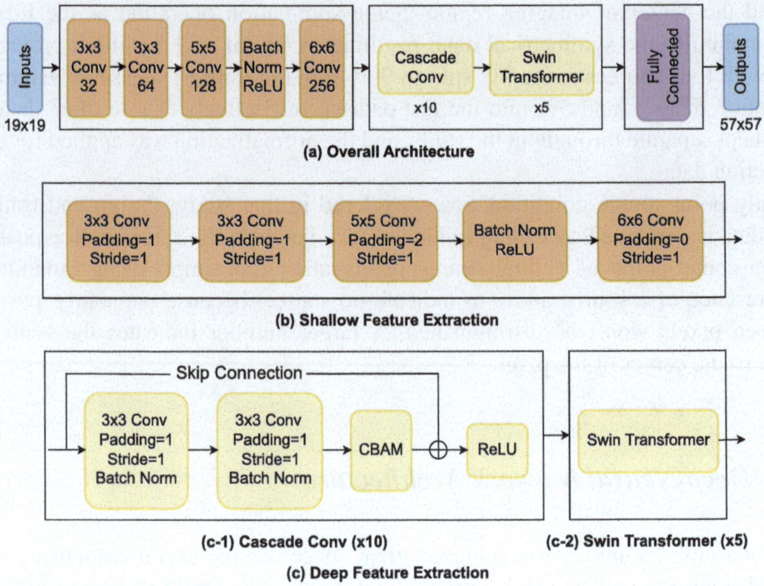

Fig. 3 (**a**) The overall architecture of our network; (**b**) shallow feature extraction; (**c**) deep feature extraction network structure

residual connection is employed to connect the former layer to the later layer to prevent the degradation problem and mitigate the vanishing gradient problem during training deep models. The residual connection allows the network to learn residual mapping instead of directly learning the underlying mapping from input to output and leads to fast convergence [44]. To facilitate these residual connections, the feature maps in each layer keep the same dimensions as the input data. The last stage of the deep feature extraction module is composed of a stack of five identical Swin Transformer layers. The Swin Transformer architecture is built on the core principles of processing input images hierarchically through multiple stages rather than processing the entire image as a single sequence, which allows the model to capture both local and global contextual information effectively. Each Swin Transformer layer has the dimensions of $14 \times 14 \times 256$ and with the patch size of 5 and the attention head number of 8. The window size for the sliding window self-attention mechanism is set to 7. Additionally, biases are added to the query, key, and value matrices. A total of 180 million parameters were included in the model.

3.4 *Training Procedure*

The model was trained using Pytorch 2.1.0 on a NVIDIA GeForce RTX4090 GPU. As a variant of the Adam optimization algorithm, AdamW optimization algorithm was used and introduced an additional weight decay term during updating weights, which can reduce the magnitude of weights and improve the generalization ability of the model. The learning rate was set to 0.001 with the weight decay of 0.05 and the momentum of $\beta_1 = 0.9$ and $\beta_2 = 0.999$, respectively. A warm-up period of five epochs was utilized to stabilize the learning process during the initial stages of model training and avoid instability caused by rapid weight updates. A cosine decay scheduler was used to help the model converge smoothly to the optimal solution in the later stages of training. Mean square error (MSE) loss function was used to calculate the difference between model output and ground truth. The total number of training epochs is 50, and the batch size is 256.

4 Results

4.1 Results of Test Dataset

The performance of the proposed deep learning model was first evaluated on the split test dataset, and the reconstruction results of eight randomly selected data were presented in Fig. 4. It can be shown that all the four sources within each data can be accurately reconstructed, but minor blurring of some sources observed. Both sides present hotspot in Fig. 4h, because the source was located at the boundary of the imaging region and the difference of the projection data between two positions is hard to be distinguished.

4.2 Results of Simulated Data with Different Number of Sources

Because the model was only trained using projection data with four sources, the feasibility of imaging other numbers of radioactive sources also needs to be evaluated. The source number of one, two, and three were tested, and all the sources can be accurately reconstructed as shown in Fig. 5.

4.3 Results of Experimental Data with Different Number of Sources

To validate the feasibility of the proposed deep neural network trained with the simulation data in real-world scenarios, we conducted an experiment using a 37 MBq ^{137}Cs point source placed on a mobile platform. Different source positions with a 2-m source-to-mask distance were measured by moving the mobile platform and controlling its height.

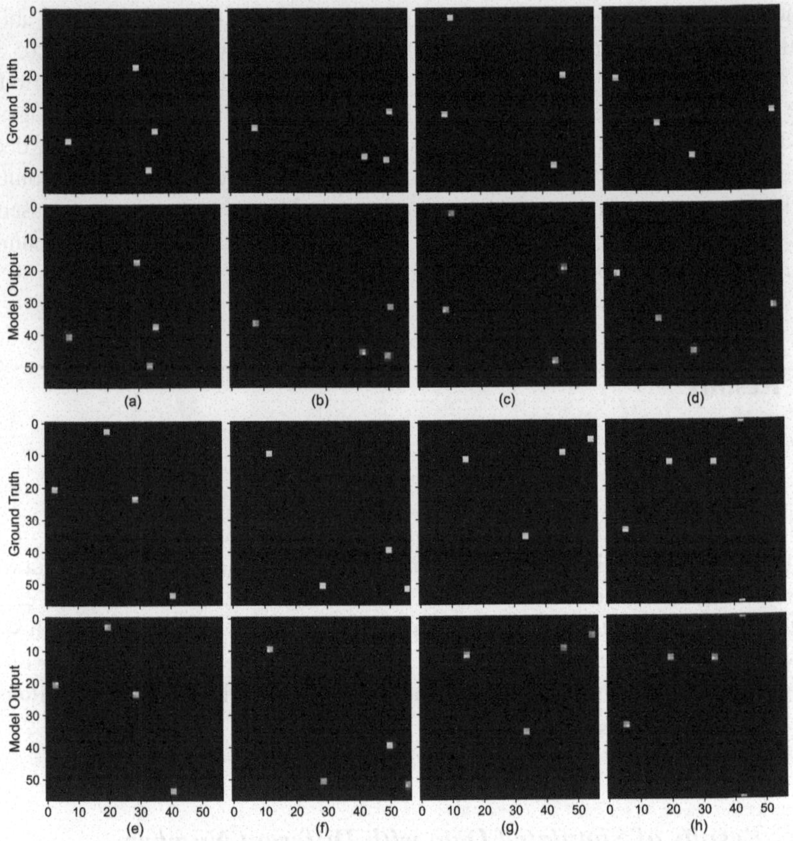

Fig. 4 Reconstruction results of eight randomly selected data evaluated on the split test dataset

At each source position, the projection data were acquired with ~180×10^3 and ~ 18×10^3 counts, respectively. The acquisition time of the 180×10^3 projection data range from ~600 s to ~1000 s, because the photons emitted from sources suffer more attenuation from the off-axis positions than the central positions. The projection of multiple sources was generated by summing the projection data of multiple single sources.

Figure 6a, b show the reconstructed images of the single source at (0, 0) and (600, 0), respectively. All source positions are in mm, showing relative offset from the centre of the FOV. The actual source position can be accurately reconstructed with great performance. Figure 6c, d represent the reconstructed images of two sources at (−600, 0) and (600, 600), (−750, 300), and (−600, 0), respectively, and similar performance can be observed as the simulation results. The artifact on the right side of Fig. 6d at 18 k counts is caused by the statistics fluctuations that make the projection data at (−750, 300) a little difficult to be distinguished with the projection data at (750, 300). In all cases, all the radioactive sources can be

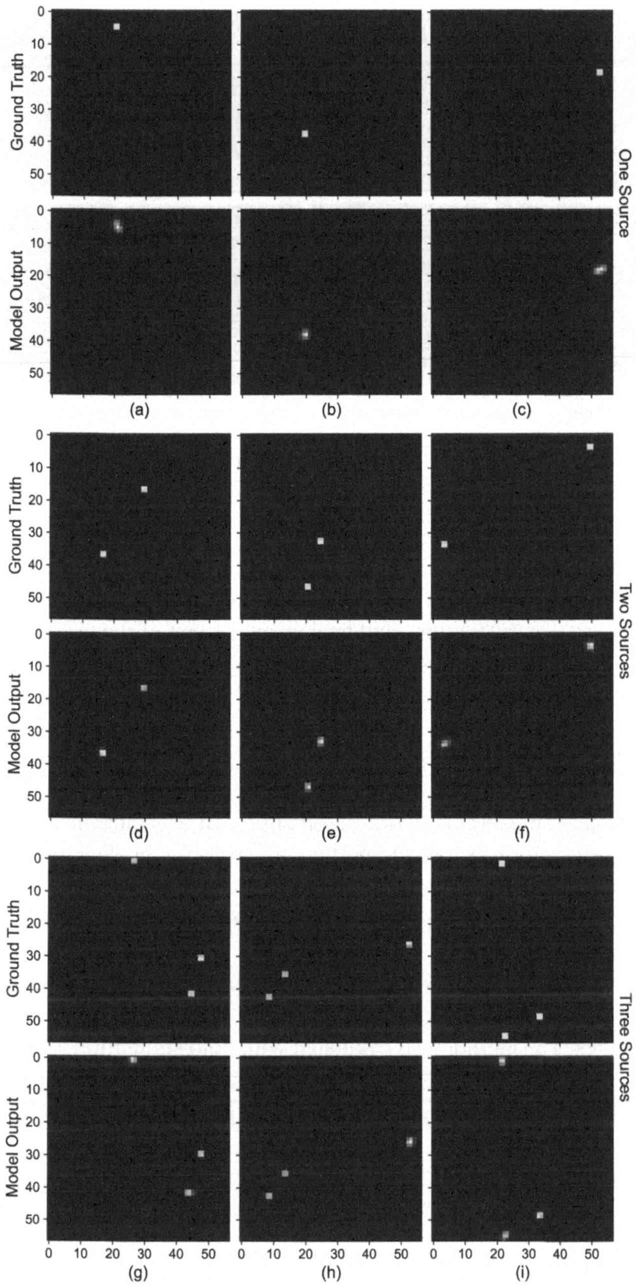

Fig. 5 Reconstruction results of simulation data with different number of sources

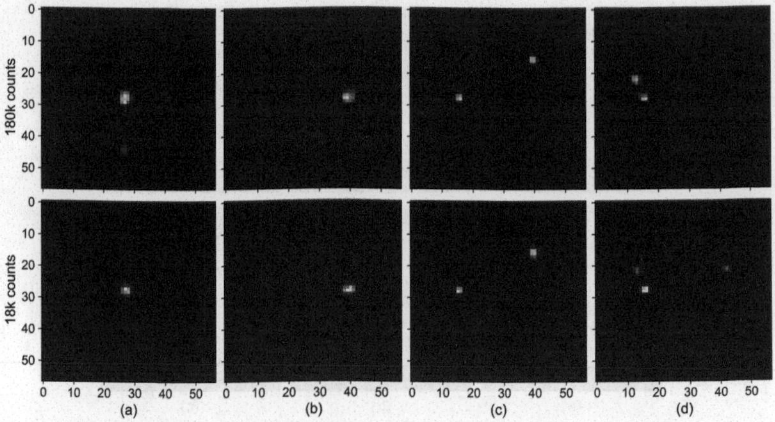

Fig. 6 Reconstruction results of real test acquisition with a ^{137}Cs source at (**a**) (0, 0), (**b**) (600, 0), (**c**) (−600, 0) and (600, 600), (**d**) (−750, 300) and (−600, 0) under 180 k and 18 k counts, respectively

accurately reconstructed even at 18 k count level, which represents the robustness to the counting statistics fluctuations and background noise of the proposed method.

5 Conclusion

In this chapter, a deep neural network-based approach was introduced to implement the image reconstruction in coded-aperture imaging for nuclear security applications. Not only is the CNN network with skip connections used to extract the compact feature maps from the input project data, but the transformer structure is also employed to capture long-term dependencies within the projection data. Both the PCFOV and FCFOV were considered during the image reconstruction, and only 0.6 million training data were computed using Monte Carlo simulations with four different energies. The method was evaluated with both simulation and experimental data, and the results show that this method can accurately reconstruct the actual source position rather than a single hot pixel when point sources lie between pixels.

References

1. IAEA, *Energy, Electricity and Nuclear Power Estimates for the Period up to 2050* (International Atomic Energy Agency, Vienna, 2023)
2. International Atomic Energy Agency, IAEA Incident and Trafficking Database 2023 Factsheet (2023)
3. E. Enlow, S. Abbaszadeh, State-of-the-art challenges and emerging technologies in radiation detection for nuclear medicine imaging: A review. Front. Phys. **11**(April), 1–19 (2023). https://doi.org/10.3389/fphy.2023.1106546

4. K. Amgarou, M. Herranz, State-of-the-art and challenges of non-destructive techniques for in-situ radiological characterization of nuclear facilities to be dismantled. Nucl. Eng. Technol. **53**(11) (2021). https://doi.org/10.1016/j.net.2021.05.031
5. L. Caballero et al., Gamma-ray imaging system for real-time measurements in nuclear waste characterisation. J. Instrum. **13**(3) (2018). https://doi.org/10.1088/1748-0221/13/03/P03016
6. O. Gal, C. Izac, F. Jean, F. Lainé, C. Lévêque, A. Nguyen, CARTOGAM – A portable gamma camera for remote localization of radioactive sources in nuclear facilities. Nucl. Instrum. Methods Phys. Res. Sect. A **460**(1), 138–145 (2001). https://doi.org/10.1016/S0168-9002(00)01108-6
7. P.T. Durrant, M. Dallimore, I.D. Jupp, D. Ramsden, The application of pinhole and coded aperture imaging in the nuclear environment. Nucl. Instrum. Methods Phys. Res. Sect. A **422**(1–3), 667–671 (1999). https://doi.org/10.1016/S0168-9002(98)01014-6
8. O. Gal et al., Development of a portable gamma camera with coded aperture. Nucl. Instruments Methods Phys. Res. Sect. A **563**(1), 233–237 (2006). https://doi.org/10.1016/j.nima.2006.01.119
9. F. Carrel et al., *GAMPIX: A new gamma imaging system for radiological safety and homeland security purposes*, in IEEE Nuclear science symposium conference record, (2011), pp. 4739–4744. https://doi.org/10.1109/NSSMIC.2011.6154706
10. M. Jeong, B. Van, B.T. Wells, L.J. D'Aries, M.D. Hammig, Scalable gamma-ray camera for wide-area search based on silicon photomultipliers array. Rev. Sci. Instrum. **89**(3) (2018). https://doi.org/10.1063/1.5016563
11. M. Jeong, M. Hammig, Development of hand-held coded-aperture gamma ray imaging system based on GAGG(Ce) scintillator coupled with SiPM array. Nucl. Eng. Technol. **52**(11) (2020). https://doi.org/10.1016/j.net.2020.04.009
12. C.G. Wahl et al., The Polaris-H imaging spectrometer. Nucl. Instrum. Methods Phys. Res. Sect. A. **784**, 377–381 (2015). https://doi.org/10.1016/j.nima.2014.12.110
13. L. Sinclair, P. Saull, D. Hanna, H. Seywerd, A. MacLeod, P. Boyle, Silicon photomultiplier-based Compton telescope for safety and security (SCoTSS). IEEE Trans. Nucl. Sci. **61**(5), 2745–2752 (2014). https://doi.org/10.1109/TNS.2014.2356412
14. A.M.L. MacLeod, P.J. Boyle, D.S. Hanna, P.R.B. Saull, L.E. Sinclair, H.C.J. Seywerd, Development of a Compton imager based on bars of scintillator. Nucl. Instrum. Methods Phys. Res. Sect. A. **767** (2014). https://doi.org/10.1016/j.nima.2014.08.012
15. B.R. Kowash, D.K. Wehe, A unified near- and far-field imaging model for rotating modulation collimators. Nucl. Instrum. Methods Phys. Res. Sect. A. **637**(1), 178–184 (2011). https://doi.org/10.1016/j.nima.2010.12.037
16. B.R. Kowash, D.K. Wehe, J.A. Fessler, A rotating modulation imager for locating mid-range point sources. Nucl. Instrum. Methods Phys. Res. Sect. A. **602**(2), 477–483 (2009). https://doi.org/10.1016/j.nima.2008.12.233
17. S.T. Brown, D. Goodman, J. Chu, B. Williams, M.R. Williamson, Z. He, Time-encoded gamma-ray imaging using a 3-D position-sensitive CdZnTe detector array. IEEE Trans. Nucl. Sci. **67**(2), 464–472 (2020). https://doi.org/10.1109/TNS.2019.2953182
18. X. Liang et al., Self-supporting design of a time-encoded aperture, gamma-neutron imaging system. Nucl. Instrum. Methods Phys. Res. Sect. A. **951**(July 2019), 162964 (2020). https://doi.org/10.1016/j.nima.2019.162964
19. N.P. Shah, P. Marleau, J.A. Fessler, D.L. Chichester, D.K. Wehe, Improved localization precision and angular resolution of a cylindrical, time-encoded imaging system from adaptive detector movements. IEEE Trans. Nucl. Sci. **68**(4), 410–425 (2021). https://doi.org/10.1109/TNS.2021.3060071
20. J. Brennan et al., Demonstration of two-dimensional time-encoded imaging of fast neutrons. Nucl. Instrum. Methods Phys. Res. Sect. A. **802**, 76–81 (2015). https://doi.org/10.1016/j.nima.2015.08.076
21. J.G. Ables, Fourier transform photography: A new method for X-ray astronomy. Publ. Astron. Soc. Aust. **1**(4), 172–173 (1968). https://doi.org/10.1017/s1323358000011292
22. R.H. Dicke, Scatter-hole cameras for X-rays and gamma rays. Astrophys. J. **153**, 101–106 (1968). https://doi.org/10.1086/180230

23. E.E. Fenimore, T.M. Cannon, Coded aperture imaging with uniformly redundant arrays. Appl. Opt. (1978). https://doi.org/10.1364/ao.17.000337
24. R. Accorsi, F. Gasparini, R.C. Lanza, Optimal coded aperture patterns for improved SNR in nuclear medicine imaging. Nucl. Instrum. Methods Phys. Res. Sect. A. **474**(3), 273–284 (2001). https://doi.org/10.1016/S0168-9002(01)01326-2
25. K. Byard, P.M.E. Shutler, Singer product apertures – A coded aperture system with a fast decoding algorithm. Nucl. Instrum. Methods Phys. Res. Sect. A **856**(June 2016), 147–153 (2017). https://doi.org/10.1016/j.nima.2016.12.042
26. P.M.E. Shutler, K. Byard, Singer quad apertures and super fast decoding. Nucl. Instrum. Methods Phys. Res. Sect. A. **944**(August), 162469 (2019). https://doi.org/10.1016/j.nima.2019.162469
27. P.M.E. Shutler, K. Byard, Signal-to-noise ratio of singer product apertures. Nucl. Instrum. Methods Phys. Res. Sect. A.. **867**(January), 237–258 (2017). https://doi.org/10.1016/j.nima.2017.05.037
28. S. Sun et al., Development of a panorama coded-aperture gamma camera for radiation detection. Radiat. Meas. **77**(June), 34–40 (2015). https://doi.org/10.1016/j.radmeas.2015.04.014
29. M.L. Galloway et al., *Status of the high efficiency multimode imager,* in IEEE nuclear science symposium conference record (2011), pp. 1290–1293. https://doi.org/10.1109/NSSMIC.2011.6154328
30. D. Hellfeld, P. Barton, D. Gunter, A. Haefner, L. Mihailescu, K. Vetter, Real-time free-moving active coded mask 3D gamma-ray imaging. IEEE Trans. Nucl. Sci. **66**(10), 2252–2260 (2019). https://doi.org/10.1109/TNS.2019.2939948
31. D. Hellfeld, P. Barton, D. Gunter, L. Mihailescu, K. Vetter, A spherical active coded aperture for \$4π\$ gamma-ray imaging. IEEE Trans. Nucl. Sci. **64**(11), 2837–2842 (2017). https://doi.org/10.1109/TNS.2017.2755982
32. G. Amoyal, V. Schoepff, F. Carrel, M. Michel, N. Blanc de Lanaute, J.C. Angélique, Development of a hybrid gamma camera based on Timepix3 for nuclear industry applications. Nucl. Instrum. Methods Phys. Res. Sect. A. **987**(October), 164838 (2021). https://doi.org/10.1016/j.nima.2020.164838
33. Q. Liu et al., Image reconstruction using multi-energy system matrices with a scintillator-based gamma camera for nuclear security applications. Appl. Radiat. Isot. **163**(September) (2020). https://doi.org/10.1016/j.apradiso.2020.109217
34. M. Jeong, G. Kim, MCNP-polimi simulation for the compressed-sensing based reconstruction in a coded-aperture imaging CAI extended to partially-coded field-of-view. Nucl. Eng. Technol. (2020). https://doi.org/10.1016/j.net.2020.02.011
35. T. Meißner, V. Rozhkov, J. Hesser, W. Nahm, N. Loew, Quantitative comparison of planar coded aperture imaging reconstruction methods. J. Instrum. **18**(1) (2023). https://doi.org/10.1088/1748-0221/18/01/P01006
36. A. Kulow et al., Comparison of three reconstruction methods based on deconvolution, iterative algorithm and neural network for X-ray fluorescence imaging with coded aperture optics. J. Anal. At. Spectrom. **35**(7), 1423–1434 (2020). https://doi.org/10.1039/d0ja00146e
37. R. Zhang et al., Reconstruction method for gamma-ray coded-aperture imaging based on convolutional neural network. Nucl. Instrum. Methods Phys. Res. Sect. A **934**(May), 41–51 (2019). https://doi.org/10.1016/j.nima.2019.04.055
38. Q. Liu et al., Neural network method for localization of radioactive sources within a partially coded field-of-view in coded-aperture imaging. Appl. Radiat. Isot. **170**(April) (2021). https://doi.org/10.1016/j.apradiso.2021.109637
39. Q. Liu, Y. Cheng, X. Tuo, Y. Yang, Radioactive source localization method for the partially coded field-of-view of coded-aperture imaging in nuclear security applications, in *Advanced X-Ray Radiation Detection: Medical Imaging and Industrial Applications*, (2022). https://doi.org/10.1007/978-3-030-92989-3_12
40. G. Daniel, O. Limousin, Extended sources reconstructions by means of coded mask aperture systems and deep learning algorithm. Nucl. Instrum.s Methods Phys. Res. Sect. A **1012** (2021). https://doi.org/10.1016/j.nima.2021.165600

41. A. Dosovitskiy et al., *An image is worth 16x16 words: Transformers for image recognition at scale*, in International conference on learning representations (2021). [Online]. https://openreview.net/forum?id=YicbFdNTTy
42. Z. Liu et al., Swin transformer: Hierarchical vision transformer using shifted windows. Proc. IEEE Int. Conf. Comput. Vis., 9992–10002 (2021). https://doi.org/10.1109/ICCV48922.2021.00986
43. S. Woo, J. Park, J.Y. Lee, I.S. Kweon, CBAM: Convolutional block attention module. Lect. Notes Comput. Sci **11211**, 3–19 (2018). https://doi.org/10.1007/978-3-030-01234-2_1
44. K. He, X. Zhang, S. Ren, J. Sun, Deep residual learning for image recognition. Proc. IEEE Comput. Soc. Conf. Comput. Vis. Pattern Recognit. (December), 770–778 (2016). https://doi.org/10.1109/CVPR.2016.90

Artificial Intelligence for X-Ray Photon Counting Technology: Current Status and Future Perspectives

Alison Deng, James Day, Elmaddin Guliyev, Kris Iniewski, and Magdalena Bazalova-Carter

1 Motivation

Photon counting computed tomography (PCCT) is an emerging CT modality that uses semiconductor-based photon-counting detectors (PCDs) [1–3]. It promises much better spatial resolution and material discrimination properties due to its inherent color X-ray detection capability. As Spectral Photon Counting (SPC) tomography collects at least ten times more medical imaging data than standard computed tomography (CT), it naturally lends itself to be used in conjunction with artificial intelligence (AI) techniques.

AI encompasses a broad range of technologies that enable computers to perform tasks that typically require human intelligence. Machine learning, a subset of AI, focuses on developing algorithms that enable computers to learn from and make predictions based on data [4–6]. In the broader field of computed tomography (CT), machine learning can optimize scanning parameters, improve image quality, and reduce noise. By learning from a dataset of annotated images, machine learning models can enhance the efficiency of image processing and analysis, leading to faster and more accurate diagnostics. Deep learning, a more advanced subset of machine learning, utilizes neural networks with multiple layers to analyze and generate complex data [5]. Deep learning algorithms excel at image classification, segmentation, and enhancement tasks in medical imaging [4, 5].

This chapter surveys four areas of applying AI techniques to PCCT. Noise reduction is a critical area where AI can substantially improve PCCT. In PCCT, noise can obscure anatomical features despite the high resolution of images, making

A. Deng · J. Day · M. Bazalova-Carter
Physics & Astronomy Department, University of Victoria, Victoria, BC, Canada

E. Guliyev · K. Iniewski (✉)
Redlen Technologies, Saanichton, BC, Canada

K. (K.) Iniewski, L. (K.) Cai (eds.), *Deep Learning for Advanced X-ray Detection and Imaging Applications*, https://doi.org/10.1007/978-3-031-75653-5_12

it challenging to achieve accurate diagnoses. AI algorithms are adept at identifying and filtering out noise, thus preserving the integrity of the image details [6, 7]. Material decomposition in PCCT is an advanced application that benefits from AI's capabilities. By differentiating between various materials based on the energy levels of counted photons, AI can provide unprecedented insights into the composition of tissues and other bodily substances [8, 9]. Metal artifact reduction is one area where AI-driven methodologies can help to mitigate the impact of metal artifacts on CT imaging, thus substantially elevating both the diagnostic precision and the visual clarity of images [10]. Image reconstruction, another domain where AI significantly contributes, holds immense potential. Traditional image reconstruction methods may only partially exploit the potential of the data collected by PCCT. However, AI-driven techniques can reconstruct images to maximize resolution and clarity, opening exciting possibilities for more precise interpretations [11]. These areas of research are beneficial in detecting and characterizing diseases. Enhancing overall image quality is one of the most straightforward yet profound benefits of applying AI in PCCT. Through deep learning, AI can improve images' clarity, contrast, and detail, making it easier for clinicians to diagnose, monitor disease progression, and plan treatments.

In essence, the combination of AI and PCCT has the potential to revolutionize medical imaging. By addressing key challenges and significantly enhancing image quality and interpretation, AI empowers PCCT to unlock new potentials in clinical research, diagnostics, and patient care, setting a new standard in the precision and efficacy of medical imaging practices.

2 AI, Machine, and Deep Learning

2.1 Historical Perspective

From movie recommendations on Netflix to conversations with ChatGPT, we hear about artificial intelligence (AI) on many occasions. Movie selection is not a big deal, but ChatGPT might someday become an incredible expert medical system, providing rapid diagnostic and therapeutic information to physicians at the point of care. It can warn about interactions and side effects, remind physicians of best practices, and generally improve the care delivered.

The application of AI is exploding. For example, it has taken ChatGPT just 5 days to reach a million users, while Spotify took 5 months and Netflix 3.5 years. There are hundreds of millions of users now, and all big corporations are spending their resources heavily on AI. Figure 1 shows the market capitalization of the top seven AI companies in the world; the first four are also the most valuable in the world.

While it can be argued whether Apple or Tesla is an AI company (the former produces phones and laptops while the latter produces cars), it is undeniable that all these seven have invested very heavily in AI development. Many other start-

Fig. 1 Market capitalization of the largest seven AI companies in the world (Apple, Microsoft, Nvidia, Alphabet (Google), Amazon, Meta (Facebook), Tesla, and IBM). Data collected on August 14, 2024; the stock market valuation changes daily. The entire stock market worldwide is worth about $100 T, so AI companies have about 20% market share already

ups, private companies, research institutions, and government research centers are developing AI algorithms. Despite the huge investment, AI technology has not yet produced any significant revenue, is still exploratory, and is not very practical. AI can also hallucinate at times, surprising with a very high degree of confidence, which is quite dangerous. That is expected to change very soon. AI techniques are getting better daily, and many jobs are ripe for displacement as AI can do the work at hand much more efficiently at a fraction of the cost of a human being. For example, 90% of the Internet content is expected to be AI or AI assisted by 2030.

As big tech companies invest billions of dollars into AI developments, countries like the USA and China vie for worldwide AI supremacy. Whoever develops the best AI system might win it all. One of the seven companies discussed, Nvidia, produces computer chips for AI and has a virtual market monopoly in that business, justifying a 3 trillion-dollar market capitalization. The largest semiconductor company in the world, TSMC, based in Taiwan, that produces all Nvidia chips is worth less than Nvidia, and the valuation of the previous king of the chips, INTEL, which employs over 100,000 people and runs 15 huge silicon fabs, is 30 times lower than Nvidia!

Computer science is pushing forward the ability of computers to handle complex calculations needed to help discover new medical drugs, identify trends and risks in financial modeling, handle large database searches, and perform massive calculations needed in many fields of science. From writing new books to fixing computer programs, AI techniques are expected to eventually proliferate in all aspects of our lives. The question is how did we get here? Figure 2 illustrates a quick history of AI development but let us discuss details a bit.

2.2 *Deep Blue and Chess*

Let us start our little AI journey with chess. Chess is a well-known board game between two players. The current form of the game emerged in Spain and the rest of Southern Europe during the second half of the fifteenth century after evolving from chaturanga, a similar but much older game of Indian origin. Unlike bridge

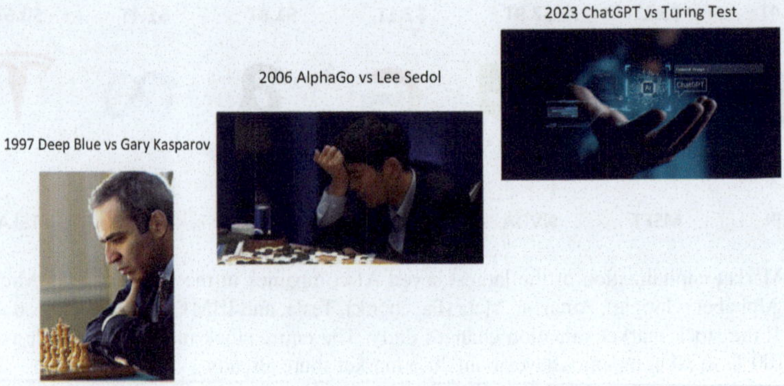

Fig. 2 Schematic illustration of the achievements of AI technology. In 1997, Gary Kasparov was beaten by Deep Blue; in 2006, Lee Sedol was beaten by AlphaGO; and in 2023, Turing test was beaten by ChatGPT

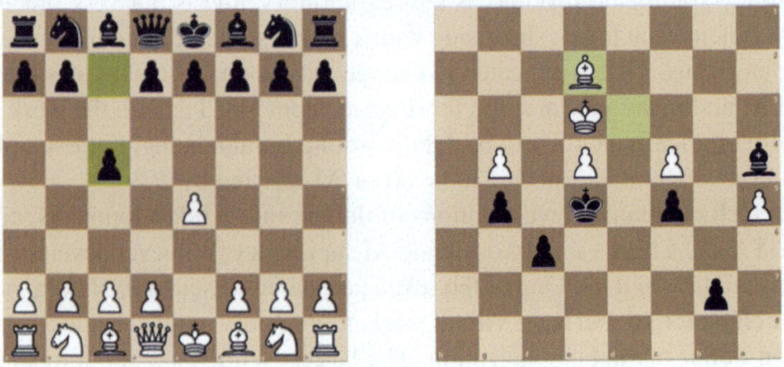

Fig. 3 Chess board game. Examples of the typical opening with 32 pieces (Sicilian defense, E2-E4, C7-C5) and the end game with only 12 pieces left

or poker, chess involves no hidden information. It is played on a chessboard with 64 squares arranged in an eight-by-eight grid with 16 pieces by each player; see Fig. 3 for illustration of the beginning of the game and its ending. The number of possible moves in chess is large, but not astronomically large; on average, every player might have 20–30 reasonable moves to choose from, and the average game lasts 40–50 moves. The complexity of the game is high, with millions of people playing it and thousands of books written about it.

It was only natural that one of the goals of early computer scientists was to create a chess-playing machine. In 1997, IBM Deep Blue became the first computer to beat the reigning World Champion Gary Kasparov. The game was close, with two wins for Deep Blue, one for Kasparov, and three draws. The match lasted several days and received massive media coverage around the world. It was the classic plot line of man vs. machine. Once the machine won that game, the game was over; machines will always be better than humans forever. Today's chess engines are significantly

Fig. 4 Go board game. Initial phase of the game (left) and final (right)

stronger than the best human players and have deeply influenced the development of chess theory.

The win for IBM was very impressive, but the Deep Blue program was quite limited in its capability; it could just play chess, nothing else. Since its "intelligence" was quite narrow, critics said at the time that the machine was simply calculating all possible move combinations. Deep Blue was supported by 480 special very large scale of integration (VLSI) silicon chips that could see 12 moves ahead, while Kasparov arguably could see about 10 moves ahead. Deep Blue could search and rank possible outcomes of 12 moves ahead, an advantage that piled on and led to the victory. That strategy was fine for chess but would never work for more complex games such as Go.

2.3 AlphaGO and Go

Go is an abstract strategy board game for two players in which the aim is to surround more territory than the opponent. The game was invented in China more than 2500 years ago and is believed to be the oldest board game continuously played to the present day. Two players, using either white or black stones, take turns placing their stones on a board; see Fig. 4. The goal is to surround and capture their opponent's stones or strategically create spaces of territory. Once all possible moves have been played, both the stones on the board and the empty points are tallied. The highest number wins.

Despite decades of work, the strongest Go computer programs could only play at the level of human amateurs. Standard AI methods, which test all possible moves and positions using a search tree, could not handle the sheer number of possible Go moves or evaluate the strength of each possible board position. As simple as the rules may seem, Go is profoundly complex. There are an astonishing 10 to the power of 170 possible board configurations—more than the number of atoms in the known universe. This makes the game of Go ten trillion (10^{10}) times more complex than chess. Go is known as the most challenging classical game for artificial intelligence because of its complexity.

In 2016, Lee Sedol, arguably the strongest Go player in history, was the first Go world champion to be defeated by a computer program. This program was called AlphaGo and was a computer program that combined advanced search trees with deep neural networks. These neural networks took a description of the Go board as an input and processed it through several different network layers containing millions of neuron-like connections. These layers were hidden deep in the network, hence the term "deep learning" used for such software. One neural network, called the "policy network," selects the next move to play. The other neural network, called the "value network," predicts the winner of the game. At the beginning, AlphaGo was introduced to numerous amateur games to help develop an understanding of reasonable human play. Then, the developers had it played against different versions of itself thousands of times, each time learning from its mistakes. Over time, AlphaGo improved and became increasingly stronger and better at learning and decision-making. This process is known as reinforcement learning. Once AlphaGo conquered the game, it was expected to move on to something more productive and beneficial to society. But things turned unexpectedly in a different direction.

2.4 ChatGPT and Turing Test

The impressive speed at which AI has evolved has never been more apparent than it is now, with ChatGPT making headlines and the dramatic evolution of large language models (LLMs) ever present in the media cycle. Millions of people worldwide have wasted no time adopting conversational AI tools in their day-to-day existence. These tools have not only enamored but also terrified audiences with their striking capabilities, efficiency, and potentially dangerous implications if not regulated well.

The idea of LLMs was first floated with the creation of the Eliza program in the 1960s, so this is not new technology. But it has taken more than 60 years to bring it to the market, why? Simply because there was not enough computer power available to process the new ideas. The significant turning point was the Google Brain concept developed in 2011. Google Brain was a deep learning artificial intelligence research team under the umbrella of Google AI, a research division at Google dedicated to artificial intelligence. It combined open-ended machine learning research with information systems and large-scale computing resources. It was merged into the former Google sister company DeepMind to form Google DeepMind in 2023.

Due to Google Brain and similar AI developments, researchers focused on building increasingly larger models. It was in 2019 that researchers from Google introduced BERT, the two-directional, 340-million parameter model that could determine context allowing it to adapt to various tasks. By pretraining BERT on a wide variety of unstructured data via self-supervised learning, the model was able to understand the relationships between words. In no time at all, BERT became the go-to tool for natural language processing tasks. In fact, it was BERT that was behind every English-based query administered via Google Search.

As BERT became more refined, OpenAI's GPT-2, at 1.5 billion parameters, successfully produced convincing prose. Then, in 2020, GPT-3 at 175 billion parameters was released, which set the standard for LLMs and formed the basis of ChatGPT. It was with the release of ChatGPT in November 2022 that the public really began to take notice of the impact of LLMs. Even nontechnical users could prompt the LLM, receive a quick response, and carry on a conversation, causing a stir of both excitement and apprehension.

Most recently, OpenAI introduced GPT-4, which is estimated at one trillion parameters—five times the size of GPT-3 and approximately 3000 times the size of BERT when it first came out, and as they say, "life has never been the same since." GPT-4 features significant enhancements, such as the ability to use computer vision to interpret visual data. GPT-4 accepts both text and images as input. To the delight of students and the consternation of teachers, GPT-4 can pass multiple standardized tests. Finally, it has managed to beat the famous Turing test (the Turing test originally developed by Alan Turing in 1950 is a test of a machine's ability to exhibit intelligent behavior equivalent to, or indistinguishable from, that of a human).

At its core, a large language model is a machine learning model that can understand and generate human language via deep neural networks. The main job of a language model is to calculate the probability of a word following a given input in a sentence. For example, "The color of the sun is ____" with the most likely answer being "yellow." The model can predict the next word in a sentence after being given a large text dataset. It learns to recognize different patterns in the words. From this process, you get a pretrained language model. Although LLMs have primarily been trained on simple tasks, like predicting the next word in a sentence, it is amazing to see how much of the structure and meaning of language they have captured, not to mention the vast number of facts they can pick up. A word of caution here: although LLM-based GPT products can learn and communicate like humans, our learning styles and behavioral patterns are still vastly more complex.

2.5 AI Classification

As the above short chess/go story illustrates, artificial intelligence has changed enormously over the last 27 years since Deep Blue victory in 1997. Various terms like AI, deep learning, and machine learning are being used in literature and popular science so it is important to explain these terms in more detail.

Machine learning approaches use training samples to discover patterns and ultimately achieve an accurate prediction of future data or trends. With the development of machine learning approaches and the increasing scale and difficulty of data analysis, traditional machine learning methods have often become unsuitable. As a result, deep learning methods have been developed because they have more com-

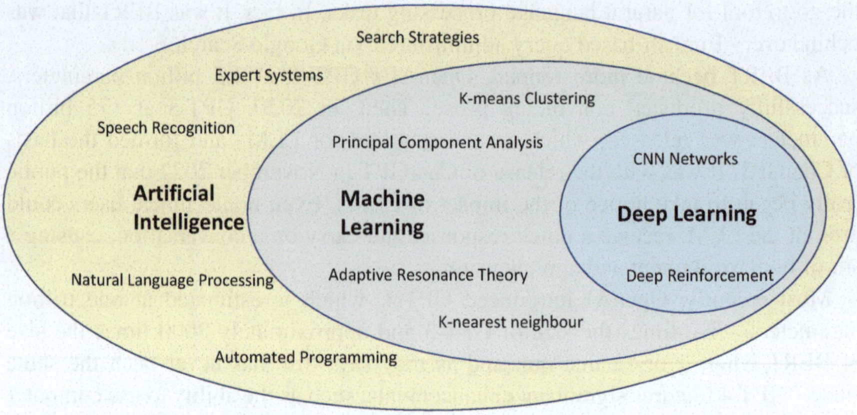

Fig. 5 Schematic illustration of the relationship between deep learning, machine learning, and artificial intelligence

plex architectures and improved data analysis capacity. The relationship between artificial intelligence, machine learning, and deep learning can be understood as three circles, deep learning being the smallest and fully contained within machine learning, which is contained within the artificial intelligence circle; see Fig. 5.

Traditional machine learning involves manually extracting features on small sample sets to balance the validity of the learning results with the interpretability of the learning model and to provide a framework for solving the learning problem when limited samples are available. Traditional machine learning can be subdivided into supervised learning, unsupervised learning, and reinforcement learning, based on the training method depending on whether the training data have labels. Statistics is one of the important theoretical foundations of traditional machine learning methods, and the main analysis includes pattern classification, regression analysis, probability density estimation, and other analyses.

To close this section, let us briefly discuss the difference between supervised learning and unsupervised learning approaches, as that would become important when discussing the application of AI in photon counting computed tomography (PCCT). Supervised learning is a machine learning approach that is defined by its use of labeled datasets. These datasets are designed to train or "supervise" algorithms to classify data or predict outcomes accurately. Using labeled inputs and outputs, the model can measure its own accuracy and learn over time. Supervised learning can be separated into two types of problems when data mining: classification and regression. Classification problems use an algorithm to accurately assign test data into specific categories, such as separating apples from oranges. Regression is another type of supervised learning method that uses an algorithm to understand the relationship between dependent and independent variables. A simple example of regression algorithms is linear regression.

Unsupervised learning uses machine learning algorithms to analyze and cluster unlabeled data sets. These algorithms discover hidden patterns in data without the need for human intervention, hence their name. Unsupervised learning models are used for two main tasks: clustering and association. Clustering is a data mining technique for grouping unlabeled data based on their similarities or differences. For example, K-means clustering algorithms assign similar data points into groups, where K represents the size of the grouping and granularity. This technique is helpful for market segmentation and image compression. Association is another type of unsupervised learning method that uses different rules to find relationships between variables in each dataset. An example would be Netflix suggesting a certain movie to watch based on previous viewing.

What is the main difference between supervised and unsupervised learning? The main distinction between the two approaches is the use of labeled datasets. Supervised learning uses labeled input and output data, while an unsupervised learning algorithm does not. In supervised learning, the algorithm "learns" from the training dataset by iteratively making predictions on the data and adjusting for the correct answer. While supervised learning models tend to be more accurate than unsupervised learning models, they require upfront human intervention to label the data appropriately. In contrast, unsupervised learning models work independently to discover the inherent structure of unlabeled data. The main difference between supervised and unsupervised learning algorithms is whether the labeled data are used or not. In each category, there are simple and complex models.

There are serious risks and drawbacks to using both approaches. The labeling process of supervised learning can be time-consuming and often requires domain expertise, especially in fields like medical imaging or natural language processing. However, the advantage is that once labeled, supervised learning models can achieve high accuracy and generalize well to new data. Meanwhile, unsupervised learning methods do not require labeled data, making them less resource-intensive in terms of data preparation. On the contrary, it can have wildly inaccurate results unless you have human intervention to validate the output variables.

Classifying large data sets can be challenging in supervised learning, but the results are highly accurate and trustworthy. In contrast, unsupervised learning can handle large volumes of data in real time. However, there is a lack of transparency into how data is clustered and a higher risk of inaccurate results. The only reliable way to select supervised vs. unsupervised training programs is to try to use both [12, 13].

A very interesting example of deep reinforcement learning was recently presented in the field of biology. In the study, published in *Nature Machine Intelligence* (www.nature.com/articles/s42256-024-00854-2), researchers trained an AI agent to direct one- elegans worms toward food in a dish, and AI algorithm performed better than the worms themselves without AI intervention. A nearby camera recorded the location and orientation of every worm's body, providing information for the previous several frames, giving it a sense of the past and present at each moment. The agent could also turn on or off a light aimed at the dish. The worms were optogenetically engineered so certain neurons would become active or inactive in response to the light, prompting movement as directed.

Future work to test this method would be of interest in larger animals with larger nervous systems than worms as the nervous system of the *Caenorhabditis elegans* is composed of only 302 neurons that are organized in several ganglia in the head and tail and into a spinal cord-like ventral nerve cord. The study opens new avenues for using deep reinforcement learning to understand neural circuits, train in biologically relevant real-world environments and modulate animal behavior.

3 Photon-Counting Technology

3.1 Photon-Counting Detectors (PCDs)

Photon-counting detectors (PCDs) are direct converters where the X-ray photon energy is directly translated into an electronic signal by the semiconductor sensor. The height of this signal is proportional to the photon energy. If the signal duration is short enough, the PCD is can detect almost every photon impinging on the sensor signal. High-Z materials, such as cadmium telluride (CdTe) and cadmium zinc tellurium (CdZnTe—CZT), are typically used for sensor fabrication to absorb X-ray photons effectively [1–3].

Photon-counting detectors (PCDs) offer several advantages over energy-integrated detectors (EIDs) both those that utilize an in-direct conversion process of X-ray conversion to visible light by a scintillating material followed by light to electronic signal conversion that uses standard photodiodes, as well as EIDs that utilize direct conversion similarly to PCDs. First, PCDs are not subject to direct signal contributions from detector intrinsic noise, also called electronic noise since their lowest energy threshold is set above the electronic noise floor. In an EID, electronic noise adds to the measured signal and becomes relevant in low flux detection. For that reason, PCDs are frequently referred to as "noise-free"; although that is highly inaccurate as a PCD, like every electronic device, suffers from noise effects, it is just that the level of electronic noise contribution in PCDs is much lower than in EIDs regardless of conversion type.

Second, PCDs can resolve photon energy, colorizing an X-ray image. This can offer a significant advantage over EID techniques particularly for material decomposition where an image is decomposed into the materials responsible. While energy integrating detectors have image techniques, such as dual energy using two perpendicular X-ray sources simultaneously, these techniques suffer from spectral overlap and misregistration reducing image quality. Spectral overlap on its own precludes EID techniques from effectively preforming the multi-energy imaging required for the decomposition of more than two materials. By binning the photons into two or more energy bins, their signals are energy-selective, and thus, a PCD intrinsically offers a spectral imaging capability, colorizing the X-ray spectrum.

Photon-counting detectors overcome the challenges with EIDs through colorization of the X-ray spectrum, allowing them to exceed two material decompositions without significant spectral overlap or misregistration. Spectral imaging allows these detectors to decompose an image into material components without motion misregistration issues.

Third, direct conversion PCDs have excellent spatial resolution as they are built using small sensor pixels. The small pixel size also helps avoiding or minimizing pulse-pileup effect, that is, counting of two photons that arrive very close in time as one single photon. In contrast, in EID, smaller pixel size requires cutting the sensor into smaller pixels and thus creating more dead space around the pixels. PCDs do not require such cutting because the pixel dimensions are determined by the pixel metal electrodes defined in the lithography process during sensor fabrication.

X-ray detectors are the major and arguably the most important component of a clinical CT system and have a substantial impact on image quality as well as patient radiation dose, which consequently affects a wide range of clinical applications. Compared to the conventional scintillator-based CT detectors using an indirect conversion technology that integrates photon energy deposited by all received X-ray photons, photon-counting detectors possess many inherent advantages, which can be largely attributed to their direct conversion technology and energy discrimination capability. Although PCD technology has been widely utilized in nuclear medicine in the last several years, its adoption in X-ray CT systems has long been hampered by the high photon flux effects encountered in clinical CT imaging. The recent significant advances in cadmium telluride (CdTe) and cadmium zinc tellurium (CZT) technology have overcome these problems and high-Z-based photon-counting computed tomography (PCCT) is now a clinical reality.

One of the main advantages of PCD is the much better spatial resolution. Detectors with high spatial resolution enable clinicians to see smaller details like plaque in coronary arteries much better and is one of the most important selling features of PCCT scanners. Spatial resolution is much better in photon-counting detectors than in energy-integrated detectors due to much smaller pixels, typically by a factor of 10 in pixel area, used in PCCT. In conventional EID detectors, a reflective layer (septa) of a finite width thickness is incorporated between each scintillating detector pixel, along the detector row and channel directions. Since scintillating detectors create visible light in the process of X-ray detection, the septa is designed to prevent the leakage of visible light between adjacent detector pixels. Figures 6 and 7 illustrate the contrast between EID and PCCT detectors.

While it is possible to create smaller EID pixels, there is a physical limit to the width of the septa. Smaller detector pixels with a finite-width septa lead to reduced geometric dose efficiency, a consequence of the reduced detector fill factor. Today's EID scanners typically utilize 1 mm pixel pitch, which translates into roughly 580 um spatial resolution at the ISO center. PCCT scanners typically utilize 330 um pixels, which translates into roughly 190 um spatial resolution at the ISO center, 3x improvements in spatial resolution.

Further improvement in EID spatial resolution can be obtained using comb filters placed near the detector to reduce the pixel aperture, resulting in smaller pixel

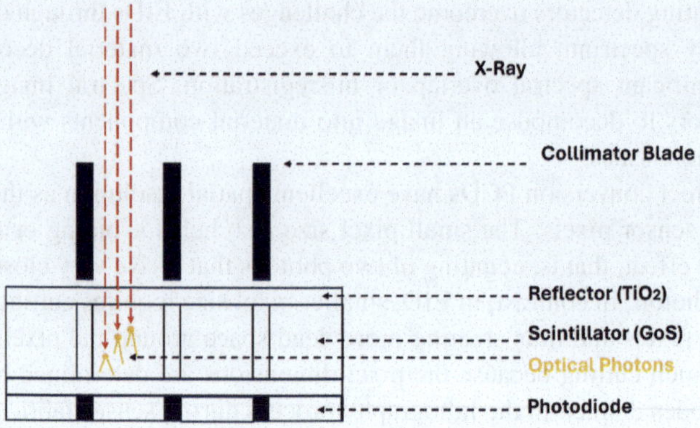

Fig. 6 Schematic cross section though EID scintillation indirect detector

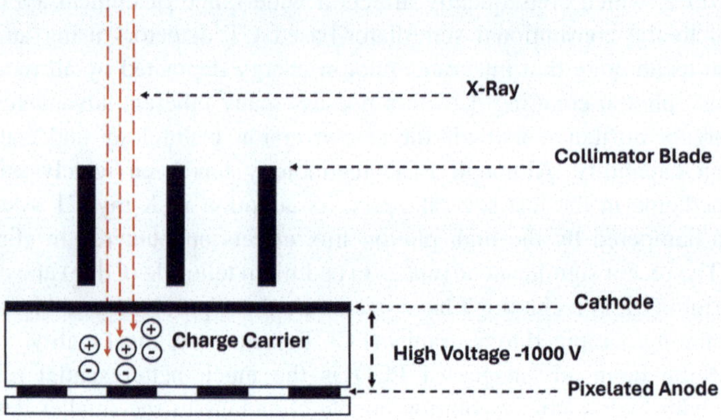

Fig. 7 Schematic cross section through PCCT direct conversion detector. Typical detector material is 2 mm thick CdTe or CdZnTe with high voltage (HV) applied between its cathode and pixelated anode

sizes for ultra-high-resolution imaging. Though this approach is routinely used for specific imaging tasks such as the CT scan of the inner ear, it is a highly dose-inefficient approach as the comb filter is placed in front of the detector array. X-rays blocked by the filter have already passed through the patient, contributing to patient dose.

This spatial resolution limitation is overcome in direct conversion PCCT technology, where no wall between detector pixels is necessary. This enables the design of smaller pixel sizes without losing fill factor and dose efficiency. Figure 7 shows a schematic cross section through a PCCT direct conversion detector. The literature reports typical anode pixel sizes ranging from 300 μm to 500 μm (pixel pitch) used in PCCT applications.

3.2 Photon-Counting Computed Tomography (PCCT)

Photon-counting detectors have several potential clinical benefits in CT imaging [14–24]. Among these high spatial resolution, increased iodine contrast-to-noise ratio and reduced electronic noise are among the more prominent. However, these detectors also have intrinsic properties that allow for increased dose efficiency and the ability to reduce beam-hardening artifacts such as those from metal objects improving clinical diagnosis.

High electronics noise is a permanent feature of the conventional EID technology as the energy integrated detectors by their principle of operation integrate all noise into the signal. The first obvious advantage of PCCT detectors is lower electronic noise. Lower noise leads to higher image quality and lower radiation dose.

Image noise in CT mainly originates from two sources: quantum noise and electronic noise. The quantum noise, which is determined by the number of detected photons, can be traced back to the random nature of X-ray photon interactions. The electronic noise comes from the detector's electronic circuits within the X-ray detection system and is, therefore, to the first degree, independent of the number of photons reaching the detector. The relative significance of quantum noise and electronic noise is determined by the incident photon flux and clinical parameters. For example, in large patients, the number of photons reaching the detector can be very small.

Figure 8 illustrates the difference in electronic noise between PCCT and EID detectors. In the PCD case, a threshold (typically at 20 keV, but a clinician can adjust it) just above the level of electronic noise can be set, making PCCT almost "noise-free." However, in the EID case, the low-energy electronic noise is integrated together with the useful signal, thereby reducing the signal-to-noise ratio (SNR).

To increase contrast-to-noise ratio (CNR), CT tests frequently utilize contrast agents. The most common one is iodine that has a K-edge position at 33.2 keV. Iodinated contrast is a form of water-soluble, intravenous radiocontrast agent containing iodine, which enhances the visibility of vascular structures and organs during radiographic procedures. Many pathologies, such as cancer, have particularly improved visibility with iodinated contrast. The radiodensity of iodinated contrast is

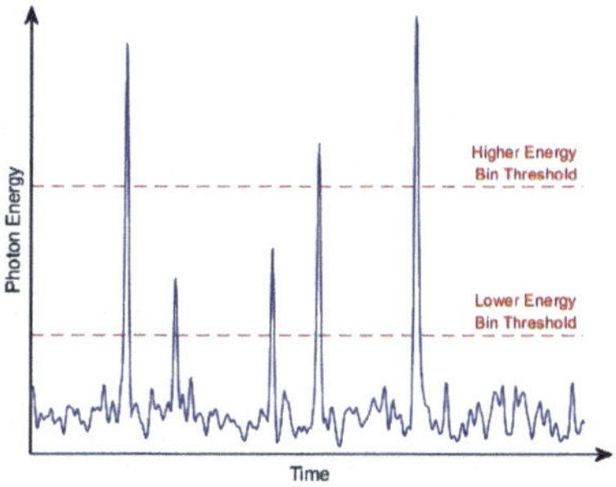

Fig. 8 Schematic illustration of photon electron pulses processed by PCCT detector. For example, with two thresholds shown, extension to a larger number of thresholds is straightforward. Typical clinical scanners utilize 4–5 energy bins

25–30 Hounsfield units (HU) per milligram of iodine per milliliter at a tube voltage of 100–120 kVpp.

PCCT detectors show great advantages over EIDs in iodine detection. In the diagnostic X-ray energy range, the X-ray photons primarily interact with scanned objects through two physical mechanisms: the photoelectric effect and Compton scattering. Theoretically, the attenuation due to the photoelectric effect is proportional to the effective atomic number of a material and is inversely proportional to the energy of incident X-ray photon. On the other hand, the energy dependence of the Compton effect is relatively flat within the diagnostic energy range, and the choice of material has little effect other than its density. Consequently, the photoelectric effect contributes more to the attenuation of low-energy photons, while Compton scattering is the dominant effect accounting for high-energy photons. As a result, high-Z materials, such as iodine, have higher photon attenuation in the low-energy range due to the photoelectric effect and therefore demonstrate higher signal or contrast in CT images.

In conventional EID detectors, the detector signal is proportional to the total energy of all detected X-ray photons. This gives more weight to higher energy photons which carry less contrast information than low energy photons. This in turn means that information content from high-Z materials such as iodine is reduced as the improved contrast of iodine is mainly explained by photoelectric interaction, which manifests its effect in the low-energy range. The efficiency of a device to convert dose to signal is referred to as dose efficiency. Consequently, the underweighting of the signal produced by lower energy photons reduces the CNR of the iodine signal in a CT system using EID. On the other hand, PCCT detectors count each individual photon equally regardless of the photon energy, without the energy weighting as in the case of EID. This results in the lower energy photons having the same weight as the higher energy photons. Relative to EIDs, the dose efficiency of a PCD is inherently improved. Hence, the low-energy photons have a greater contribution to the image contrast for PCD-CT compared to EID-CT, which improves the iodine contrast and CNR. For similar reasons, radiation dose in PCCT can be reduced without affecting image quality. This advantage of PCCT is frequently referred to as the better Swank factor and has been shown to produce 30–40% gains in CNR values and which can be, practice, turned into a similar reduction in radiation dose.

Beam hardening is a well-known effect in X-ray imaging. As photons pass through the scanned object, low-energy photons are preferentially attenuated compared to high-energy photons. Since poly-energetic beams are used in CT, this causes the effective photon energy to shift toward the higher end of the spectrum. This makes the beam more resilient to attenuation and, thus, harder to shift/change further. This effect introduces artifacts, typically dark areas adjacent to highly attenuating objects such as cortical bone, affecting the image appearance and CT number accuracy for nearby soft tissues. In the presence of foreign objects such as metal implants, severe image streaking artifacts together with dark/bright regions can be observed across the image.

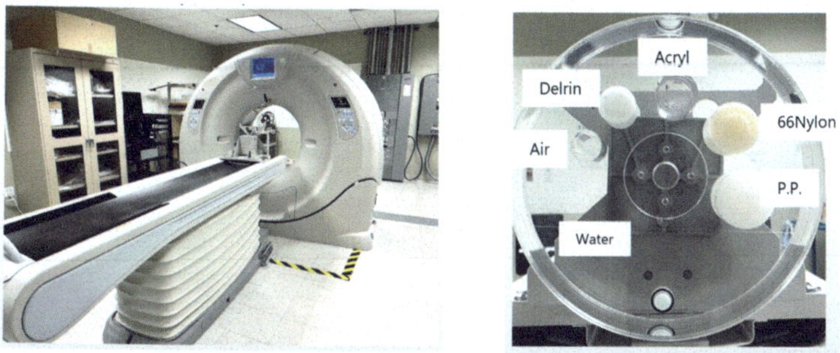

Fig. 9 Canon's first engineering prototype CZT based PCCT system (left) and image of one of the phantoms used in the study (right) [15]

These metal artifacts have a characteristic appearance and are caused by multiple physical mechanisms, including X-ray scattering, photon starvation, as well as beam hardening effect. The severity of metal artifacts can vary according to the atomic number, density, and shape of the metal, as well as the surrounding anatomy and scan acquisition/reconstruction parameters. In many cases, the resulting metal artifacts can dramatically obscure critical structures and results in significantly reducing diagnostic confidence in various clinical tasks, such as distinguishing pathologic findings from normal structures.

Since each individual photon is sorted according to its energy in a PCCT acquisition, an energy-bin image can be reconstructed using only higher-energy photons. Compared to the conventional EID-CT image or the low-energy threshold image in a PCCT acquisition, the high-energy bin image is more immune to beam hardening effect in areas around dense bones.

3.3 *Phantom Case Study*

The superior PCCT spatial resolution and lower radiation dose is illustrated using the following performance evaluation of a full-size photon-counting CT system, shown in Fig. 9, through various imaging phantom studies [14, 15]. This prototype system supports a 500 mm field-of-view (FOV) and 10 mm z-coverage at isocenter. Phantom scans were acquired using 120 kVpp from 50 to 400 mAs to assess the imaging performance on CT number accuracy, uniformity, noise, spatial resolution, and material differentiation and quantification.

Image reconstruction workflow is shown in Fig. 10. Both qualitative and quantitative evaluations show that PCCT has superior imaging performance with lower noise and improved spatial resolution compared to conventional energy integrating detector (EID)-CT, under the tested conditions. Using a projection domain material decomposition approach with multiple energy bin measurements, PCCT virtual monoenergetic images (VMIs) have lower noise and good accuracy in quantifying

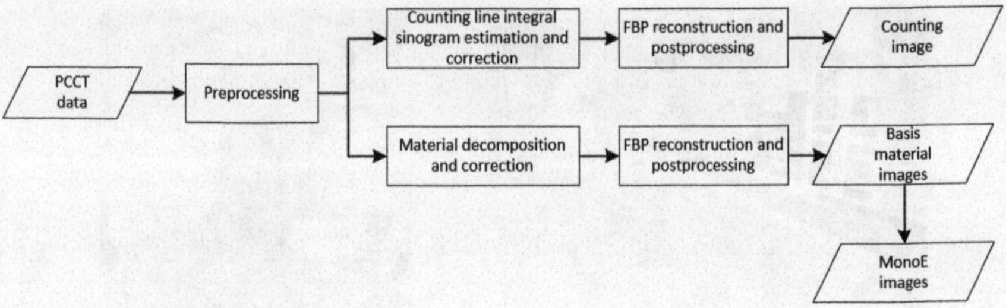

Fig. 10 PCCT image reconstruction data flow used in the study [15]

iodine and calcium concentrations. These results lead to an increased contrast-to-noise ratio (CNR) for both high- and low-contrast phantom objects compared to EID-CT at a matched dose and spatial resolution. PCCT can also generate super-high resolution (SHR) images using a much smaller detector pixel size than EID-CT, greatly improving image spatial resolution.

Improved spatial resolution and quantification accuracy with reduced image noise in PCCT images may lead to better diagnoses at reduced radiation doses compared to conventional EID-CT. The increased CNR achieved by PCCT suggests a potential reduction in iodine contrast media load, resulting in better patient safety and reduced cost. The original research covers details that are beyond the scope of this book. The curious reader is encouraged to read the publication by Zhan et al. [15] for complete details.

4 Using AI in PCCT

4.1 Noise Reduction

The intersection of AI and PCCT has led to significant advancements in noise reduction techniques, enhancing image clarity and diagnostic value. Studies have explored various network architectures and their effectiveness across different domains, indicating a potential for broader application and improved image quality in medical diagnostics.

Baffour et al. [7] investigated the application of deep learning for denoising photon-counting CT images and focused on whether denoising is more effective in the projection or image domain [7]. Of three common neural network architectures (U-Net, Dilated U-Net, and ResNet), the authors indicated that U-Net and Dilated U-Net architectures performed best in the image domain for the specific phantoms trained on.

However, denoising in the projection domain generally yielded better results for different phantoms, suggesting a potential for broader generalization. They suggest that the choice of domain for implementing deep learning denoising can significantly impact the quality of photon-counting CT images, with implications for optimizing CT imaging protocols.

Further expanding on the utility of PCCT when combined with deep learning, another study focused on noise reduction for detecting multiple myeloma lesions (blood cancer which shows symptoms in bones), benchmarking it against the conventional energy-integrating detector (EID) CT. Twenty-seven adults with multiple myeloma were scanned and compared using both a PCD CT and an EID CT system. Combined with deep learning denoising processes, the PCD CT outperformed the EID system, providing superior visualization of multiple myeloma, lytic (bone cancer) and intramedullary lesions (cancers attached to the spinal cord) as well as related pathological fractures. This suggests that PCD CT could significantly enhance the detection and diagnosis of multiple myeloma, potentially influencing treatment and management strategies for this condition without increasing radiation dose.

4.2 Material Decomposition

In material decomposition, deep learning also plays a crucial role in advancing capabilities in PCCT, enabling more precise differentiation of materials based on photon energy levels. This precision is critical for diagnosing and characterizing diseases, further underscoring AI's transformative impact on medical imaging.

A deep learning post-processing approach was introduced by Eguizabal et al. [8] to enhance the maximum likelihood estimate of material decomposition in photon-counting spectral CT. It leverages convolutional residual blocks, blending model-based and data-driven strategies for noise reduction and accuracy improvement. Their method showed promising results in simulations using Shepp-Logan phantoms and human numerical phantoms, demonstrating its potential to improve CT image quality efficiently. The approach suggests a balance between computational efficiency and the quality of material decomposition, indicating its promise for clinical application.

4.3 Metal Artifact Reduction

AI-driven methods have shown promising results in reducing artifacts caused by metal objects in CT images, improving diagnostic accuracy and image clarity. These advancements highlight the potential of integrating spectral information and sophisticated algorithms to enhance CT imaging quality.

Schmidt et al. presented a photon-counting normalized metal artifact reduction (PCNMAR) method to enhance diagnostic CT imaging by minimizing distortion caused by metal objects. They introduced a technique that leverages spectral information intrinsic to photon-counting CT alongside a sophisticated prior-based metal artifact reduction algorithm. By integrating these approaches, PCNMAR effectively reduced metal artifacts across the various imaging bins and bin combinations, showcasing a notable improvement over the conventional method with the potential to significantly enhance the clarity of CT images, offering promising implications for clinical diagnostics [10].

4.4 Image Reconstruction Methods

AI-based image reconstruction techniques offer advantages over traditional methods, including improved image quality, resolution, and reduced computation time. These innovations represent a significant leap forward in utilizing PCCT data more effectively, enhancing the diagnostic process.

Crosstalk between a detector's imaging bins is one complicated issue that requires AI to address properly. The incorporation of a learned crosstalk model was required to sufficiently address the challenge of efficiently modeling this effect. The model by Strom et al. effectively reduced crosstalk with a notable increase in image quality (signal-to-noise ratio) while benefiting from reducing computation time by as much as 50% over previous methods. Their approach is applicable to other situations involving different data sets and detector settings, suggesting that partially learned forward operators could be a computationally advantageous method for modeling data generation in photon-counting CT [11].

Machine learning models have permeated many aspects of photon-counting advancements and will continue to expand and has become indispensable for improving these already complex systems. Noise was already a complicated issue in the previous generation of these detectors. Already many denoising algorithms are under research to refine PCD CT imaging and characterize the optimal method of enhancing image quality, improving signal to noise ratio and maximize diagnostic imaging quality despite limited photon statistics [16–24].

5 Case Study: Using AI for Cardiac X-Ray Applications

Detection of cardiac anomalies from X-ray images in a fast and accurate way is important in diagnostic medicine. Despite advancements in medical imaging, the complexity of cardiac structures and variability among patients pose significant challenges to traditional analysis methods. This section introduces an example of using convolutional neural network (CNN)-based framework developed to enhance cardiac detection accuracy and efficiency from X-ray imaging. The proposed CNN model demonstrated better performance in detecting cardiac anomalies compared to traditional imaging analysis methods. In the future, current applications of AI have great potential to be extended to PCCT.

This research offers an example of the potential of CNNs in medical imaging, particularly for cardiac detection. By offering a more accurate, efficient, and accessible means of diagnosing heart conditions, the findings pave the way for enhanced patient care and management. Moreover, the study's insights into CNN architecture optimization for medical imaging can inform future developments in the field, extending beyond cardiac diagnostics to other complex imaging tasks.

The advent of digital imaging has revolutionized medical diagnostics, providing radiologists with an invaluable tool for detecting and treating diseases. Among these diseases, cardiac-related diseases stand as a leading cause of morbidity and mortality worldwide [25]. The early detection of cardiac anomalies is paramount in mitigating their progression and improving patient outcomes [26]. X-ray imaging, due to its widespread availability and cost-effectiveness, is often the first diagnostic step in evaluating heart health. However, the interpretation of cardiac X-ray image is challenging, requiring specialized expertise, and often resulting in variability in diagnostic accuracy [27, 28].

The complexity of the heart's anatomy, coupled with its variable appearance due to different patient physiologies and the presence of disease, necessitates a diagnostic approach that is both sophisticated and nuanced. Traditional methods of X-ray image analysis rely heavily on the subjective interpretation of radiologists, leading to a demand for more objective, reliable, and efficient diagnostic tools. This need is further amplified by the increasing volume of medical imaging data, which strains healthcare resources and underscores the necessity for automation in image analysis.

In this context, convolutional neural networks (CNNs) have emerged as a groundbreaking technology in the field of computer vision, demonstrating exceptional ability in recognizing patterns and features in images. The application of CNNs to medical image analysis has shown promising results, offering a new paradigm in the detection and characterization of diseases [25]. Specifically, in cardiac diagnostics, CNNs can leverage their pattern recognition capabilities to identify subtle anomalies in X-ray images that may be indicative of cardiac conditions, thereby enhancing the accuracy and efficiency of diagnostic processes.

Early attempts to automate the analysis of medical images date back several decades, with initial methodologies focusing on rule-based systems that mimicked expert decisions. However, these systems were limited by their inability to learn from data and adapt to new patterns not encoded in their rules. The advent of machine learning, and subsequently deep learning, shifted this paradigm by enabling models to learn directly from the data, improving their performance as more data became available. CNNs have distinguished themselves in image recognition tasks among deep learning architectures due to their ability to extract features automatically and hierarchically from images.

Recent studies have demonstrated the efficacy of CNNs in detecting a wide range of pathologies in chest X-rays, including pulmonary conditions, fractures, and neoplasms. For cardiac detection, Liao et al. [29] presented a model that accurately identified cardiomegaly from chest X-rays, achieving an area under the curve (AUC) of 0.95 in receiver operating characteristic (ROC) analysis. Similarly, Zhang and

Fig. 11 Chest X-ray image used in this test case

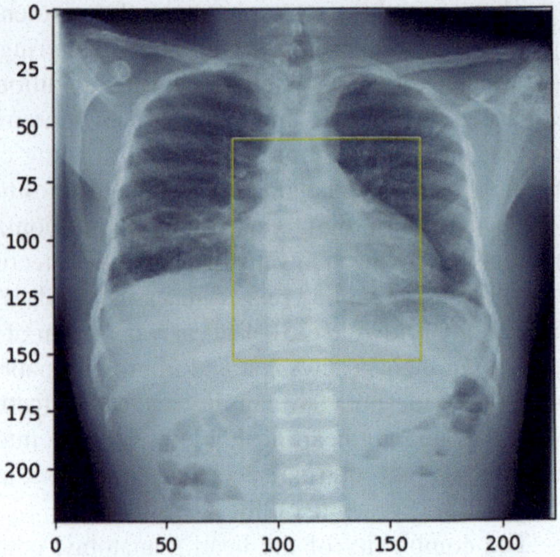

colleagues (2021) developed a CNN [30] that outperformed traditional imaging analysis in detecting coronary artery disease, highlighting the potential of deep learning in cardiac diagnostics.

Despite these advancements, several gaps remain in the literature, particularly concerning the generalizability of models across diverse patient populations and imaging modalities. Many existing studies are constrained by the homogeneity of their datasets, which are often collected from single institutions and lack representation of varied demographics and pathologies. This limitation raises concerns about the models' performance in real-world settings, where variability in patient characteristics and image quality is the norm.

The dataset, retrieved from a comprehensive Kaggle repository (www.kaggle.com/datasets), includes chest X-ray images such as the one displayed in Fig. 11. This image is representative of the data going to be used to train the convolutional neural network (CNN) for cardiac abnormality detection. The image depicts a standard posteroanterior (PA) chest X-ray, providing a clear view of the thoracic cavity, inclusive of the heart, lungs, and major blood vessels. A blue bounding box is superimposed on the X-ray, likely indicating the region of interest (ROI) that the CNN is configured to analyze. This ROI seems to encapsulate the cardiac silhouette, which is crucial for detecting conditions such as cardiomegaly, structural anomalies, or other pathologies that may affect the size, shape, or position of the heart.

Our model uses a customized convolutional neural network (CNN) tailored explicitly for cardiac feature extraction from X-ray images. At the heart of the model lies a sequence of convolutional layers, each meticulously designed to parse through the intricacies of cardiac anatomical structures within the X-ray images. The architecture commences with a convolutional layer adapted to handle single-channel (grayscale) input, which is crucial for the grayscale nature of X-ray images.

This initial layer uses a 7×7 kernel to capture the broad features of the image with a stride of 2 for spatial reduction.

As the architecture deepens, subsequent convolutional layers with varying kernel sizes and filters refine the feature maps, distilling the essence of cardiac shapes and potential anomalies. Each convolutional layer is followed by a batch normalization step to accelerate training and a rectified linear unit (ReLU) activation to introduce nonlinearity, essential for capturing complex patterns. The culmination of the convolutional base is a series of fully connected layers that condense the high-level features into a four-dimensional output. This output corresponds to the coordinates of the cardiac region's bounding box, thus focusing the model on the localization task at hand.

Distinct from models adopted in literature, the model benefits from several key advantages. By constructing a custom CNN from the ground up, we ensure that each layer is optimized for the specific task of cardiac detection, allowing for finer control over the feature extraction process. Transitioning from classification to localization, the model's output layer is fine-tuned to provide precise bounding box coordinates, aligning with the clinical need for accurate cardiac region identification. Integral to the model is the capability to visualize the learning process. The logging function overlays predicted coordinates on X-ray images, enabling immediate visual assessment of the model's accuracy.

Utilizing an Adam optimizer with a carefully tuned learning rate ensures that the model benefits from adaptive learning rates, crucial for the convergence of a custom-built CNN. The employment of mean squared error loss directly minimizes the distance between predicted and actual bounding box coordinates, a direct measure of the model's performance in the localization task. As we proceed, the ensuing sections will elucidate the empirical validation of the model. The performance metrics and visual evidence provided will illuminate the model's efficacy and delineate its potential as a diagnostic tool in clinical settings.

The series of chest X-ray images presented in Fig. 12 represents the model's performance at various stages of training: early (epoch 10), middle (epoch 50), and upon completion (epoch 100). At epoch 10, the bounding boxes, indicative of the model's predictions, are broadly placed, encapsulating the cardiac region but with evident imprecision. This initial stage reflects the model's developing understanding of the cardiac structures, where it is beginning to learn the relevant features but has yet to refine its localization accuracy.

Advancing to epoch 50, there is a marked improvement in the precision of the bounding boxes. The model, benefiting from the iterative learning process, starts to home in on the cardiac silhouette with greater accuracy. The progression from epoch 10 to 50 demonstrates the model's ability to adjust its parameters effectively in response to the complex patterns inherent in X-ray images.

The images from epoch 100 depict a notable enhancement in localization performance. The bounding boxes are tightly aligned with the cardiac region, suggesting that the model has achieved a sophisticated level of feature recognition and spatial understanding. The accuracy of these predictions, closely hugging the contours of the heart, signifies a matured learning that aligns with the intended outcomes of the study.

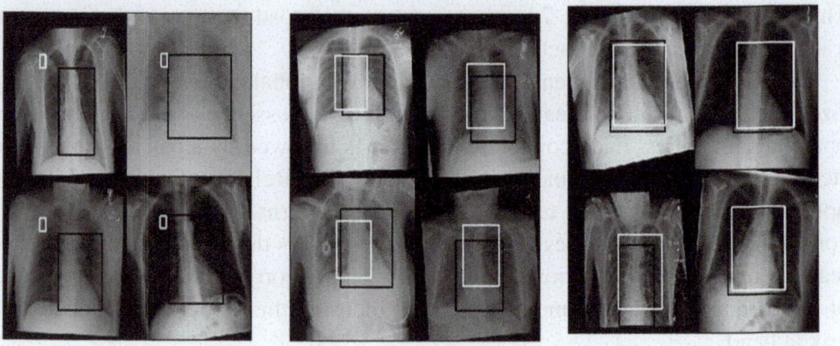

Fig. 12 X-ray images after model's performance at various stages of training left—early (epoch 10), middle (epoch 50), and right—upon completion (epoch 100)

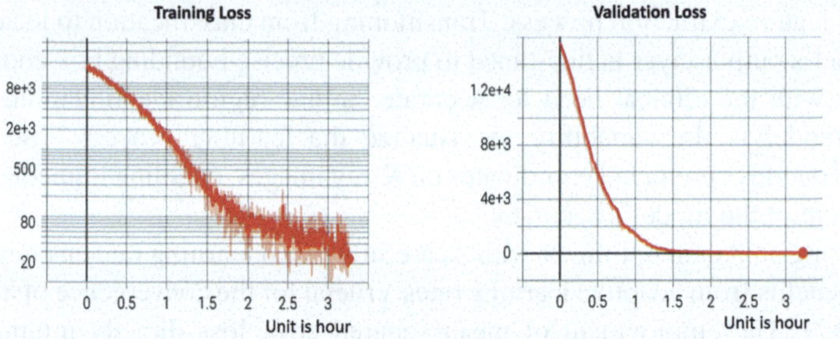

Fig. 13 Training new data. It shows a similar trend of sharp decline in the initial phases, affirming that the model's early learning is robust and not merely overfitting to the training data. Crucially, the validation loss curve plateaus and maintains stability, without an uptick, suggesting that the model has not begun to overfit to the training data—a common concern in deep learning applications

The training and validation loss plots, shown in Fig. 13, are instrumental in understanding the learning dynamics of the model. The training loss plot exhibits a steep decline in the initial epochs, indicating rapid learning and feature extraction by the model. This sharp decrease is expected as the model initially adjusts from a high entropy state, starting with little to no knowledge about the task at hand, toward a more ordered state reflecting an understanding of the cardiac structures in the X-ray images.

As training progresses, the decrease in loss tapers off, suggesting that the model is approaching a state of convergence where it has largely adapted to the patterns within the training data. The flattening curve toward the later epochs indicates a diminishing return on loss reduction for each additional epoch, characteristic of the learning plateau typically reached by well-tuned models.

The sustained low level of validation loss toward the later epochs is particularly encouraging. It implies the model has achieved a generalizable understanding of cardiac anomalies detectable in X-ray images. This is evidenced by its consistent performance on unseen data, a strong indicator of the model's potential efficacy in a clinical setting.

A key outcome of this study, as demonstrated by the validation loss plot, is that the model seems well-poised for clinical translation. The low variance in loss across epochs and the convergence of training and validation losses suggest that the model is likely to perform reliably when deployed in a real-world diagnostic scenario, maintaining accuracy and consistency.

The evidence presented by the loss plots leads to several important conclusions. First, the model has demonstrated the capacity for rapid and sustained learning, a desirable characteristic for any deep learning application in medicine. Second, the model's generalization abilities suggest it is well-equipped to handle the variability inherent in medical images across different patient populations. Future research directions may include further exploration into model interpretability and the deployment of the model in a live clinical environment to assess its performance in real-time diagnostics. Additionally, studies could investigate the integration of the model into existing healthcare systems, evaluating its impact on diagnostic workflows and patient outcomes.

6 AI Future in Healthcare

6.1 Evolution of Healthcare Needs

There is a lot of confusion and excitement surrounding whether and how artificial intelligence (AI) will transform healthcare. Effectively, AI is a collection of algorithms. The effectiveness of any AI algorithm can be tested in practice. However, the problem might not be a specific algorithm's capability but humans' inability to clearly identify the clinical problem that the algorithm can solve. To truly flex AI in healthcare, problems should be identified and grounded in specific patient needs.

Let us consider first what humans need and want. The current trend in the developed world is that consumers are taking their health into their own hands. Inspired by a new wave of longevity, concierge medicine, and diagnostic companies, people are getting extensive preventative testing—from comprehensive blood tests, CT scans, to cancer screenings and genetic panels—to fully understand what is happening in their bodies. The question is who is going to deliver these services and what role AI will play.

Typically, the companies that change business are founded by outsiders and operate outside the bounds of traditional health systems. History teaches us that major step changes in consumer experiences are generally driven by outsiders who fundamentally reimagine how things could work. In retail space, Amazon, not Walmart, enabled anyone to have anything shipped to their home. As a result, the typical shopping experience has changed dramatically. In music, it was Spotify, not Universal, that gave us unlimited access to music for a low price. In silicon chips, it was Nvidia, not Intel, that gave us computer chips for AI. In payments, it was PayPal, not Bank of America, that revolutionized online payments and digital wallets. In Internet search, it was Google, not Microsoft, that gave us unlimited search capabilities. Finally, it was Uber, not a taxi service, that gave us instantaneous access to rides.

New healthcare services will be provided by new, innovative companies, that change the way things have been done so far, likely using AI as a leverage for efficiency and speed of processing. The question is whether people want to interface with AI directly or whether the medical doctor will still need to act as an intermediary.

Imagine that you are suffering from anxiety. To solve the problem, you turn to the Internet. You describe your anxiety symptoms in a few words, hit return, and get a million results in a fraction of a second. You find some relaxation and mindfulness suggestions and sign up for an anxiety app. In other words, you identified and described your symptoms to a search engine, found candidate solutions, and tested those solutions.

If the app and meditation do not help, you schedule an appointment with your clinician. But what is it you hope your clinician will do for you? You already have all the available data about your specific anxiety and mood as well as access to a wealth of medical information. Yet a search engine alone has not solved your problem, which is your inability to interpret and act on the data you have. You still need a doctor to prescribe you medication unless we reach the stage of human development where we will use AI to write our medication prescription (or have the meds delivered directly to your home, bypassing the "unneeded" stage of writing the actual prescription).

Taking that into account, AI scientists might consider that a doctor solves a patient's problem by gathering information and helping them interpret and act on their data. The clinical examination is simply a structured data collection procedure, just an old-fashioned data exchange, usually in the form of talking. It is not clear that people will be comfortable talking to AI directly, although recent reports about people's attachment to AI tools seem to indicate that sooner or later that will be the case.

Framed as a data problem, it is clear where and why AI can benefit healthcare. Before, during, and between clinical encounters, AI algorithms can help gather, interpret, and act on clinical data. With the exciting new developments of foundation models and intelligent agents, an AI system could tailor the action based on a specific patient, clinician, or hospital system.

The problems and risks that the AI tool must manage depend on which stage of data problem it is meant to solve. Any tool that might help doctors more efficiently gather your history, interpret that history in the context of growing clinical research, and act on the problem would benefit health care. To enter clinical practice, AI tools must be verified, tested, and be cost efficient, a process that will require greater specificity to the way physicians gather, interpret, and act to benefit a patient.

The combination of AI and PCCT is a promising area of research, with ongoing advancements expected to further improve diagnostic capabilities. AI can leverage the detailed data from PCCT to predict disease progression and treatment outcomes, contributing to personalized medicine.

6.2 AI Challenges in Healthcare

To illustrate AI challenges in healthcare, we will use X-ray computed tomography as an example, as that is the topic of this chapter. The challenges are expected to be very similar in other medical imaging modalities and healthcare in general. The prerequisite of large datasets for training deep learning models poses a significant challenge, amplified by concerns over privacy and restrictions on data sharing. This issue has been prevalent in medical imaging in general; and PCCT, unfortunately, cannot be excused from these concerns. These hurdles complicate the data acquisition process and raise ethical issues that must be navigated carefully. Ensuring data diversity to train robust and generalizable models across various demographics and conditions adds another layer of complexity. Innovations in synthetic data generation offer potential pathways to mitigate these challenges, enabling the use of AI in settings where data are scarce or sensitive [23–25].

The opaque nature of deep learning models often referred to as "black boxes" poses significant challenges to clinician trust and technology adoption for PCCT. The inability to understand the decision-making process of AI models for medical imaging has constantly been discussed in clinical research and practice, where explainability and accountability are paramount. Developing explainable AI methods that provide insights into the models' reasoning processes is critical to overcoming this barrier.

Algorithm and model development is crucial in overcoming the barriers of model complexity and data scarcity that currently hinder broader AI integration in medical imaging, including PCCT. Current research endeavors can be directed toward creating powerful and understandable deep learning frameworks that are both data and resource efficient. Innovations in algorithm design focus on models that can learn from smaller datasets without compromising their predictive accuracy, thus democratizing AI technology for a broader range of healthcare institutions, regardless of size or resource availability. Moreover, advancements in making these models more interpretable aim to bridge the gap between AI capabilities and clinical applicability, ensuring that healthcare professionals can understand and trust AI-generated insights.

Clinical validation is another critical area of research in applying AI techniques to PCCT. The path to integrating deep learning–enhanced PCCT into routine clinical practice requires rigorous validation through extensive clinical trials. One pioneering example is using a PCCT simulator to generate projection images of computational phantoms, given the geometry and physics of the scanner and imaging parameters for specific clinical tasks [24]. Clinical validation studies are essential to ascertain the diagnostic precision and effectiveness of AI-augmented PCCT across a spectrum of medical fields, including but not limited to oncology and cardiology. By systematically evaluating the performance of these AI-enhanced imaging modalities in diverse clinical scenarios, researchers can identify areas of improvement, adapt AI algorithms to meet clinical needs better, and ultimately demonstrate the practical value of this technology in enhancing patient outcomes. The successful clinical validation of PCCT diagnostic accuracy will pave the way for its adoption in standard medical protocols, revolutionizing patient care with cutting-edge precision diagnostics.

6.3 Future of AI

Despite the many challenges the AI is facing, we should not be afraid of using AI in medicine. Despite the wide scale automatization and digitalization, humans will always be needed for specific tasks. But the fear of technology is there, from artificial intelligence taking the place of doctors in various specialties, and robots surpassing surgeons' skills to taking jobs in pharma.

AI will be bigger than all other tech revolutions and will rival those related to computer and Internet revolutions. Many white color jobs will disappear, but many new jobs will also get created. Robots, supported by AI, are likely to replace most labor jobs within 10–20 years. Machines will substitute many doctors in the future in a healthcare scene driven by entrepreneurs. Dermatologists already use AI to detect skin cancer with near-perfect accuracy. It is likely we will not be needing radiologists in the future as AI algorithms will read medical images with better precision at a fraction of the cost. However, the human brain is complex and able to oversee such a vast scale of knowledge and data that it might take several years developing an AI that takes over the brain's job. For now, the task of AI is to program repetitive, data-based tasks, and leave the complex analysis/decision to people.

The perceived threat of AI is often exaggerated and can take unexpected directions. Consider chess. Many believed the game would lose its popularity after Garry Kasparov's defeat by an AI in 1997. Instead, nearly a billion people today engage in chess daily, with AI becoming an indispensable tool for players, enthusiasts, and coaches. Similarly, when the generative AI boom began with ChatGPT in 2022, there was a vision of superintelligent AI tools that would revolutionize productivity and economic gains. While that vision has not fully materialized, AI tools are gradually integrating into our lives. Some people have even formed relationships with AI systems, interacting with them as friends, therapists, or teachers. While initially surprising, these developments offer potential solutions to challenges. AI language models, which are based on large language models (LLMs), predict the next likely word in a sentence. This predictive ability can result in interactions that feel engaging and even lifelike. However, these systems also raise important ethical considerations. For instance, while AI can provide companionship, it can sometimes generate inaccurate or misleading information. This underscores the need for careful oversight and responsible use of AI, ensuring that it complements human interaction without replacing the authenticity and trust that only real human relationships can provide.

AI has demonstrably enhanced chess and Go by creating a more competitive and popular, game environment. AI can transform healthcare as well, where its adoption promises similar advancements. Perhaps the biggest gains can be achieved when people's creativity and problem-solving skills are combined with the huge AI computing power and cognitive resources of technology.

6.4 PCCT and AI Outlook

CdTe and CZT-based PCCT detectors present a variety of advantages over conventional EID detectors. In this chapter, we have highlighted the primary benefits of this technology for computed tomography applications: higher spatial resolution, absence of electronic noise, lower radiation and contrast dose, and the possibility of using advanced contrast agents. The great value of photon-counting technology in CT is that it incorporates all these elements into a single acquisition platform. For many routine applications, PCCT scanners using CZT detectors will provide great improvement in clinical care. Scans with metal hardware will become increasingly easier to read because of better artifact suppression. Scans with iodine contrast will show better detectability, leading to lower imaging dose and contrast dose, especially at high kVpp.

New capabilities of PCCT scanners include simplification and consolidation of a standardized universal imaging protocol and the ability to classify small high-contrast lesions such as lung tumors and kidney stones. For examinations of high-resolution structures such as the temporal bone or sinus, radiation dose will be reduced. Truly transformative capabilities may occur when targeted molecular contrast agents become clinically available. PCCT technology is only in its infancy, with the first human scanners currently being announced and deployed. The next several years are sure to reveal a wealth of information on the potential applications of PCCT.

PCCT has the potential to dramatically alter the clinical use of CT in the coming years. By using energy-resolving detectors instead of EID, these new CT systems can substantially reduce image noise, increase the spatial resolution, and use K-edge imaging to measure the concentration of specific elements with reduced radiation doses compared to current standards. These properties can be useful in the evaluation of multiple anatomical districts, where it is necessary to identify small physiological and pathological structures, as well as for the use of different contrast agents, even concurrently, as shown in our review.

In addition, the synergy between AI and PCCT holds immense promise for revolutionizing medical imaging. The fusion of these cutting-edge technologies enables overcoming longstanding limitations in traditional imaging methods, offering clearer, more detailed images and facilitating the early detection and accurate characterization of a wide range of medical conditions. Although powerful, applying

AI techniques to PCCT should still need to be navigated carefully. Again, this overview of the current research trends and challenges between AI and PCCT does not intend to cover all the ongoing research in the field. As the field continues to evolve, the continued exploration and integration of AI within PCCT promises not only to redefine the boundaries of medical imaging but also to enhance patient care and treatment outcomes.

References

1. Spectral, Photon Counting Computed Tomography Technology and Applications, Edited by Katsuyuki Taguchi, Ira Blevis, and Krzysztof Iniewski (CRC Press, 2020), https://www.routledge.com/Spectral-Photon-Counting-Computed-Tomography-Technology-and-Applications/Taguchi-Blevis-Iniewski/p/book/9780367490119
2. T. Schmidt, Future prospects of spectral CT: Photon counting, in *Computed Tomography: Approaches, Applications, and Operations*, ed. by E. Samei, N.J. Pelc, (Springer, 2020), pp. 269–286. https://doi.org/10.1007/978-3-030-26957-9
3. S. Hsieh, K.K. Iniewski, *Photon Counting Computed Tomography: Clinical Applications, Image Reconstruction and Material Discrimination* (Springer 2023), https://link.springer.com/book/10.1007/978-3-031-26062-9
4. P. Ongsulee, *Artificial intelligence, machine learning and deep learning*, in 2017 15th International conference on ICT and knowledge engineering (ICT&KE), pp. 1–6 (2017). https://doi.org/10.1109/ICTKE.2017.8259629 . ISSN: 2157-099X
5. I. Castiglioni, L. Rundo, M. Codari, G. Di Leo, C. Salvatore, M. Interlenghi, F. Gallivanone, A. Cozzi, N.C. D'Amico, F. Sardanelli, AI applications to medical images: From machine learning to deep learning. Phys. Med. **83**, 9–24 (2021). https://doi.org/10.1016/j.ejmp.2021.02.006
6. H. Karlsson, V. Moro, A. Eguizabal, M. Eriksson, A. Agren, D. Akerstrom, M.U. Persson, Deep-learning-based denoising for photon-counting CT: Image domain or projection domain? in *Medical Imaging 2022: Physics of Medical Imaging*, vol. 12031, (SPIE, 2022), pp. 698–715. https://doi.org/10.1117/12.2612480
7. F.I. Baffour, N.R. Huber, A. Ferrero, K. Rajendran, K.N. Glazebrook, N.B. Larson, S. Kumar, J.M. Cook, S. Leng, E.R. Shanblatt, C.H. McCollough, J.G. Fletcher, Photon-counting detector CT with deep learning noise reduction to detect multiple myeloma. Radiology **306**(1), 229–236 (2022). https://doi.org/10.1148/radiol.220311
8. A. Eguizabal, M.U. Persson, F. Gronberg, A deep learning post-processing to enhance the maximum likelihood estimate of three material decomposition in photon counting spectral CT, in *Medical Imaging 2021: Physics of Medical Imaging*, vol. 11595, (SPIE, 2021), pp. 1080–1089. https://doi.org/10.1117/12.2581044
9. V.D. Trapani, L. Brombal, F. Brun, Multi-material spectral photon-counting micro-CT with minimum residual decomposition and self-supervised deep denoising. Opt. Express **30**(24), 42995–43011 (2022). https://doi.org/10.1364/OE.471439. Optica Publishing. Accessed 2024-03-29
10. T.G. Schmidt, B.A. Sammut, R.F. Barber, X. Pan, E.Y. Sidky, Addressing CT metal artifacts using photon-counting detectors and one-step spectral CT image reconstruction. Med. Phys. **49**(5), 3021–3040 (2022). https://doi.org/10.1002/mp.15621
11. E. Strom, M. Persson, A. Eguizabal, O. Oktem, Photon-counting Ct reconstruction with a learned forward operator. IEEE Trans. Comput. Imaging **8**, 536–550 (2022). https://doi.org/10.1109/TCI.2022.3183405
12. J. O'Connell, K. Murphy, S. Robinson, K. Iniewski, M. Bazalova-Carter, Unsupervised learning methods in X-ray spectral imaging material segmentation. JINST, P06022 (2019) https://iopscience.iop.org/article/10.1088/1748-0221/14/06/P06022 & https://arxiv.org/abs/1904.03701

13. D. Richtsmeier, E. Guliyev, K. Iniewski, M. Bazalova-Carter, Contaminant detection in non-destructive testing using a CZT photon counting detector. JINST, P01011 (2021) https://iopscience.iop.org/article/10.1088/1748-0221/16/01/P01011/meta
14. Start of Clinical Research of the First Japanese Produced Next-Generation Photon-Counting Computed Tomography, Canon Medical, www.global.medical.canon/News/PressRelease/Detail/131535-834 (2023)
15. X. Zhan, R. Zhang, X. Niu, I. Hein, B. Budden, W. Shuoxing, N. Markov, C. Clarke, Y. Qiang, H. Taguchi, K. Nomura, Y. Muramatsu, Y. Zhou, T. Kobayashi, R. Thompson, H. Miyazaki, H. Nakai, Comprehensive evaluations of a prototype full field-of-view photon counting CT system through phantom studies. Phys. Med. Biol. **68**, 175007 (2023). https://doi.org/10.1088/1361-6560/acebb
16. J. Hsieh, T. Flohr, Computed tomography recent history and future perspectives. J. Med. Imaging **8**(5), 052109 (2021). https://doi.org/10.1117/1.JMI.8.5.052109
17. T. Flohr, M. Petersilka, A. Henning, S. Ulzheimer, J. Ferda, B. Schmidt, Photon-counting CT review. Phys. Med. **79**, 126–136 (2020). https://doi.org/10.1016/j.ejmp.2020.10.030
18. C.H. McCollough, S. Leng, L. Yu, J.G. Fletcher, Dual- and multi-energy CT: Principles, technical approaches, and clinical applications. Radiology **276**(3), 637–653 (2015). https://doi.org/10.1148/radiol.2015142631. Radiological Society of North America
19. D. Richtsmeier, P.-A. Rodesch, K. Iniewski, W. Siu, M. Bazalova-Carter, The feasibility of accurate stent visualization with photon-counting detector CT and K-edge imaging. IEEE Trans. Radiat. Plasma Med. Sci. **7**(7), 712–718 (2023). https://doi.org/10.1109/TRPMS.2023.3287566. https://pubmed.ncbi.nlm.nih.gov/31882698/
20. D. Richtsmeier, C. Dunning, K. Iniewski, M. Bazalova-Carter, Multi-contrast K-edge imaging on a bench-top photon-counting CT system: Acquisition parameter study. J. Instrum. **15** (2020). https://doi.org/10.1088/1748-0221/15/10/P10029. https://iopscience.iop.org/article/10.1088/1748-0221/15/10/P10029
21. L. Shen, Y. Xing, L. Zhang, Joint reconstruction and spectrum refinement for photon-counting-detector spectral CT. IEEE Trans. Med. Imaging **42**(9), 2653–2665 (2023). https://doi.org/10.1109/TMI.2023.3261999
22. Toyoda, T., Sato, S., Kiji, H., Kataoka, J., Kotoku, J., Taki, M.: Application of machine-learning models to improve the image quality of photon-counting CT images. J. Instrum. 16(05), 05021 (2021) https://doi.org/10.1088/1748-0221/16/05/P05021. IOP
23. E. Abadi, C. McCabe, B. Harrawood, S. Sotoudeh-Paima, W.P. Segars, E. Samei, Development and clinical applications of a virtual imaging framework for optimizing photon-counting CT. Proc. SPIE Int. Soc. Opt. Eng. **12031**, 120311 (2022). https://doi.org/10.1117/12.2612079
24. S. Wang, Y. Yang, Z. Yin, A.S. Wang, Noise2Noise for denoising photon counting CT images: Generating training data from existing scans, in *Medical Imaging 2023: Physics of Medical Imaging*, vol. 12463, (SPIE, 2023), pp. 15–19. https://doi.org/10.1117/12.2653975
25. Decline in deaths from heart disease and stroke–United States, 1900–1999 Centers for Disease Control and Prevention (CDC)
26. S.A. Alowais, S.S. Alghamdi, N. Alsuhebany, T. Alqahtani, A.I. Alshaya, S.N. Almohareb, A. Aldairem, M. Alrashed, K.B. Saleh, H.A. Badreldin, M.S. Al Yami, S. Al Harbi, A.M. Albekairy, Revolutionizing healthcare: The role of artificial intelligence in clinical practice. BMC Med. Educ. **23**, 689 (2023)
27. S. Wann, J. Tunio, Current challenges and future directions in cardiac imaging. J. Saudi Heart Assoc. **22**(3), 121–124 (2010)
28. J.D. Dormer, M. Halicek, L. Ma, C.M. Reilly, E. Schreibmann, B. Fei, Convolutional Neural Networks for the Detection of Diseased Hearts Using CT Images and Left Atrium Patches
29. C. Liao et al., A deep-learning-based framework for identifying and localizing multiple abnormalities and assessing cardiomegaly in chest X-ray. Nat. Commun. **15**, 1347 (2024)
30. Y. Zhang, S. Liu, Z. He, Y. Zhang, C. Wang, A CNN model for cardiac arrhythmias classification based on individual ECG signals. Cardiovasc. Eng. Technol. **13**(4), 548–557 (2022)

Correction to: Enhanced 3D X-Ray Tomography: Deep Learning-Based Advanced Algorithms for Fiber Instance Segmentation

Vijay Kumar Pediredla, Mengqi Fang, Yankai Cao, Andre Phillion, Mark Martinez, and Bhushan Gopaluni

Correction to:
Chapter 9 in: K. (K.) Iniewski, L. (K.) Cai (eds.), *Deep Learning for Advanced X-ray Detection and Imaging Applications*, https://doi.org/10.1007/978-3-031-75653-5_9

The book was inadvertently published without the co-authors' names (Andre Phillion and Mark Martinez) in Chapter 9. This has now been corrected now.

The updated version of this chapter can be found at
https://doi.org/10.1007/978-3-031-75653-5_9

K. (K.) Iniewski, L. (K.) Cai (eds.), *Deep Learning for Advanced X-ray Detection and Imaging Applications*, https://doi.org/10.1007/978-3-031-75653-5_13

Index

K. (K.) Iniewski, L. (K.) Cai (eds.), *Deep Learning for Advanced X-ray Detection and Imaging Applications*, https://doi.org/10.1007/978-3-031-75653-5

The manufacturer's authorised representative in the EU is Springer Nature Customer Service Centre GmbH, Europaplatz 3, 69115 Heidelberg, Germany. If you have any concerns regarding our products, please contact ProductSafety@springernature.com

Printed and bound by CPI Group (UK) Ltd, Croydon, CR0 4YY
30/09/2026
02225562-0007